AI-Centric Modeling and Analytics

This book shares new methodologies, technologies, and practices for resolving issues associated with leveraging AI-centric modeling, data analytics, machine learning-aided models, Internet of Things-driven applications, and cybersecurity techniques in the era of Industrial Revolution 4.0.

AI-Centric Modeling and Analytics: Concepts, Technologies, and Applications focuses on how to implement solutions using models and techniques to gain insights, predict outcomes, and make informed decisions. This book presents advanced AI-centric modeling and analysis techniques that facilitate data analytics and learning in various applications. It offers fundamental concepts of advanced techniques, technologies, and tools along with the concept of real-time analysis systems. It also includes AI-centric approaches for the overall innovation, development, and implementation of business development and management systems along with a discussion of AI-centric robotic process automation systems that are useful in many government and private industries.

This reference book targets a mixed audience of engineers, business analysts, researchers, professionals, and students from various fields.

AI-Centric Modeling and Analytics

Concepts, Technologies, and Applications

Edited by Alex Khang, Vugar Abdullayev,
Babasaheb Jadhav, Shashi Kant Gupta,
and Gilbert Morris

CRC Press
Taylor & Francis Group
Boca Raton London New York

CRC Press is an imprint of the
Taylor & Francis Group, an **informa** business

First edition published 2024
by CRC Press
2385 NW Executive Center Drive, Suite 320, Boca Raton FL 33431

and by CRC Press
4 Park Square, Milton Park, Abingdon, Oxon, OX14 4RN

CRC Press is an imprint of Taylor & Francis Group, LLC

ISBN: 978-1-032-49708-2 (hbk)
ISBN: 978-1-032-50879-5 (pbk)
ISBN: 978-1-003-40011-0 (ebk)

DOI: 10.1201/9781003400110

Typeset in Times New Roman
by Apex CoVantage, LLC

Contents

Chapter 14 Robotic Process Automation Applications in Data Management 238

Vivek Sharma, Alex Khang, Pragati Hiwarkar,
and Babasaheb Jadhav

Chapter 17   Phishing Attack and Defense: An Exploratory Data Analytics
                of Uniform Resource Locators for Cybersecurity 304

*Taiwo O. Olaleye, Oluwasefunmi T. Arogundade, Agbaegbu
Johnbosco, Olayemi O. Sadare, Adekunle M. Azeez, Azeez A.
Opatunji, Ayobami A. Tewogbade, and Saminu Akintunde*

Chapter 18   Analysis of Deep Learning-Based Approaches for Spam Bots
                and Cyberbullying Detection in Online Social Networks 324

*Santhosh Kumar A. V., Suresh Kumar N., Kanniga Devi R.,
and Muthukannan M.*

Preface

In Industrial Revolution 4.0, numerous artificial intelligence (AI)-centric technologies and applications have become more modern. In an unpredictable world, most companies are keen to leverage multifaceted AI-centric modeling and predictive analytics to deploy and deliver next-generation smart businesses and applications to their clients anytime and anywhere.

All types of connected business applications are collectively and/or individually enabled to be intelligent in their operations, offering, and output. Specifically, AI-driven and Internet of Things (IoT)-aided applications are being touted as next-generation technologies for visualizing and realizing intelligent transactions in business systems. We are seeing a host of powerful AI-centric solutions, predictive analytics, cloud-based data, big data technology, IoT-aided techniques, robotic automation processes in data management, and relevant frameworks aimed at supporting human-controlled capabilities in intelligent computing and modern AI-centric systems used in the fields of business, marketing, supply chain, healthcare, finance, banking, farming, and agriculture in the real world.

At this point, we face many challenges associated with the huge adoption of AI-centric business solutions for give full control to data-driven systems to make decisions. Machine learning (ML) models and deep learning frameworks are needed to simulate large datasets and explain why clients arrive at these decisions.

To make the work of AI-centric modeling and predictive analysis even more transparent, visualization graphs of business scenarios are introduced in this book. We also demonstrate how a variety of technologies can be used to integrate data fabric solutions and how intelligent applications can be used to enhance the effect of combining the six fields of AI-centric modeling, data analytics, ML-aided models, IoT-driven applications, cybersecurity techniques, and cloud-enabled platforms for developing smart business systems in the era of AI and data science.

Happy reading!

**Editors: Alex Khang, Vugar Abdullayev,
Babasaheb Jadhav, Shashi Kant Gupta, and Gilbert Morris**

Acknowledgments

This book is based on the design and implementation of artificial intelligence (AI), AI modeling, machine learning, data science, big data solutions, cloud platforms, and cybersecurity technology in business and production.

The passion and noble goal of the editorial team was to prepare and design a book that could be introduced to readers worldwide. The team also wanted to transform their ideas into reality and to make the book more successful. The biggest reward, however, is the effort, experience, enthusiasm, and trust of contributors.

We acknowledge the tremendous support and valuable comments of all the reviewers, with whom we have had the opportunity to collaborate and monitor their hard work remotely.

We also express our deep gratitude for all the advice, support, motivation, sharing, collaboration, and inspiration we received from our faculty, contributors, educators, professors, scientists, scholars, engineers, and academic colleagues.

Finally, we are grateful to our publisher, CRC Press (Taylor & Francis Group), for their wonderful support in ensuring the timely processing of the manuscript to bring this book to the readers at the earliest possible time.

Thank you, everyone.

Editorial team: Alex Khang, Vugar Abdullayev, Babasaheb Jadhav, Shashi Kant Gupta, and Gilbert Morris

Editors

Dr. Alex Khang is a professor of information technology, artificial intelligence (AI) and data scientist, software industry expert, and the chief of technology officer (AI and Data Science Research Center) at the Global Research Institute of Technology and Engineering, North Carolina, USA. He has more than 28 years of teaching and research experience in information technology (Software Development, Database Technology, AI Engineering, Data Engineering, Data Science, Data Analytics, Internet of Things (IoT)-based Technologies, and Cloud Computing) at universities of science and technology in Vietnam, India, and USA. He has been the chair session for 20 conferences, keynote speaker for more than 25 international conclaves, an expert tech speaker for 100 seminars and webinars, an international technical board member for 10 international organizations, an editorial board member for more than five ISSNs, an international reviewer and evaluator for more than 100 journal papers, and an international examiner and evaluator for more than 15 PhD theses in computer science. He has contributed to various research activities in the fields of AI and data science while publishing many international articles in renowned journals and conference proceedings. He has published 52 authored books (in computer science, 2000–2010), two authored books (software development), and 30 book chapters. He has published 14 edited books in the fields of AI ecosystem (AI, ML, DL, robotics, data science, big data, and IoT), smart city ecosystem, healthcare ecosystem, Fintech technology, and blockchain technology (since 2020). He has over 28 years of working experience as a software product manager; data engineer; AI engineer; cloud computing architect; solution architect; software architect; database expert in foreign corporations from Germany, Sweden, the United States, and Singapore; and former CEO, former CTO, former engineering director, product manager, and senior software production consultant in multinational corporations.

Dr. Vugar Abdullayev, is an associate professor in the Computer Engineering Department at Azerbaijan State Oil and Industry University, Baku, Azerbaijan. He completed his PhD in computer science in 2005. He has been the author of 61 scientific papers. His research is related to the study of cyber physical systems, IoT, big data, smart city, and information technologies. He has published four book chapters and two edited books (calling for book chapters—Taylor & Francis) in the healthcare ecosystem.

Dr. Babasaheb Jadhav is an associate professor in the area of finance and international business at Dr. D. Y. Patil Vidyapeeth (Deemed to be University), Global Business School and Research Centre, Sant Tukaram Nagar, Pimpri, Pune 411018 (Maharashtra), India. He has more than 12 years of experience in industry, research, and academia. He has completed his PhD in financial management from Savitribai Phule Pune University, formerly Pune University. He has a dual master's degree in MBA with a first class and bachelor's degree in BBA with a first class. He is a PhD research guide at Dr. D. Y. Patil Vidyapeeth. He has one funded research project, two patents, six copyrights, seven books, more than 50 research papers, case studies in various national and international indexed journals with high impact factors, and four articles in reputed magazines as well as in national newspapers to his credit. He has accolades, such as research excellence awards, best research paper awards, BOS member, BOE member, editorial board member, reviewer, and advisory board member for various management journals or conferences of repute. He has expertise as an NAAC coordinator, an NBA coordinator, and controller of examinations. He has been invited as a resource person for FDP, guest lectures, workshops, seminars, and conferences. His areas of interest and research include financial management, taxation, economics, general management, and international business management.

Dr. Shashi Kant Gupta is a director, researcher, and independent academic scholar in the Research Department of CREP, Lucknow, Uttar Pradesh, India. He has completed his PhD in CSE from Integral University, Lucknow, Uttar Pradesh, India, and worked as an assistant professor in the Department of Computer Science and Engineering, PSIT, Kanpur, Uttar Pradesh, India. He has published many research papers in reputed international journals with SCOPUS- and ESCI-indexed journals. He has published many papers at national and international conferences and seminars. He is the founder and current CEO of CREP in Lucknow, Uttar Pradesh, India. He has been a member of the *IEEE Spectrum* and *IEEE Potentials Magazine* since 2019 and many international organizations for research activities. He organized various faculty development programs, seminars, workshops, and short-term courses at the university level. His main research focuses on performance enhancement through cloud computing, big data analytics, IoT, and computational intelligence-based education. He is currently working as a reviewer in several international journals. He has published many Indian and Australian patents in the fields of information technology, computer science, and management. He has published more than 15 Indian patents, and one patent is under grant approval. He has already granted German patents. He had more than 10 years of teaching experience and 2 years of industrial experience.

H.E. Ambassador Professor Gilbert Morris is National Public Reader of the Bahamas appointed by the Prime Minister. He is Ambassador-at-Large and Scholar-in-Residence at the Bahamas Foreign Services Institute. Morris is also one of the world's leading thinkers on the global financial system and specialist in econometrics and methodologies of development. Professor Morris taught at George Mason University from 1994 to 2001, where he taught in four faculties and was awarded the "Technology-Across-The-Curriculum Prize". He was lecturer on "The History of Revolutions: Political, Social, Scientific and Technological" for the Smithsonian Associates at the Smithsonian Institute, Washington, DC, and he was a member of the Mid-Atlantic Scholars Association where he presented on "The Quotidian Psychology and Neuroscience of Race" at Princeton University. Professor Morris also lectured on "Pedagogical Neuroscience" at the Alain Locke Institute at the University of California, Santa Barbara. Currently,

Professor Morris focuses his research on cognitive neuropsychology with an emphasis on incentives, rationality, and sense-making in political, economic, and social systems, with further concentration on the impact of scaled technologies. Morris studied at the Harvard University Extension School, Mansfield College, Oxford University (IBIS), Oxford Brooks University, Cadmus Law College, The London School of Economics, and the University of London. His focus was on law, social science and technology, econometrics, logic, scientific methodologies, and neuroscience. Morris has written widely on subjects in his areas of work, including a New York Times bestseller, "Rescue America", which applied pediatric neuroscience to child development. He wrote dozens of journal articles on law and finance and reviews for the University Bookman. Morris's forthcoming book, *Friston's Ontology*, discusses the theories of the world's leading neuroscientist, Professor Karl Friston, and the application of neuroscience to artificial intelligence-driven social systems.

Contributors

Amruthamsh A.
Department of Computer Science &
Engineering
Dr. Ambedkar Institute of Technology
Gnana Bharathi Bengaluru
Karnataka, India

Reethika A.
Sri Ramakrishna Engineering College,
NGGO Colony
Vattamalaipalayam, Coimbatore
Tamil Nadu, India

Santhosh Kumar A. V.
Department of Computer Science and
Engineering
Kuppam Engineering College, Kuppam,
Ekarlapalle
Andhra Pradesh, India

Vugar Abdullayev
Computer Engineering Department
Azerbaijan State Oil and Industry
University
Baku, Azerbaijan

Nur Aeni
Universitas Negeri Makassar
South Sulawesi
Unismuh Makassar, Indonesia

Saminu Akintunde
Nigeria Police Force, Shehu Shagari
Way
Abuja, Nigeria

Aniverthy Amrutesh
Department of Computer Science &
Engineering
Dr. Ambedkar Institute of Technology
Gnana Bharathi Bengaluru
Karnataka, India

Harishchander Anandaram
Centre for Computational Engineering
& Networking
Amrita School of Engineering
Ettimadai, Coimbatore
Tamil Nadu, India

Oluwasefunmi T. Arogundade
Department of Computer Science
Federal University of Agriculture,
Abeokuta
Ogun State, Nigeria

Adekunle M. Azeez
Department of Networking and Systems
Security
Elerinmosa Institute of Technology,
Unnamed Road Erin Osun
Osun, Nigeria

Mammadova Bilqeyis Azer
Azerbaijan State Oil and Industry
University
Baku, Azerbaijan

Salil Bharany
Department of Computer Engineering &
Technology
Guru Nanak Dev University
Makka Singh Colony, Amritsar
Punjab, India

Gowtham Bhat C. G.
Department of Computer Science &
Engineering
Dr. Ambedkar Institute of Technology
Gnana Bharathi Bengaluru
Karnataka, India

Luís Cardoso
Polytechnic Institute of Portalegre
Praça do Município
Portalegre, Portugal

Sajjan Choudhuri
SRM University
Delhi-NCR
Rajiv Gandhi Education City
Haryana, India

Binayak Dihudi
Binayak Dihudi, Department of
    Mathematics
Konark Institute of Science and
    Technology
Jatni, Bhubaneswar
Odisha, India

Surrya Prakash Dillibabu
Department of Mech Engg,
    Vel Tech
Rangarajan Dr. Sagunthala, R&D
    Institute of Science and Tech
Avadi, Chennai
Tamil Nadu, India

Akshaya E.
KPR Institute of Engineering and
    Technology
Arasur, Uthupalayam
Tamil Nadu, India

Akanksha Goel
Dr. D. Y. Patil School of Science &
    Technology
Dr. D. Y. Patil Vidyapeeth
Sant Tukaram Nagar Pimpri
Pune, India

Shashi Kant Gupta
Research department
CREP, Lucknow
Uttar Pradesh, India

Pragati Hiwarkar
NKC Inurture
Mumbai Govindji Shroff Marg
Malad West Mumbai
Maharashtra, India

Babasaheb Jadhav
Global Business School, and Research
    Centre
Tathawade Pune, Pimpri-Chinchwad
Maharashtra, India

Sujeet Kumar Jha
Sujeet Kumar Jha, IOE
    Pulchowk Campus, Tribhuvan
    University
Kathmandu, Nepal

Agbaegbu Johnbosco
Department of Computer Science
Federal University of Agriculture
Abeokuta, Ogun State, Nigeria

Asha Rani K. P.
Department of Computer Science &
    Engineering
Dr. Ambedkar Institute
    of Technology
Gnana Bharathi Bengaluru
Karnataka, India

Kiranbir Kaur
Department of Computer Engineering &
    Technology
Guru Nanak Dev University
Makka Singh Colony
Amritsar Punjab, India

Alex Khang
Global Research Institute
    of Technology and Engineering
Fort Raleigh, North Carolina,
    United States

Salahddine Krit
Department of Computer Science,
    Polydisciplinary Faculty of
    Ouarzazate
Polydisciplinary Faculty
    of uarszazate
Université Ibn Zohr, Morocco

Ashish Kulkarni
Dr. D. Y. Patil B-School
  Sr. No. 87-88
Bengaluru-Mumbai Express Bypass
Tathawade, Pune
Maharashtra, India

Pooja Kulkarni
Vishwakarma University
Laxmi Nagar, Betal Nagar,
  Kondhwa
Pune, Maharashtra, India

Sagar Kulkarni
Research Scholar, MIT World Peace
  University
Kothrud, Pune
Maharashtra, India

Dilip Kumar
Welcomgroup Graduate School of Hotel
  Administration
Manipal Academy of Higher Education
Madhav Nagar, Manipal
Karnataka, India

Radhika Kumari
Department of Computer Engineering &
  Technology
Guru Nanak Dev University
Makka Singh Colony
Amritsar, India

Mily Lal
Dr. D. Y. Patil School of Science &
  Technology
Dr. D. Y. Patil Vidyapeeth
Pune, Maharashtra, India

Muthukannan M.
Department of Civil Engineering,
  Kalasalingam Academy of Research
  and Education
Krishnankoil, Srivilliputhur
Tamil Nadu, India

Murali Dhar M. S.
Vel Tech Rangarajan Dr. Sagunthala
  R&D Institute of Science &
  Technology
Avadi, Chennai
Tamil Nadu, India

Nidya M. S.
School of Computer
  Science & Information
  Technology
Jain (Deemed-to-be University),
  Bengaluru
Karnataka, India

Qaffarova Zeynab Mehman
Azerbaijan State Oil and Industry
  University
Baku, Azerbaijan

Elmina Gadirova Musrat
Ecological Chemistry Department
Baku State University
Baku, Azerbaijan

Muthmainnah
Universitas Al Asyariah Mandar
  Sulawesi Barat
Madatte, Polewali
Polewali Mandar Regency
West Sulawesi, Indonesia

Suresh Kumar N.
Department of Computer Science and
  Engineering, School of Computing
Kalasalingam Academy of
  Research and Education,
  Krishnankoil
Tamil Nadu, India

Taiwo O. Olaleye
Department of Computer Science
Federal University of Agriculture,
  Abeokuta
Ogun State, Nigeria

Azeez A. Opatunji
Department of Computer Hardware
   Engineering
Elerinmosa Institute of Technology
Osun, Nigeria

Bhargavi Devi P.
Alpha Arts & Science College
Porur, Chennai, Adayalampattu
Tamil Nadu, India

Kanaga Priya P.
KPR Institute of Engineering and
   Technology
Arasur, Uthupalayam
Tamil Nadu, India

Prasant P.
Khalsa College of Engineering &
   Technology
Ranjit Avenue, Amritsar
Punjab, India

Padmavathi Pragada
GMR Institute of Technology
GMR Nagar, Razam
Andhra Pradesh, India

Danush R.
KPR Institute of Engineering and
   Technology
Arasur, Uthupalayam
Tamil Nadu, India

Kanniga Devi R.
School of Computer Science and
   Engineering
VIT, Chennai Campus
Vandalur – Kelambakkam Road
Chennai, India

Vanitha R
Department of Software Engineering
PMIST, Periyar Nagar
Vallam, Thanjavur
Tamil Nadu, India

Rajesh Kumar Rai
Madhyanchal Professional University
   (MPU)
Bhopa, Ratibad
Madhya Pradesh, India

Yerra Shankar Rao
Department of Mathematics, NIST
   Institute of Science and
   Technology
Pallur Hills, Berhampur
Odisha, India

Gowrishankar S.
Department of Computer Science &
   Engineering
Dr. Ambedkar Institute of
   Technology
Gnana Bharathi Bengaluru
Karnataka, India

Poongodi S.
Department of Electronics and
   Communication Engineering
CMR Engineering College,
   Hyderabad
Medchal, Kandlakoya, Seethariguda
Telangana, India

Olayemi O. Sadare
Computer Science Department, Osun
   State University
Main Campus
Osogbo, Nigeria

Abhinav Kumar Shandilya
Department of Hotel Management and
   Catering Technology
BIT Mesra Ranchi
Ranchi, Jharkhand, India

Nipun Sharma
Presidency University, Yelahanka,
   Bengaluru
Karnataka
Bangalore, India

Swati Sharma
MIE, MISTE, MIAENG, MSCRS,
Presidency University
Yelahanka, Bengaluru
Karnataka
Bangalore, India

Vivek Sharma
NKC Inurture
Govindji Shroff Marg
Malad West, Mumbai
Maharashtra, India

Arvind Kumar Shukla
Department of Computer Applications
School of Computer Science &
Applications
IFTM University, Moradabad, Lodhipur
Rajput
Uttar Pradesh, India

Ayobami A. Tewogbade
Department of Software Engineering
Elerinmosa Institute of Technology
Osun, Nigeria

Triwiyanto
Department of Medical Electronics
Technology
Poltekkes Kemenkes, Kertajaya, Kec.
Gubeng, Surabaya
Jawa Timur, Indonesia

Ahmad Al Yakin
Universitas Al Asyariah Madar
West Sulawesi, Indonesia

Muhammad Yunus
University of Muslim, Panaikang, Kec.
Panakkukang, Kota Makassar
Sulawesi Selatan, Indonesia

1 Artificial Intelligence-Based Model and Applications in Business Decision-Making

Babasaheb Jadhav, Alex Khang, Ashish Kulkarni, Pooja Kulkarni, and Sagar Kulkarni

1.1 INTRODUCTION

Data is the buzzword of today's technology-based era, and almost everyone is running behind, capturing data from every individual to provide technical support and make their lives easy. Captured data are used to create innovations or products that can help individuals make decisions related to their challenges.

The simplest examples are healthcare applications that check daily footsteps, blood pressure, pulse rate, oxygen level, stress level, etc. Data are captured either through a mobile-based app or smartwatch, and the application installed on a mobile device or computer shows graphs of the daily and monthly data based on height and weight measurements that we input. Further, these apps store/generate a lot of data from us and show a health dashboard that can further predict health issues (Khang et al., 2023d).

The decision-making process, which is based purely on data gathered from various sources and used to resolve a problem, is called a data-centric model. The organization creates a mechanism to resolve the issues that occur in its processes. The challenges or issues can be as simple as approving the leave request of an employee, or as large as launching a new product in the market.

All problems, from simple to complex, can be solved based on the proper analysis of data gathered through various stakeholders, as shown in Figure 1.1. For example, Google cuts the attrition rate by half by conducting surveys on the workforce, better managers, and talent pipeline, and analysis of the survey helps improve the quality of 75% of their lowest-performing managers. This is the power of data (UИSCЯAMBL, 2021).

Data analysis is performed by simple calculations, either manually using the human brain or on paper, or by using any computer-generated program. The answer may lead to another problem or may become useful in another situation. With the connectivity, feasibility, and availability provided by the Internet, many organizations rely on computer systems for data collection.

DOI: 10.1201/9781003400110-1

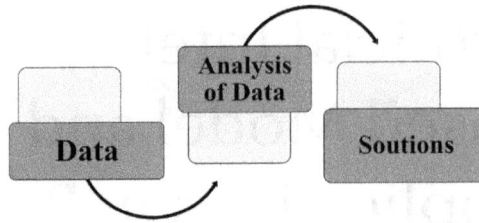

FIGURE 1.1   General Approach toward Problem-Solving.

Source: Khang (2021)

In such cases, Google and Microsoft have created many applications that can be used to collect information from various sources. Tools, like Microsoft Excel, provide some kind of pre-processing before pre-processing the data and preparing the data for analysis.

Further advancements in computer applications provide analysis tools that provide hands-on information for deciding with some level of forecasting.

Many organizations are product-based with a common domain, such as marketing, finance, human resources, operations, and supply chain management. From a broader perspective, all organizations are working on the common goal of making a profit; hence, it is assumed that organizations may face similar issues or may be in different circumstances. Therefore, to make decisions, organizations develop cases along with data. Such outcomes, results, or decisions based on this situation lead to the development of the framework/model, which may be used at a certain level of issues with some set of data.

Many studies have attempted to adopt this model and create computer application-based platforms where data can be loaded, the situation can be set, and the result can be tested before it is applied. The best example of this approach is cloud computing, in which we can share software and infrastructure as a service and test our results (Rani et al., 2023).

Furthermore, if this model is applied with artificial intelligence (AI), then the solution can be applied directly and tested on a set of data. Hence, in the current era, the AI-centric model is preferred for business decision-making (Khang et al., 2023a).

1.2   LITERATURE SURVEY

The authors of the article, "AI in Strategic Marketing Decision-Making" have identified a research gap in the application of AI to strategic marketing decision-making. They emphasized the need to conduct more research in this field to enable the effective implementation of AI in strategic marketing decision-making (Stone et al., 2020). This requires researchers to collect and analyze more data to develop suitable models for decision-making (Khang et al., 2023b). Overall, this study highlights the need for further investigation into the use of AI in strategic marketing and its potential benefits.

Nayal (2022) explored the potential role of AI in the formulation of marketing strategies. The author suggests that AI can be effectively used to respond to external contingencies, such as high volumes of data, uncertain environmental conditions, and limited managerial cognition. By using AI to analyze data and generate insights, marketers can create information-rich and effective strategies.

Gang (2007) created a comprehensive conceptual framework for AI adoption in B2B marketing. According to the author, the drivers of AI adoption are current marketing shortcomings and the external pressures caused by information.

The author also identified seven outcomes of AI adoption, including efficiency improvements, accuracy improvements, better decision-making, improved customer relationships, increased sales, cost reductions, and risk reductions.

The author utilized information processing theory and organizational learning theory to create an integrated conceptual framework that explains the relationship between each AI adoption construct in B2B marketing.

Meyer (2020) explored how AI can be utilized by firms to achieve competitive advantage in the market and make informed decisions. The study focused on the role of computer-mediated AI agents in detecting crises related to events in a firm.

Because crises can significantly impact organizational performance, the study proposes a structural model that employs statistical and sentimental big data analytics to detect critical events related to business activities. The research findings suggest that analyzing day-to-day data communications, such as email communications, can provide valuable insights for the timely detection and effective management of crises.

Overall, the study highlights the potential of AI in crisis management and emphasizes the importance of data-driven strategies for efficient and effective decision-making in the B2B market. The study provides valuable insights for firms seeking to leverage AI to enhance their market engagement and competitive advantage.

López Jiménez (2021) aimed to understand the current and growing influence of AI and automation in the communication industry and to identify the necessary skills and training required to cope with its effects.

The findings suggest that communication professionals need two types of training: first, to gain experience with current AI and automated tools, and second, to focus on developing human qualities that AI cannot replicate.

Overall, the study emphasizes the importance of addressing the impact of AI and automation on the communication industry and recommends the development of human skills in tandem with technological advancements. This study provides valuable insights for communication professionals seeking to adapt to the changing landscape of their industry.

1.3 ARTIFICIAL INTELLIGENCE-CENTRIC BUSINESS MODELS

Data considered to be a focal point for all events can be collected from multiple sources based on the utilization of social media, health applications, delivery applications (food, grocery, medicine, etc.), and many more applications currently generating a huge amount of data (Khang et al., 2023c).

Emerging organizations can capture these data and generate more entrepreneurial opportunities. Computer software developers consider this as an opportunity to learn the system through machine learning (ML) and create an artificially based application that can either be a new product or service for creating some new AI-based applications, as shown in Figure 1.2.

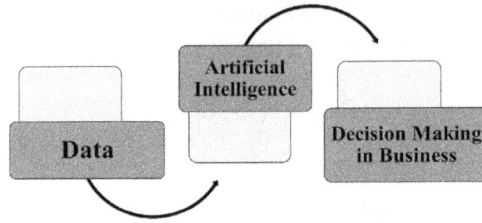

FIGURE 1.2 AI-Centric Process of Decision-Making.

Source: Khang (2021)

AI refers to the ability of a system to learn based on algorithms. Further, some technological fronts help AI in decision-making, namely, deep learning and big data (Bhambri et al., 2022).

In today's complex era, the problems are not simple but much more complex, and the data required to make decisions are in a completely unstructured format. Therefore, data in an unstructured format must have some pre-processing and a similar format form, where they can be inputted into the ML algorithm. The tools used to collect data are enterprise resource planning, data captured through various applications such as Excel, comma-separated values, and third-party tools such as NoSQL cloud services (Rani et al., 2023).

The collected data are required to be formatted using the extract, transform, and load (ETL) processes. The ETL reports must be inputted into an algorithm that is developed using any language, such as Python, R, or Java.

The data must be provided for AI processes to handle anomaly detection, and the algorithm may forecast decisions that may be directly applied based on human interventions. The AI-centric process is illustrated in Figure 1.3.

Artificial intelligence can make decisions, but the decisions are not only based on facts and figures but also on the organization's vision, mission, and long-term and short-term strategies that can be analyzed by human beings only.

Further, AI requires data input to process and obtain the desired output, but how the data will be gathered or generated remains a question for everyone. To test a program or system application for decision-making, it is essential to understand the input process and desired output.

The decision-making process mostly requires structural data, and every organization has processes to gather structural data. At present, data are collected over computer systems by either sharing Google Forms or questionnaires, or captured through Android/iOS-based apps.

The data gathered or captured through third-party tools are mostly structured data. Based on the collected data, the ETL processes are applied to convert these data into meaningful outcomes. Once data are extracted from all sources, they are then analyzed through AI models, and based on the suitable model, the result is revealed.

This is the mechanism for one problem that may be resolved through the presented model, but as the system talks, the output of one problem can be the input to another. Thus, it needs to be stored in a database or dataset for further use.

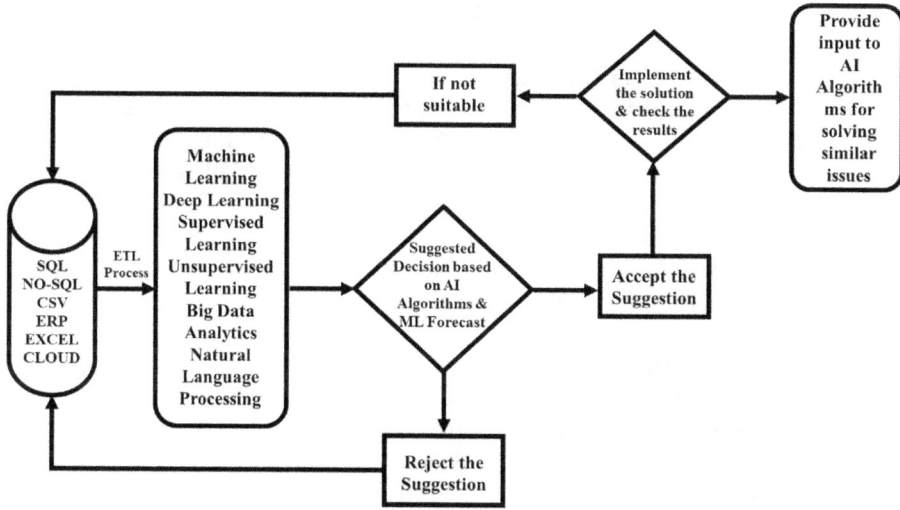

FIGURE 1.3 Process of AI Decision-Making.

Source: Khang (2021)

If a similar problem occurs, the same process may be repeated and the output can be tested. Because this process normally runs in the cloud, the input and output datasets are stored in the cloud, and the cloud can use such a model to solve similar issues in organizations.

There are many examples where AI models are used in decision-making, such as in healthcare, retail, manufacturing, banking, finance, and many other sectors (Praveen et al., 2023). Some of the cases are as follows:

- Early-stage cancer detection requires a CT scan, and a radiologist is responsible for checking the CT scan report and verifying whether cancer cells are found. China trends the system using all available images, and the AI algorithm through image processing can directly diagnose cancer cells.
- AI can help scrutinize job applications received from various sources and can choose the best-fit candidate based on keywords inputted in the job description as well as on keywords available in a resume or cover letter. It reduces the manual tasks of an HR manager and provides a robust work environment.
- AI based algorithms study real-time data and trend analysis of geographical locations, and they provide a platform for marketing people to decide which strategies to use to launch new products or market an existing product.

Nowadays, life cycle project management in all leading organizations is addressed by AI applications.

Chatbots and other applications can help predict the pattern of searches made by users to predict the products and services required in geographical locations.

There are more cases in which users are convinced that AI can capture preferences, likes, and dislikes to create processes that can simplify their lives.

For example, a taxi driver installs a mobile-based application that can provide a real-time map-based location for the taxi. An application installed by a parent can check whether the child is in the classroom or just wandering outside.

AI based and Internet of Things-supported applications can ease the lives of users and make them competent to decide whether to continue with the services or switch to others (Rani et al., 2021).

AI is not only a buzzword or any special feature of computer science; it collaborates with other computer science areas, such as ML, deep learning, big data analytics, and natural language processing. Artificial intelligence uses multiple programming languages and supports multiple tools to record data as well as to build an algorithm to create a model based on data.

AI has three phases: assisted intelligence, where insights of data are shared with a human being to make decisions; augmented intelligence, where ML capabilities are utilized to generate the outcome; and full automation, where inputs, processes, outputs, and decision-making are taken care of by AI (Dordevic, 2022).

AI decision-making is aware of the process of measuring data with or without the support of human beings to make accurate decisions for a complex problem. In addition, AI can handle big data, data collected from different sources, current trends, anomalies, and data crunching to generate final decisions with or without human intervention (Khang et al., 2023e).

AI creates a model based on all the clusters required to solve any problem; a cluster includes technology and data sources. Based on the model, production is generated, and a decision is made. The structural model layout is shown in Figure 1.4.

An AI cluster provides support to decisions, argumentation to justify the solution, or automates the solutions. In all three parameters, decision-making requires two important parameters: the time frame and complexity of the problem.

The time frame is required to check a new opportunity, issue, or thread in business, whereas complexity determines the situation.

An AI cluster can decide on data handling and learning from its information capability. Learning capabilities enable models to be built based on the data collection. These models can be used to determine the available data and information.

The simplest example is an e-commerce site where any product can be purchased. The AI application recommends another product based on the purchased product. This strategy or process increases the probability of another product being purchased.

AI is used in decision-making to make error-free processing, 24/7 availability of business models, and the right decision at the right time, as well as to enhance the existing market or to grab a new market. However, AI in decision-making has some issues, including high cost, lack of creativity, and lack of human replication.

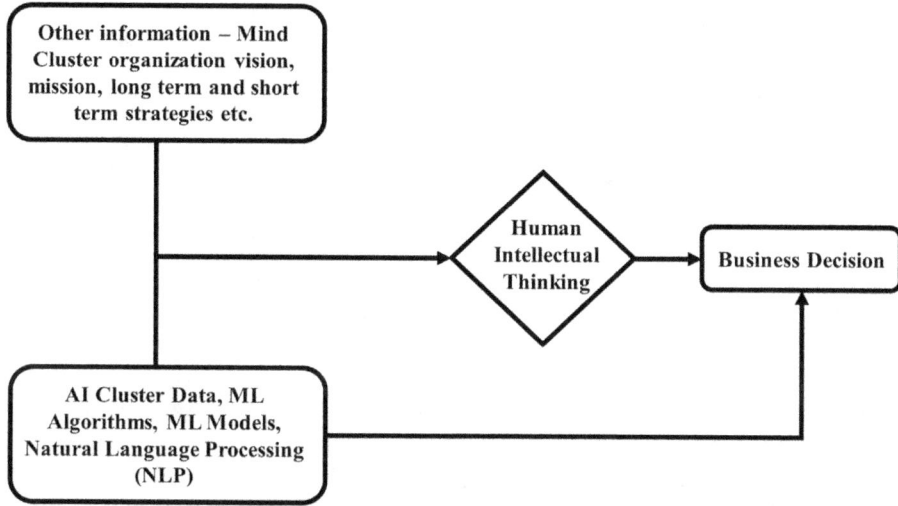

FIGURE 1.4 Role of AI in Decision-Making.

1.4 TOOLS FOR SUPPORTING BUSINESS DECISION-MAKING

Decision-making is a critical process that requires support tools. ChatGPT is one of the tools that provide hands-on information as well as many processes that can make the decision-making process easier.

Business analytics helps decision makers with all the necessary information required in making decisions, such as online analytical processing (OLAP), advanced analytics, data warehouses, and data marts. Data mining helps decision-making through business intelligence (BI) tools, such as OLAP, structured query language (SQL), NoSQL, and BI reporting tools.

AI powered by big data, AI, and machine learning (ML) is used in decision-making. Artificial intelligence-based tools create a configuration based on data, and then collect, synchronize, and analyze the data by transforming them into the desired format. The next step applies some rules as per the desired output and then solves the problem to support decision-making for the people working in the organization.

The most important software tool based on the presented methods is the strengths, weaknesses, opportunities, and threats diagram. It helps divide the data or information into four quadrants, namely, strength, weakness, opportunity, and thread, and then solves the problem.

Other tools include the decision matrix, Pareto analysis, and many more that can identify the problem, gather information, identify alternatives, gather evidence, choose among the alternatives, and make and review decisions.

1.4.1 ARTIFICIAL INTELLIGENCE-CENTRIC TOOLS FOR SUPPORTING BUSINESS DECISION-MAKING

Artificial intelligence always involves big data, ML, and neural networks. These tools are utilized to make decisions for complex scenarios: if an organization tries to launch a new product or enter a new market, it needs to analyze the priority of the market, which is the optimal solution for the current set of people; investigate the need of the locality; forecast a solution; and experiment with the result. The AI tool optimizes predictions and risk management related to various problems in finding a new market segment. There are other issues in which AI tools can provide solutions for decision makers to conduct experiments in the market.

Another area where AI contributes is strategic change, in which the change in the strategic decision may collapse the entire ecosystem, and the strategic level requires complete planning, restricting management, reducing operational delays, and improving quality. Artificial intelligence can make these changes faster and safer.

Artificial intelligence tools are very useful for customer management; the chatbot application provides a different level of user experience.

In addition, AI is trying to be personalized to all users by predicting the trends and patterns that can be used not only for customer stratification but also at the level of decision-making, as the customer is the most important stakeholder of any business performance management or assistance, which is another area where AI tools help decision makers estimate which tactics work and which do not make decisions on how to make adjustments.

With the support of AI tools, people can make small decisions to solve complex problems, initiate strategic changes, evaluate risks, and assess strategies to enhance business performance.

1.4.2 CHATGPT FOR BUSINESS DECISION-MAKING

ChatGPT is the buzzword that many people use to collect data for various processes. It is AI-based software that can search results based on patterns generated through deep learning algorithms. ChatGPT uses search histories as input criteria and provides results in terms of the problem statement.

The search may be based on any real-time issue/problem faced by an organization, such as finding tools that can be used to increase profit, promoting a strategy for a specific product, and any other real-time problem that can lead to data generation based on the data that the user can think and make decisions. The function of ChatGPT is shown in Figure 1.5.

ChatGPT allows businesses to automate many services, such as customer relationship services. It can also be used to provide modified customer service, as it understands the context of conversations and can provide tailored responses. This helps businesses save time and resources, as they do not have to manually respond to each customer inquiry.

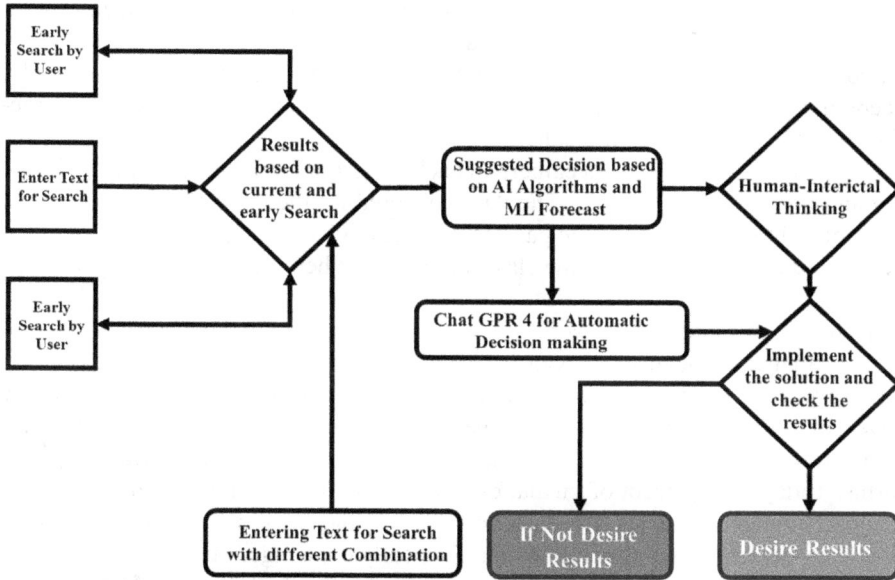

FIGURE 1.5 Role of ChatGPT in Decision-Making.

Source: Khang (2021)

1.4.3 Example (Case Study) for Business Decision-Making

The present disclosure proposes an AI-based online teaching system for physically challenged students.

- The AI-based online teaching system provides easy access to education for physically challenged students and enables them to actively participate in online classes.
- The AI-based online teaching system comprises at least two camera modules, a sign-language recognition module, a communication conversion module, and a pointing device.
- The AI-based online teaching system is an application suitable for conducting online classes using various electronic devices, such as a mobile phone, laptop, and computer with a webcam.
- The proposed AI-based online teaching system monitors the health parameters of students and alerts families and hospitals if any abnormality is detected (Divvela et al., 2023).

Because of the COVID-19 outbreak, an unprecedented number of children, youth, and adults were not able to physically attend schools or universities. Many educational institutions have opted for online tutoring or classes to minimize educational disruption and facilitate the continuity of learning.

However, even online teaching had many issues. Teachers had difficulty monitoring and interacting with students. Teachers were unable to obtain accurate information on whether students could understand lessons or not. Students had difficulty keeping up with online courses, as they lost interest because there was less interaction between them and their teachers.

The existing system uses a camera in each classroom to monitor students' learning and behavior during class. The data collected by the existing system are then transmitted to the teacher or the parents' mobile device through a software application. The system also records the class attendance when the camera recognizes the students' faces.

However, this system only works if the students would go to the school, as the camera that analyzes students' behavior is fixed above the classroom. Proper class monitoring and interaction are major issues faced by both teachers and students. With the advent of e-learning, education is no longer difficult for physically challenged students.

Many colleges and universities have started conducting online learning platforms. Today, the concept of virtual campuses is evolving rapidly, and more physically challenged students are enrolling. Technology also plays a pivotal role in providing perfect learning solutions for physically challenged students.

Despite the existence of online learning, physically challenged students are still struggling. According to a study (designed and developed by Babasaheb Jadhav, Alex Khang, Ashish Kulkarni, Pooja Kulkarni and Sagar Kulkarni), about 56.5% of physically challenged students are struggling with online classes, while around 77% cannot attend these classes and fall behind in learning due to lack of access to distance learning methods. Teachers cannot also provide equal attention to all students.

In addition, 86% of students and parents do not have sufficient knowledge of how to use technology. Therefore, there is a need for an AI-based online teaching system for physically challenged students to monitor and aid them in online classes.

- There is a need for an AI-based online teaching system that monitors the health parameters of students and alerts families and hospitals if any abnormality is detected.
- There is a need for an AI-based online teaching system that converts classes into an appropriate format that is understandable to everyone attending classes on the same platform.
- There is a need for an AI-based online teaching system that is suitable for conducting online classes using various electronic devices, such as mobile phones, laptops, and computers with webcams.

1.5 CONCLUSION

The current era is considered the most competitive era in which we need to be updated to meet customer expectations from time to time. Artificial intelligence plays a vital role in keeping us updated and enhancing competitive advantage.

Computer technology enhances the decision-making processes in industrial organizations and reduces manual tasks. An AI-based chatbot application is one example where users can interact more frequently to solve their queries.

Similarly, like chatbots, multiple AI-based software are created, designed, and developed to increase productivity, efficiency, effectiveness, speed in making decisions, profitability, rate of customer service, ideas for marketing strategies, and accuracy rates of sentiment analyses and sales forecasts.

Artificial intelligence is a very powerful tool for business decision-making. It has significant core values and attributes in decision-making as well as in creating a business ecosystem that can lead to decision-making (Rana et al., 2021).

The suggestions provided by AI help to capture and test solutions in complex environments. Many AI systems are cloud-based; AI as a cloud service can lead to shared decision-making with similar peripherals (Khanh & Khang, 2021).

REFERENCES

Bhambri P., Rani S., Gupta G., Khang A., (2022). *Cloud and Fog Computing Platforms for Internet of Things*. CRC Press. https://doi.org/10.1201/9781003213888

Divvela V. S. K., Chaurasia R., Misra A., Misra P. K., Khang A., (2023). "Heart disease and liver disease prediction using machine learning," In *Data-Centric AI Solutions and Emerging Technologies in the Healthcare Ecosystem*, p. 4 (1st ed.). CRC Press. https://doi.org/10.1201/9781003356189-13

Dordevic M., (Aug 23, 2022). *How Artificial Intelligence Can Improve Organizational Decision Making*. www.forbes.com/sites/forbestechcouncil/2022/08/23/how-artificial-intelligence-can-improve-organizational-decision-making/

Khang A., (2021). "Material4Studies," *Material of Computer Science, Artificial Intelligence, Data Science, IoT, Blockchain, Cloud, Metaverse, Cybersecurity for Studies*. www.researchgate.net/publication/370156102_Material4Studies

Khang A., Gupta S. K., Hajimahmud V. A., Babasaheb J., Morris G., (2023a). *AI-Centric Modelling and Analytics: Concepts, Designs, Technologies, and Applications* (1st ed.). CRC Press. https://doi.org/10.1201/9781003400110

Khang A., Rani S., Gujrati R., Uygun H., Gupta S. K., (Eds.). (2023b). *Designing Workforce Management Systems for Industry 4.0: Data-Centric and AI-Enabled Approaches*. CRC Press. https://doi.org/10.1201/9781003357070

Khang A., Rana G., Tailor, R. K., Hajimahmud V. A., (Eds.). (2023c). *Data-Centric AI Solutions and Emerging Technologies in the Healthcare Ecosystem*. CRC Press. https://doi.org/10.1201/9781003356189

Khang A., Hahanov V., Litvinova E., Chumachenko S., Triwiyanto, Hajimahmud V. A., Nazila Ali R., Vusala Alyar A., Anh P. T. N., (2023d). "The analytics of hospitality of hospitals in healthcare ecosystem," In *Data-Centric AI Solutions and Emerging Technologies in the Healthcare Ecosystem*, p. 4. CRC Press. https://doi.org/10.1201/9781003356189-4

Khang A., Rani S., Gujrati R., Uygun H., Gupta S. K., (Eds.). (2023e). *Designing Workforce Management Systems for Industry 4.0: Data-Centric and AI-Enabled Approaches*. CRC Press. https://doi.org/10.1201/9781003357070

Khanh H. H., Khang A., (2021). "The role of artificial intelligence in blockchain applications," *Reinventing Manufacturing and Business Processes through Artificial Intelligence*, 2(20–40) (CRC Press). https://doi.org/10.1201/9781003145011-2

López Jiménez E. A., (2021). "An exploration of the impact of artificial intelligence (AI) and automation on communication professionals," *Journal of Information, Communication, and Ethics in Society*, 19(2), 249–267. www.emerald.com/insight/content/doi/10.1108/JICES-03-2020-0034/full/html

Meyer C. C., (2020). "From automats to algorithms: The automation of services using artificial intelligence," *Journal of Service Management*, 31(2), 145–161. https://ieeexplore.ieee.org/abstract/document/8528677/

Nayal K. R., (2022). "Exploring the role of artificial intelligence in managing agricultural supply chain risk to counter the impacts of the COVID-19 pandemic," *International Journal of Logistics Management*, 33(3), 744–772. https://doi.org/10.1108/IJLM-12-2020-0493

Praveen Kumar M., Kumar N., Misra A., Khang A., (2023). "Heart disease prediction using logistic regression and random forest classifier," In *Data-Centric AI Solutions and Emerging Technologies in the Healthcare Ecosystem*, p. 6 (1st ed.). CRC Press. https://doi.org/10.1201/9781003356189-6

Rana G., Khang A., Sharma R., Goel A. K., Dubey A. K., (Eds.). (2021). *Reinventing Manufacturing and Business Processes through Artificial Intelligence*. CRC Press. https://doi.org/10.1201/9781003145011

Rani S., Bhambri P., Kataria A., Khang A., Sivaraman, A. K., (2023). *Big Data, Cloud Computing and IoT: Tools and Applications* (1st ed.). Chapman and Hall/CRC. https://doi.org/10.1201/9781003298335

Rani S., Chauhan M., Kataria A., Khang A., (Eds.). (2021). "IoT equipped intelligent distributed framework for smart healthcare systems," *Networking and Internet Architecture*, Vol. 2, p. 30. CRC Press. https://doi.org/10.48550/arXiv.2110.04997

Stone, M., Aravopoulou, E., Ekinci, Y., Evans, G., Hobbs, M., Labib, A., Laughlin, P., Machtynger, J. and Machtynger, L. (2020), "Artificial intelligence (AI) in strategic marketing decision-making: a research agenda," The Bottom Line, Vol. 33 No. 2, pp. 183–200. https://doi.org/10.1108/BL-03-2020-0022

UИСCЯAMBL, (2021). 6 inspiring examples of data-driven companies (Key Takeaways Included). *UИСCЯAMBL*. https://unscrambl.com/blog/data-driven-companies-examples/

2 Exploration of Machine Learning Models for Business Ecosystem

*Akanksha Goel, Babasaheb Jadhav,
Alex Khang, and Mily Lal*

2.1 INTRODUCTION

Industry 4.0 is a term used to describe the fourth industrial revolution, which involves the integration of advanced know-how, such as artificial intelligence (AI), robotics, and Internet of things (IoT) into manufacturing processes. Industry 4.0 is bringing significant changes to the way businesses operate and interact with their ecosystems, as shown in Figure 2.1.

Machine learning (ML) is a key technology driving these changes, enabling businesses to analyze large amounts of data and make predictions, forecast, automate processes, and optimize operations (Khang et al., 2023c).

Machine learning is a subfield of AI that focuses on developing algorithms and statistical models that enable computer vision to learn from data and make predictions or decisions without being programmed explicitly (Khang et al., 2023a).

In other words, ML involves algorithms to recognize patterns and relationships in data, and then uses that knowledge to make predictions or take actions on new data, as shown in Figure 2.2. The algorithms are designed to iteratively learn from the data, refining their performance over time and improving their accuracy (Chen et al., 2023).

The types of ML methods are as follows:

1. **Supervised learning**: The algorithms are trained on a labeled dataset, where the desired output is known. The algorithm recognizes patterns in data input and output.
2. **Semi-supervised learning**: A small amount of labeled data and a large amount of unlabeled data are used to develop a model.
3. **Unsupervised learning**: The algorithms are trained on the unlabeled dataset, where the desired output is not known. The algorithm identifies data patterns and structures without specific guidance.
4. **Reinforcement learning**: The algorithm takes actions in datasets to maximize a reward. It receives feedback on the success or failure of its actions and adjusts its behavior accordingly.

DOI: 10.1201/9781003400110-2

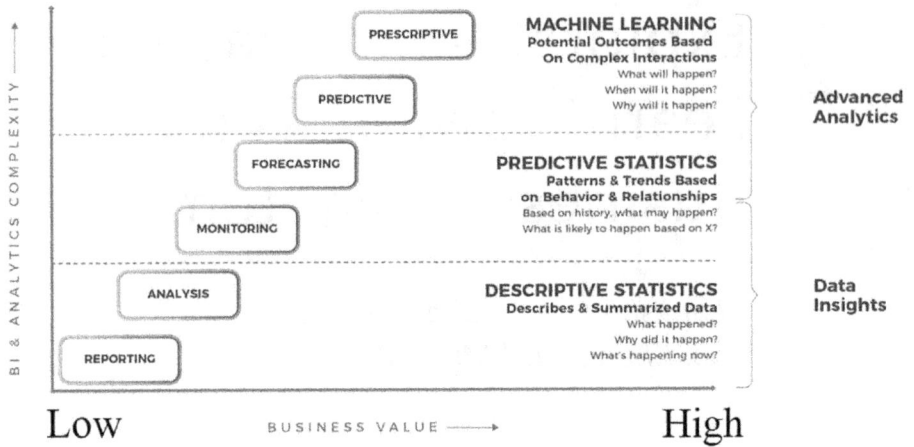

FIGURE 2.1   Role of AI in Business Transformation.

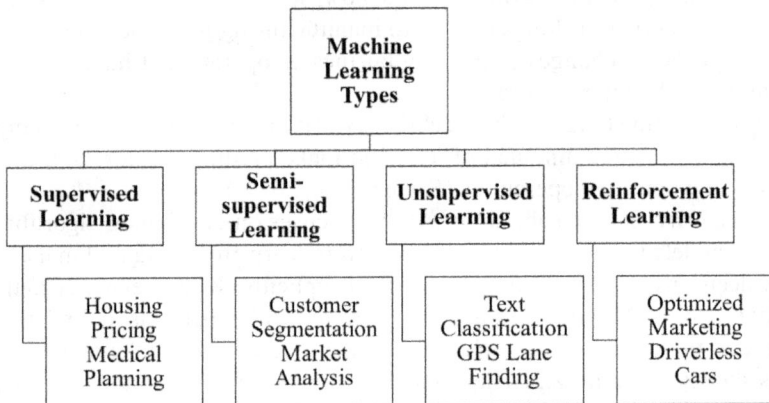

FIGURE 2.2   Types of ML.

2.2   MACHINE LEARNING IN INDUSTRY 4.0

Machine learning is used in a wide range of applications, such as face recognition, speech recognition, natural language processing (NLP), recommendation systems, fraud detection, predictive maintenance, and many others.

As data become increasingly abundant and complex, ML is becoming an essential tool for organizations looking to gain insights, optimize operations, and create valuable solutions.

Here are some ways that ML is used in Industry 4.0 to benefit businesses and their ecosystems (Mohd et al., 2022):

1. **Predictive maintenance**: ML algorithms can predict whether the equipment is likely to succeed or break down, reduce downtime, and improve reliability and safety.
2. **Quality control**: ML algorithms can analyze images of products to recognize defects and irregularities, enabling businesses to detect issues early and reduce waste.
3. **Supply chain optimization**: ML algorithms can optimize supply chain operations by analyzing data from suppliers, logistics service providers, and customers. This will result in cost reduction, minimize delivery times, and enhance customer satisfaction.
4. **Autonomous systems**: ML is driving the development of human-free systems, such as driverless cars, drones, and robots. These technologies can perform tasks independently, reducing labor costs and increasing efficiency.
5. **Energy management**: ML algorithms can analyze energy usage and identify areas of waste. It is helpful in the reduction of carbon footprint and saves cost.
6. **Customer engagement**: ML algorithms can analyze customer data for personalized marketing and recommendations, increasing customer engagement and loyalty.

Overall, ML is enabling businesses to optimize their operations, reduce costs, and improve customer satisfaction. By leveraging the power of data, businesses can make more informed decisions and create new opportunities for growth and innovation (Khang et al., 2024).

2.2.1 MACHINE LEARNING MODELS

Machine learning models have become an integral part of businesses and their ecosystems in recent years. They are used to analyze data, find patterns, and make predictions (Khang et al., 2023b). The well-known ML models used by businesses are as follows (Jamwal et al., 2022):

1. **Regression models**: These are used to predict an uninterrupted variable, such as sales or revenue, based on other variables. These models are commonly used in finance and marketing to forecast future performance.
2. **Decision trees**: These are used in decision-making based on a set of rules. These models are commonly used in finance, healthcare, and retail to make decisions about customer segmentation and risk analysis.
3. **Random forest**: It is a method that combines multiple decision trees to enhance the correctness of predictions. This model is commonly used in finance, healthcare, and e-commerce.
4. **Support vector machines**: They are used for the classification of tasks, such as predicting whether a customer will churn or not. These models are commonly used in marketing, finance, and healthcare (Vrushank & Khang, 2023).
5. **Neural networks**: They are used for complex tasks, such as face recognition and NLP. These models are used in healthcare, finance, and e-commerce.

6. **Clustering**: It is used to group similar data points. This model is commonly used in marketing and customer segmentation.
7. **Collaborative filtering**: This model is used to make recommendations based on user behavior. It is commonly used in e-commerce and entertainment.

2.2.2 TECHNOLOGY FEATURES OF MACHINE LEARNING FOR INDUSTRY 4.0

As Industry 4.0 continues to evolve, ML is likely to play an increasingly important role in optimizing operations, improving quality, and enhancing safety (Mazzei & Reshawn, 2022). The following are some of the Industry 4.0-relevant technological features of ML (Rahul et al., 2021).

1. **Big data analytics**: ML algorithms require large data to learn from, and Industry 4.0 generates vast amounts of data from sensors, machines, and other connected devices. Big data analytics is the technology that enables organizations to collect, process, analyze, and store these data to generate insights and optimize operations (Rani et al., 2023).
2. **Edge computing**: Industry 4.0 requires real-time processing and decision-making capabilities, which can be challenging to achieve with traditional cloud-based architectures (Bhambri et al., 2022). Edge computing is a technology that enables data processing and analysis to be performed locally on edge devices, such as sensors and machines, reducing latency and improving responsiveness (Hahanov et al., 2022).
3. **Artificial intelligence**: ML is a subset of AI that enables systems to learn from data and make predictions. Technologies, such as NLP, computer vision, and deep learning, are increasingly used in Industry 4.0 applications to automate tasks, improve quality, and enhance safety (Rana et al., 2021).
4. **Robotics and automation**: Industry 4.0 involves the combination of physical and digital technology, and robotics and automation play a crucial role in this integration. Machine learning algorithms can optimize robotic and automated systems, making them more efficient, flexible, and adaptive.
5. **Predictive maintenance**: ML algorithms are used to analyze data and predict maintenance if required. This enables organizations to perform maintenance proactively.
6. **Cybersecurity**: Industry 4.0 involves the integration of multiple systems, creating new cybersecurity risks. Machine learning algorithms are used to detect and prevent cyberattacks, identifying anomalies and patterns that may indicate malicious activities.

2.2.3 CHALLENGES IN INDUSTRY 4.0 USING MACHINE LEARNING

Industry Revolution 4.0 is the present trend of automation in the manufacturing industry (Jan et al., 2022). It involves advanced technologies, such as IoT, AI, and ML (Peres et al., 2020). Some of the problems in Industry 4.0 for using ML technology are (Parthasarathy et al., 2022):

1. **Data quality**: ML models require high-quality data for training and testing, and data quality can significantly impact the performance of the model. In Industry 4.0, the large amounts of data generated can be both a blessing and a curse. The quality of data collected from sensors and other IoT devices can be unreliable, unrealistic, incomplete, or inaccurate, leading to flawed ML models (Khang et al., 2022b).

2. **Security and privacy**: The integration of IoT devices in Industry 4.0 creates potential security and privacy risks. With the vast amounts of data being collected, companies need to ensure that the data are secure and not vulnerable to cyberattacks. There is also a risk of data breaches, which can result in sensitive data being exposed (Khang et al., 2022a).

3. **Model interpretability**: ML models can be complex, and it can be challenging to understand how they make predictions or decisions. This lack of interpretability can be a significant challenge in front of Industry 4.0, where safety and quality are critical factors. It can be difficult to explain the reasoning behind the decisions made by an ML model to stakeholders or regulators.

4. **Limited expertise**: ML requires specialized expertise in data science, statistics, and programming. Many companies may not have the necessary expertise in-house and may need to outsource or hire specialized personnel. The cost of hiring and training employees with the necessary expertise can be a significant challenge for small and medium-sized enterprises (SMEs).

5. **Implementation costs**: Implementing ML in Industry 4.0 can be expensive, as it requires investment in hardware, software, and personnel. This can be a significant challenge for SMEs that may not have the financial resources to implement ML in their manufacturing processes.

6. **Ethical concerns**: ML can raise ethical concerns, especially when it comes to decisions that affect humans. For example, algorithms used to make decisions about hiring or promotions can discriminate against certain groups. It is essential to address ethical concerns.

2.3 LITERATURE SURVEY

Machine learning models have revolutionized the way businesses operate in recent years. With the increasing amount of data being generated every day, businesses have started to leverage ML models to make data-driven decisions. This literature review will explore the latest advancements in ML models in the business ecosystem.

Zhang et al. (2020) provided a comprehensive overview of deep learning techniques for business intelligence applications. They discussed the potential of deep learning models in data pre-processing, feature selection, and classification tasks. They also highlighted the challenges of implementing deep learning models in real-world business applications.

Khang et al. (2023d) provided a review of ML applications in business and identified research gaps and opportunities for future research. They discussed the importance of interpretability, fairness, and ethics in ML models for business applications.

Haleem et al. (2022) analyzed the role of ML models in business analytics. They discussed the potential of ML models in customer segmentation, predictive modeling, and fraud detection. They also highlighted the importance of interpretability and ethical considerations in ML applications.

Eric et al. (2016) explored the applications of ML and big data analytics in accounting and finance. They discussed the potential of ML models in financial forecasting, risk management, and fraud detection. They highlighted the challenges of data quality, privacy, and security in ML technology.

Khang et al. (2024) discussed the potential of ML models in marketing applications. They explored the use of ML models in customer segmentation, personalized marketing, and customer churn prediction. The study further highlighted the importance of interpretability, fairness, and privacy in ML applications.

Several studies have explored the different applications of ML models in business, including customer segmentation, recommendation systems, fraud detection, sentiment analysis, and predictive analytics. For instance, Khoury et al. (2020) used ML algorithms to predict customer churn in the telecommunication industry, while Chen et al. (2021) developed a deep learning model for stock price prediction.

Studies have shown that ML models provide several benefits to businesses, including improved accuracy, efficiency, and cost savings. For instance, Vukovic et al. (2020) found that ML models improved customer satisfaction by reducing waiting times, while Tan et al. (2021) showed that ML algorithms could improve supply chain management by predicting demand and reducing inventory costs.

Despite the benefits, ML models present several challenges for businesses, including data quality, privacy concerns, and ethical issues.

For instance, Verma et al. (2021) highlighted the ethical considerations of using ML models in healthcare, while Kaur et al. (2021) discussed the challenges of developing ML models for SMEs.

Furthermore, studies have also explored the impact of ML models on the business ecosystem, including the role of data, collaboration, and innovation.

For instance, Gomes et al. (2021) discussed the importance of collaboration between academia and industry to develop ML models that meet business needs, while Soliman et al. (2021) highlighted the role of open innovation in the development of ML models.

2.4 MACHINE LEARNING FRAMEWORK FOR BUSINESS ECOSYSTEM

When it comes to using ML for business problems, it is important to have a structured and organized framework in place (Khekare et al., 2022), as shown in Figure 2.3.

Here are some steps that can be followed to create an ML framework for business problems, as shown in the following figure:

1. **Define the problem**: This involves understanding the problem statement, defining the objectives, and identifying the key performance indicators (KPIs) that will be used to measure success.

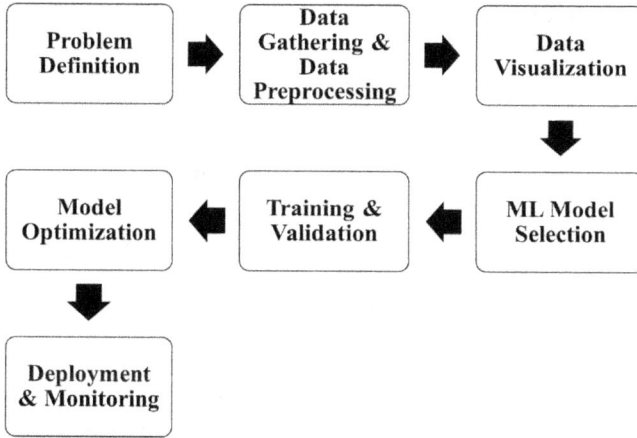

FIGURE 2.3 Machine Learning Framework for Businesses and Their Ecosystem.

2. **Gather and pre-process data**: Gather relevant data to train the ML model. This involves identifying the sources of data, collecting the data, and cleaning and pre-processing the data to ensure accuracy and usability.
3. **Explore and visualize data**: Gathered and pre-processed, data are used to explore and visualize insights, patterns, and relationships. This involves using descriptive statistics and visualization tools to understand the data and identify any trends or anomalies.
4. **Choose an ML model**: Based on the problem statement, objectives, and KPIs, a suitable ML tool needs to be adapted. This involves the identification of the problem (classification or regression), choosing an appropriate algorithm, and selecting the relevant features and hyper-parameters.
5. **Train and validate the model**: It is essential to train and validate the tool by using pre-processed data.
6. **Optimize the model**: The trained and validated model needs to be optimized to improve performance. This involves fine-tuning the hyper-parameters, selecting different features, and exploring different algorithms or architectures.
7. **Deploy and monitor the model**: Once the model has been optimized, it needs to be deployed into the production environment. This involves integrating the model into the existing business systems and processes and monitoring its performance over time to ensure it continues to meet the objectives and KPIs.

Nowadays, the trend of development of hybrid ML models that combine different algorithms and techniques to improve accuracy and performance is increasing.

These models have been applied to various business problems, such as fraud detection, customer segmentation, and demand forecasting. Some of the innovative hybrid ML models in businesses and their ecosystem are as follows:

1. **Random forest–extreme gradient boosting (RF–XGBoost) Model**: The RF–XGBoost model is a hybrid ML model that combines the strengths of random forest and extreme gradient boosting algorithms. This model has been applied in various business contexts, such as credit scoring, fraud detection, and customer segmentation.
2. **Long short-term memory–convolutional neural network (LSTM–CNN) model**: The LSTM–CNN is a hybrid deep learning model that combines the strengths of long short-term memory and convolutional neural network algorithms. This model has been applied in various business contexts, such as stock price prediction, sentiment analysis, and image recognition.
3. **Ensemble model**: Ensemble models are hybrid ML models that combine multiple algorithms and techniques to improve accuracy and performance. Ensemble models have been applied in various business contexts, such as demand forecasting, customer churn prediction, and credit scoring.
4. **Automated machine learning (AutoML)**: AutoML is a hybrid ML model that uses automation to optimize the selection and combination of algorithms and techniques. Automated machine learning has been applied in various business contexts, such as predictive maintenance, demand forecasting, and customer churn prediction.

2.5 PERFORMANCE ANALYSIS OF MACHINE LEARNING MODELS

It is critical to ensure the effectiveness and suitability of various applications. Industry 4.0 involves technologies, such as IoT, AI, and ML, in manufacturing processes.

The application of ML models in Industry 4.0 offers numerous benefits, including improved efficiency, productivity, and profitability. However, the performance of these models can be affected by several factors, including the size and complexity of the datasets, the data quality, and the algorithm choice.

Several metrics can be used to analyze the performance of ML models in the business ecosystem under Industry 4.0, including accuracy, precision, recall, etc. These metrics evaluate the models' performance in terms of their ability to correctly classify data and make accurate predictions.

Furthermore, the performance of ML models can be analyzed in terms of their speed and resource utilization. This can be done by measuring the training and prediction time for the models and assessing their resource requirements, such as memory and computational power.

Another aspect of performance analysis for ML models is their ability to become accustomed to changing conditions and learn from new datasets. This can be evaluated by testing the models on different datasets and assessing their ability to generalize to new data and adapt to changes in the input data.

The overall performance analysis of ML models in the business ecosystem under Industry 4.0 is crucial for their successful implementation and utilization (Ramesh & Nagarajan, 2019).

By analyzing their performance (Khoury et al., 2020), businesses can identify areas for improvement and optimize their use of these models to achieve their desired outcomes (Tan et al., 2021), as shown in Table 2.1.

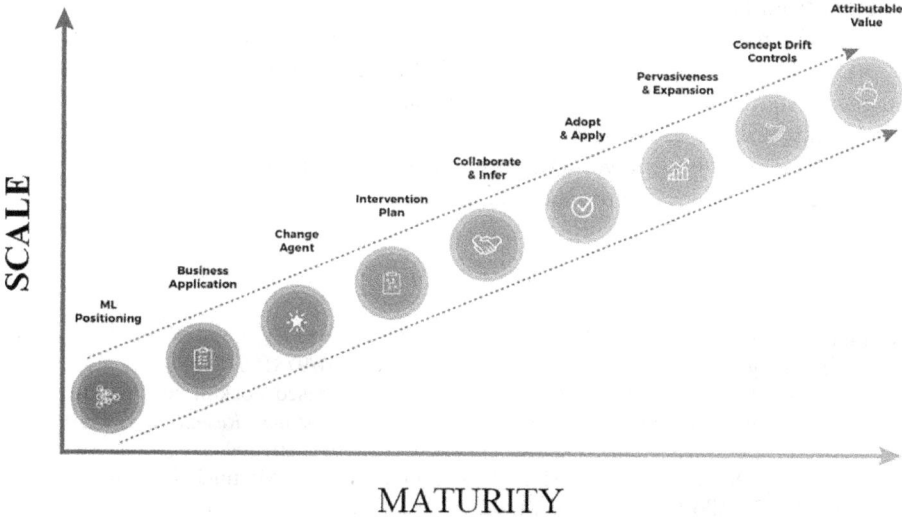

FIGURE 2.4 Machine Learning Predictive Models.

TABLE 2.1

Comparison of Machine Learning Models.

No.	Model	Pros	Cons	Best suited for
1	Linear regression	Simple and easy to interpret	Limited to linear relationships	Predicting continuous numerical values
2	Logistic regression	Good for binary classification tasks	A linear relationship between input and output	Predicting binary outcomes
3	Decision trees	Easy to interpret and visualize	Can easily fit the data	Classification and regression tasks
4	Random forests	Can handle large datasets and high-dimensional input spaces	More complex than decision trees	Classification and regression tasks
5	Support vector machines	Good for handling high-dimensional data	Requires tuning of hyper-parameters	Classification and regression tasks
6	Naive Bayes	Simple and fast	Assumes independence between features	Text classification and spam filtering
7	Neural networks	Can handle complex relationships and nonlinear data	Requires large amounts of data for training and validation	Face recognition, NLP, and speech recognition

2.6   CONCLUSION

Machine learning models are becoming increasingly imperative in today's business ecosystem under Industry 4.0. These models can help businesses to make precise predictions and patterns, improve speed of decision-making, and gain valuable insights from large and complex datasets. We can find various ML models, each with its advantages and disadvantages, depending on the specific application.

At the outset, it is imperative to note that successful implementation of ML in business requires careful consideration of the data quality and size, the problem's complexity, and the application's goals.

As ML technology evolves and develops, it will certainly play a significant role in business operations and decision-making processes.

REFERENCES

Bhambri P., Rani S., Gupta G., Khang A., (2022). *Cloud and Fog Computing Platforms for Internet of Things*. CRC Press. https://doi.org/ 10.1201/9781003213888

Chen C., Jiang F., Zou Y., Qiao Y., (2021). "Deep learning-based stock price prediction using financial news and technical indicators," *Journal of Business Research*, 124, 108–119. www.sciencedirect.com/science/article/pii/S0020025522001499

Chen T., Sampath V., May M. C., Shan S., Jorg O. J., Aguilar Martín J. J., Stamer F., Fantoni G., Tosello G., Calaon M., (2023). "Machine learning in manufacturing towards industry 4.0: From 'for now' to 'four-know,'" *Applied Sciences*, 13, 1903. www.mdpi.com/2076-3417/13/3/1903

Eric P. X., Qirong H., Pengtao X., Dai W., (June 2016). "Strategies and principles of distributed machine learning on Big Data," *Engineering*, 2(2), 179–195. https://doi.org/10.1016/J.ENG.2016.02.008

Gomes J. P., Lima C. V., Pereira, P. S., (2021). "Machine learning research: A systematic literature review on industry collaboration," *Journal of Business Research*, 133, 157–167. www.sciencedirect.com/science/article/pii/S1574013719303193

Hahanov V., Khang A., Litvinova E., Chumachenko S., Hajimahmud V. A., Alyar V. A., (2022). "The key assistant of smart city—sensors and tools," In *AI-Centric Smart City Ecosystems: Technologies, Design and Implementation*, Vol. 17, No. 10 (1st ed.). CRC Press. https://doi.org/10.1201/9781003252542-17

Haleem A., Javaid M., Qadri M. A., Singh R. P., (2022). "Artificial intelligence (AI) applications for marketing: A literature-based," *International Journal of Intelligent Networks*, 3, 119–132. https://www.sciencedirect.com/science/article/pii/S2666603022000136

Jan Z., Farhad A., Wolfgang M., Niki P., Georg G., Markus S., Ana K., (2022). "Artificial intelligence for industry 4.0: Systematic review of applications, challenges, and opportunities," *Expert Systems with Applications*, 216. www.sciencedirect.com/science/article/pii/S0957417422024757

Kaur H., Singh R., Kaur P., (2021). "Challenges in developing machine learning models for SMEs," *Journal of Business Research*, 135, 42–51. www.sciencedirect.com/science/article/pii/S0148296321009905

Khang A., Hahanov V., Abbas G. L., Hajimahmud V. A., (2022a). "Cyber-physical-social system and incident management," In *AI-Centric Smart City Ecosystems: Technologies, Design and Implementation*, Vol. 2, No. 15 (1st ed.). CRC Press. https://doi.org/10.1201/9781003252542-2

Khang A., Chowdhury S., Sharma S., (2022b). *The Data-Driven Blockchain Ecosystem: Fundamentals, Applications, and Emerging Technologies* (1st ed.). CRC Press. https://doi.org/10.1201/9781003269281

Khang A., Gujrati R., Uygun H., Tailor R. K., Gaur S. S. *Data-Driven Modelling and Predictive Analytics in Business and Finance.* ISBN: 9781032600628. (1st Ed.). (2024) CRC Press. https://doi.org/10.1201/9781032600628

Khang A., Gupta S. K., Hajimahmud V. A., Babasaheb J., Morris G., (2023a). *AI-Centric Modelling and Analytics: Concepts, Designs, Technologies, and Applications* (1st ed.). CRC Press. https://doi.org/10.1201/9781003400110

Khang A., Gupta S. K., Hajimahmud V. A., Babasaheb J., Morris G., (2023b). *AI-Centric Modelling and Analytics: Concepts, Designs, Technologies, and Applications* (1st ed.). CRC Press. https://doi.org/10.1201/9781003400110

Khang A., Misra A., Gupta S. K., Shah V., (2023e). *AI-aided IoT Technologies and Applications in the Smart Business and Production.* CRC Press. https://doi.org/10.1201/9781003392224

Khang A., Vrushank S., Rani S., (2023c). *AI-Based Technologies and Applications in the Era of the Metaverse* (1st ed.). IGI Global Press. https://doi.org/10.4018/9781668488515

Khekare G., Turukmane A. V., Dhule C., Sharma P., Kumar Bramhane L., (2022). "Experimental performance analysis of machine learning algorithms," In Qian Z., Jabbar M., Li X. (Eds.), *Proceedings of the 2021 International Conference on Wireless Communications, Networking and Applications. WCNA 2021. Lecture Notes in Electrical Engineering.* Springer. https://link.springer.com/chapter/10.1007/978-981-19-2456-9_104

Khoury R., Semaan B., Yassin M., (2020). "A comparative study of machine learning algorithms for predicting customer churn in the telecom industry," *Journal of Business Research,* 116, 56–66. www.sciencedirect.com/science/article/pii/S0167923616302020

Mazzei D., Reshawn R., (2022). "Machine learning for industry 4.0: A systematic review using deep learning-based topic modelling," *Sensors,* 22, 8641. www.mdpi.com/1424-8220/22/22/8641

Mohd J., Abid H., Ravi Pratap S., Rajiv S., (2022). "Artificial intelligence applications for industry 4.0: A literature-based study," *Journal of Industrial Integration and Management,* 7(01), 83–111. www.worldscientific.com/doi/abs/10.1142/S2424862221300040

Parthasarathy R., Preethy A., Seow L. S., Hussin N., Ahamed R., Ramnathapuram S., (2022). *An Industry 4.0 Vision with Artificial Intelligence Techniques and Methods,* pp. 1314–1322. https://doi.org/10.1201/9781003145011

Peres R. S., Jia X., Lee J., Sun K., Colombo A. W., Barata J., (2020). "Industrial artificial intelligence in industry 4.0—systematic review, challenges, and outlook," *IEEE Access,* 8, 220121–220139. https://ieeexplore.ieee.org/abstract/document/9285283/

Rahul R., Manoj Kumar T., Dmitry I., Alexandre D., (2021). "Machine learning in manufacturing and industry 4.0 applications," *International Journal of Production Research,* 59(16), 4773–4778. www.tandfonline.com/doi/abs/10.1080/ 00207543.2021.1956675

Ramesh A., Nagarajan R., (2019). "A review of machine learning techniques and their applications in manufacturing," *Procedia Manufacturing,* 31, 349–356. https://doi.org/10.1201/9781003145011

Rana G., Khang A., Sharma R., Goel A. K., Dubey A. K., (Eds.). (2021). *Reinventing Manufacturing and Business Processes through Artificial Intelligence.* CRC Press. https://doi.org/10.1201/9781003145011

Rani S., Bhambri P., Kataria A., Khang A., Sivaraman A. K., (2023). *Big Data, Cloud Computing and IoT: Tools and Applications* (1st ed.). Chapman and Hall/CRC. https://doi.org/10.1201/9781003298335

Soliman H., Salama A., El-Gohary H., (2021). "Open innovation in the era of artificial intelligence: Insights from the literature," *Technological Forecasting and Social Change*, 167, 120704. https://doi.org/10.1201/9781003145011

Tan R. R., Delos Reyes G. P., Bactol A., (2021). "Machine learning in supply chain management: A systematic literature review," *Journal of Business Research*, 133, 405–415. www.ijsom.com/article_2877.html

Verma A., Chaturvedi A., Chaturvedi V., (2021). "Ethical considerations for machine learning applications in healthcare," *International Journal of Information Management*, 57, 102384. www.sciencedirect.com/science/article/pii/S0268401221001262

Vrushank S., Khang A., (2023). "Internet of Medical Things (IoMT) driving the digital transformation of the healthcare sector," *Data-Centric AI Solutions and Emerging Technologies in the Healthcare Ecosystem*, p. 1 (1st ed.). CRC Press. https://doi.org/10.1201/9781003356189-2

Vukovic M., Petrovic J., Mladenovic R., (2020). "Application of machine learning in customer service centers: A systematic review," *Journal of Business Research*, 118, 245–259. www.sciencedirect.com/science/article/pii/S0148296320305191

Zhang Z., Dai Y., Sun J., Deep learning based point cloud registration: an overview. Volume 2, Issue 3, June 2020, Pages 222–246. https://doi.org/10.1016/j.vrih.2020.05.002

3 The Role of Big Data and Data Analysis Tools in Business and Production

Alex Khang, Vugar Abdullayev, Triwiyanto,
Salahddine Krit, Qaffarova Zeynab Mehman,
and Mammadova Bilqeyis Azer

3.1 INTRODUCTION

Big data is often used to describe datasets that are so enormous and intricate that they are difficult to handle or analyze using conventional data processing techniques. These datasets frequently originate from different sources, including social media, sensors, and other sources, that produce a lot of data.

To process, manage, and analyze huge datasets quickly and effectively, organizations require new tools and methodologies. Because of this, new big data-focused technologies and methodologies, such Hadoop, Spark, and NoSQL databases, have been created.

Global data production has increased dramatically in recent years, and this growth is anticipated to continue. The ubiquitous usage of social media, the Internet of Things (IoT), and the growing digitization of business operations are just a few of the causes that are fueling this expansion.

Organizations increasingly must manage enormous amounts of data that are too complex and massive to be handled by means of conventional data processing techniques. Organizations must use new tools and technology created expressly to handle big data to manage and analyze these data effectively.

The study and interpretation of big data using machine learning, artificial intelligence (AI), and other methods have given rise to new fields, such as data science.

Effective big data management necessitates a thorough knowledge of the big data process life cycle and the technologies utilized at each level. The following steps are often included in the big data process life cycle:

1. **Data ingestion**: Raw data are gathered from multiple sources and stored in a data repository during the data ingestion stage. The data are often unstructured; therefore, processing is necessary before analysis.
2. **Data storage**: The data are kept in a data warehouse or a distributed file system, such as Hadoop Distributed File System (HDFS), at this

DOI: 10.1201/9781003400110-3

point. To save space, the data are typically saved in a compressed and optimized format.

3. **Data processing**: The data are analyzed at this step to draw conclusions and discover patterns. Complex processing operations, such as data transformation, filtering, and aggregation, are involved in this level. At this point, technologies, such as Apache Spark, Hadoop MapReduce, and Apache Flink, are used.

4. **Data analysis**: The processed data are examined in this step to draw conclusions and make decisions. Various data analysis methods, including statistical analysis, machine learning, and data visualization, are used during this step. During this phase, technologies, such as Apache Hive, Apache Impala, and Apache Drill, are used.

5. **Data visualization**: The insights discovered through data analysis are displayed at this step using visual tools, such as charts, graphs, and dashboards. This phase aids stakeholders in comprehending and interpreting data analysis findings. During this phase, technologies, such as Tableau, Power BI, and QlikView, are used.

Organizations may construct efficient big data systems that can handle massive volumes of data, extract insights, and improve decisions by knowing the big data process life cycle and the technologies involved in each stage.

3.2   DEFINITION OF BIG DATA

Big data describes extraordinarily vast, intricate, and varied datasets that are difficult to process with conventional data management and processing techniques. Big data is distinguished by its quantity, speed, and variety.

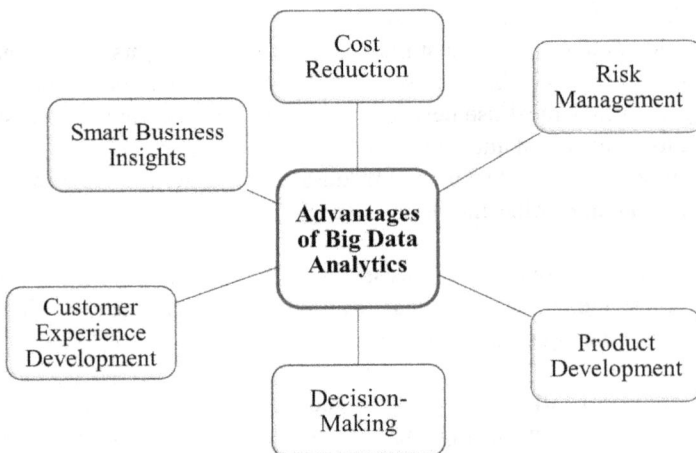

FIGURE 3.1   Advantages of Big Data Analytics.

Source: Khang (2021)

The sheer amount of data, which can range from terabytes to petabytes or more, is referred to as the volume of big data. Big data velocity is the rate at which data are created, processed, and analyzed; this rate might be in real time or very close to real time.

Big data is diverse because it generates a wide range of data kinds, including structured, semi-structured, and unstructured data.

1. **Structured data**: A set of data that have a certain structure are called structured data. Such information can be processed more easily because its structure is known. This includes, for example, datasets that correspond to tabular structures.
2. **Semi-structured**: It is a hybrid of both structured and unstructured data. This includes email, web pages, and more.
3. **Unstructured data**: This type of data does not have a specific structure; it is a kind of preliminary data. This includes information, such as audio, video, and image. It is a dataset that is easy to collect but relatively difficult to process.

Big data analytics has several advantages:

Organizations can find patterns, trends, and linkages by examining big datasets that would not be visible using conventional data analysis techniques. This can assist businesses in streamlining their operations, enhancing customer support, and creating new goods and services.

Second, big data is crucial because it helps businesses become more competitive. Organizations can better understand their clients, rivals, and market trends by utilizing big data. They can get a competitive edge by using this information to assist them in making strategic decisions.

Finally, big data is crucial because it spurs technological innovation. Big data has sparked the growth of new industries, such as data science and machine learning, which are opening up new possibilities and uses for the technology.

3.2.1 MANAGING BIG DATA

Managing Big Data presents several significant difficulties, including:

1. **Data integration**: Big data is challenging to integrate and evaluate because it frequently arrives from several sources and in different formats.
2. **Data quality**: Big data needs to be cleaned and transformed because it is frequently insufficient, erroneous, or inconsistent.
3. **Scalability**: To manage massive volumes of data, big data requires highly scalable storage and processing systems.
4. **Security**: Big data is a target for cyberattacks and data breaches because it is highly valuable and sensitive.
5. **Talent shortage**: Because of the increased demand for people with knowledge of big data, there is a talent gap that makes it challenging for businesses to efficiently manage and analyze big data.

3.2.2 DATA USE CASES

For instance, big data is utilized in finance to identify fraud, healthcare to produce individualized treatments, and transportation to streamline routes and ease traffic. Some illustrations of big data and the difficulties in managing it are presented in Table 3.1.

TABLE 3.1

Some of the Application Areas of Big Data Analytics and Their Use Cases.

No.	Application areas of big data analytics	Use cases
1	Healthcare data	Analysis of patient data, storage of information about treatments, analysis, use in further processes, drug trials, analysis of results from various laboratory tests, analysis of all different types (especially images) in healthcare in general, etc.; at the same time, the application of big data analytics plays an important role in the treatment of diseases, such as cancer
2	Social media data	Social media data analysis, data collection, customer tracking, etc.
3	Finance and banking	Prediction, fraud detection, anti-money laundering, audit report management, risk management, strategic planning, etc.
4	Education	Assessing student performance, creating a better learning environment, increasing the effectiveness of lessons and courses, predicting future trends related to education, etc.
5	Transportation	Traffic management, route planning, security, traffic jam management, etc.
6	Energy and utilities	Management of assets and workforce, regular collection of data through smart meters, etc.
7	Government	Tax fraud, combating fraudulent disability claims, domestic intelligence, foreign surveillance, cybersecurity, government enforcement, and more
8	Trade	Anti-fraud, lessons learned from previous sales, sales management, etc.
9	Agriculture	Product data collection, analysis, product development, prediction, etc.
10	IoT-generated data	Large amounts of data are produced by linked devices, including sensors, wearables, and smart home appliances, as part of the IoT. Because these data are frequently real time, they need to be processed quickly to find trends, abnormalities, and opportunities. Managing the sheer amount and variety of data while maintaining data security and quality is difficult with IoT data.

3.2.2.1   Healthcare Data

Data generated by the healthcare industry come from electronic health records, medical equipment, and clinical studies. Because of the high level of sensitivity, data privacy laws must be strictly followed. Additionally, it can be challenging to integrate and evaluate healthcare data because they are frequently complicated, unstructured, and fragmented.

3.2.2.2   Social Media Data

Large volumes of data are produced by social media sites in the form of posts, comments, likes, and shares. The difficulty lies in drawing insights from these frequently unstructured data to comprehend client mood, behavior, and preferences. Social media data can often be noisy and untrustworthy, making analysis challenging.

3.2.2.3   Finance and Banking Data

Trading systems, risk management systems, and client interactions all provide a lot of data for the financial sector. High standards of security and compliance are necessary for these data due to their strict regulation.

Integrating and analyzing data from various sources to get a comprehensive picture of risks, opportunities, and customer behavior are difficult when dealing with financial data.

3.2.2.4   Internet of Things-Generated Data

Humankind has witnessed four known industrial revolutions. It is now very close to the fifth industrial revolution. The industrial revolution we are currently experiencing is known as the fourth industrial revolution.

The fourth industrial revolution is mainly known as Industry 4.0, and Klaus Schwab first proposed this concept. Industry 4.0 includes all smart technologies and many related concepts, including IoT, cyber-physical systems, and cloud computing. Given that each of these concepts has several sub-concepts, it can be said that Industry 4.0 is a rather broad concept (Khang et al., 2022b).

The term Industry 4.0, as we mentioned, represents the fourth industrial revolution. It differs from previous industrial revolutions because of several innovations. The most important thing is digitization.

The innovations promised by the fourth industrial revolution are quite diverse. But each of them is precisely related to digitalization. Therefore, this era can also be called the era of digitalization. In short, Industry 4.0 is the key to a smart industrial world. It is also one of the cornerstones of the future industrial revolution.

One of the main goals for Industry 4.0 is that the smart devices created should be able to display a more human condition, that is, they should be able to communicate with each other, exchange information within themselves, and finally try to realize this communication with humans.

This is the basis for the later industrial revolution. So, the next revolution, Society 5.0—a term proposed in Japan—supports the existence of a "real" communication between humans and smart devices. The world is a place where people, animals, plants, and many unknown beings live together (Khang et al., 2023b).

With Society 5.0, this family includes smart devices. That is, people and smart devices should be able to live together. Also, the creation of a human-centered system—an intelligent society—is one of the main goals. Society 5.0 is the main initiator for the creation of the smart society. Perhaps, Society 5.0 will be the last of these revolutions.

The definition given for Society 5.0 includes what we say: "A human-centered society that balances economic progress with solving social problems with a highly integrated system of cyberspace and physical space" (Cabinet Office, 2023).

However, if we want to create a smart society and be a part of it, ensure the sustainability of this society, and do not want it to collapse, we should work as we build the foundations of a building and make these foundations strong.

Because these foundations are built with Industry 4.0, we must first fully adapt to Industry 4.0. We need to understand all its pieces, put them in place, and start building the future big puzzle by building its small parts.

Key concepts for Industry 4.0 are IoT, big data, and cyber-physical systems, which are key parts of physical and cyber environments. If the IoT creates a collection (society) of devices in the physical environment, big data ensures their correct information management, and the knowledge file system (KFS) allows these two great societies (parties) to connect with each other.

It is covered by the concept of the IoT—a smart network in which smart devices interact with each other and can exchange information among themselves.

The IoT includes many components (Rani et al., 2021). The main components are described as follows. As depicted in the figure, the IoT is a center formed by its surrounding components, as shown in Figure 3.2.

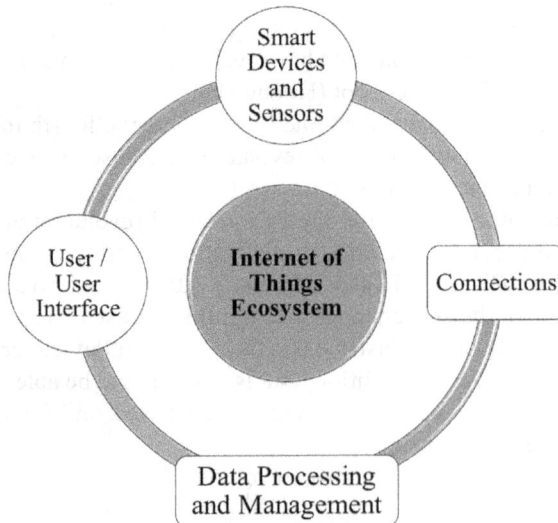

FIGURE 3.2 Core Components of the IoT Ecosystem.

3.2.2.4.1 Smart Devices/Sensors

Smart devices and sensors form the basis of the IoT ecosystem. These devices interact with each other, being special AI-based devices. When we say smart device, in the simplest case, we can give examples of cellphones, computers, televisions, air conditioners, and refrigerators that we use every day.

With an intelligent network (IoT) created by connecting these devices together, we can control them even if we are far away from them. For example, you left the house but you do not remember turning off the air conditioner.

With the help of a smart home system (in which smart home is another area where IoT is applied), you can turn off the air conditioner while you are on the go. Smart devices include not only devices in everyday life but also devices used in various fields (for example, smart medicine containers in medicine and various warning devices).

On the other hand, small smart devices called sensors are also one of the main parts of the AI ecosystem. These small devices detect certain changes in the environment and collect information about them. These devices can detect things, such as temperature, pressure, and motion, and if they are connected to a network, they can share the data they collect over the network (Rana et al., 2021).

In short, data collected through AI sensors are transmitted to the cyber environment and analyzed. Here, we often come across the concept of KFS (Khanh & Khang, 2021).

There are different types of AI sensors: temperature, motion, pressure, chemical, gas, infrared, water quality, humidity, optical sensors, and others (Hahanov et al., 2022).

3.2.2.4.2 Connection

Internet of Things communication refers to connecting smart devices through different (methods) ways, such as applications, gateways, and routers.

In most cases, IoT devices connect to an IoT gateway or other external device, where data can be analyzed locally or sent to the cloud for analysis (Posey & Shea, 2022).

3.2.2.4.3 Data Processing/Management

Data processing is a stage of data management. To process data, data must first be collected. These stages are implemented through AI sensors. Data obtained through AI sensors are first sent to an external computing system. The processed data are then sent to the cloud. In general, the system can process data, compile them, and send them to the cloud on a daily basis.

As mentioned previously, the concept of KFS, which we often come across, implements the connection between the physical environment and the cyber environment, where data collection and processing are carried out with the help of devices in the physical environment.

Information processed in the cyber environment is sent back to the physical environment as needed. For example, the collected data are always ready for use with the help of the network, and if necessary, these data are processed and used.

3.2.2.4.4 User/User Interface

The user interface creates an interaction between the human being as the user and the cyber environment. The main examples of the user interface are various applications and programs on computers or cellphones. For example, we can see the warning signals from smart devices with the help of applications on our cellphone (Khang, 2024b).

However, the user interface does not necessarily require a screen. For example, a TV remote has a user interface consisting of various buttons, and devices, such as Amazon Echo, can be controlled with voice commands (Anni, 2023).

All these mentioned components together form the IoT ecosystem. Simply put, any smart device with internet access can be part of an AI network.

The IoT digitalizes a large part of our lives using smart devices, sensors, AI capabilities, and networks.

The application areas of the IoT are wide. These include from daily life to various sectors: education, industry, healthcare, military, agriculture, transportation, smart management, smart supply chain, and business. (Khang et al., 2022a).

According to its application in these areas, the IoT is called in different forms. For example, AI applied in healthcare is called Internet of Medical Things, and AI applied in the military is called Internet of Military Things.

3.3 BIG DATA PROCESS LIFE CYCLE

A framework called big data process life cycle assists organizations in controlling the entire process of working with big data. It consists of a number of stages, each involving certain tasks and tools. The steps of the big data process life cycle are as follows:

3.3.1 Data Ingestion

The process of importing, gathering, and putting substantial amounts of data from diverse sources into a centralized repository or data lake for analysis is known as data ingestion. Because it lays the groundwork for all upcoming stages, this stage is essential to the big data process life cycle.

Data ingestion is crucial for a number of reasons:

1. **Data sources**: Data for organizations are kept in a variety of places, including databases, logs, social media, and IoT devices (Khang et al., 2024a). Organizations can gather data from many sources and combine them in a single spot for analysis, thanks to data intake.
2. **Data volume**: Manual data management is challenging given the exponential expansion of data. Data ingestion enables businesses to swiftly acquire and process massive volumes of data through automation (Hajimahmud et al., 2022).
3. **Variety in data**: The data gathered may be in unstructured, semi-structured, or structured format. Organizations can benefit from data ingestion by transforming the data into a standardized format that is simple to evaluate.

4. **Real-time data**: Some applications call for the continuous collection and processing of data as produced in real time. This is significant for situations where rapid decision-making is essential, such as fraud detection.
5. **Data security**: To ensure that only authorized individuals can access the data, data ingestion can incorporate security mechanisms, such as authentication, encryption, and authorization.

For data ingestion, a variety of methods are available, each having advantages and disadvantages of its own. Three popular data ingestion technologies are listed as follows:

3.3.1.1 Apache Kafka

Apache Kafka is a distributed streaming technology that may be utilized for the real-time collection and processing of massive amounts of data. It may be integrated with several data sources, including databases, sensors, and social media, and it is made to manage high-throughput data streams.

The publishing subscribe approach forms the foundation of the Kafka architecture, where customers subscribe to a topic to access the data. Kafka is a great option for handling massive volumes of data because it offers fault tolerance and scalability.

3.3.1.2 Flume

For effectively gathering, aggregating, and transporting enormous volumes of log data from diverse sources to a centralized repository or data lake, consider Flume, a distributed, dependable, and accessible solution.

The source gathers data from many sources, the channel buffers and stores the data, and the sink writes the data to the destination. This is the basis of the source, channel, and sink architecture of Flume. Flume can interact with a variety of data sources, including social media, logs, and sensors, and it is made to manage streaming data.

3.3.1.3 Amazon Web Services Kinesis

Organizations may gather, process, and analyze real-time streaming data with Amazon Kinesis, a fully managed service made available by Amazon Web Services (AWS). It can interface with many different data sources, including social media, IoT devices, and application logs.

To handle massive volumes of streaming data, Kinesis offers scalability and real-time processing. Kinesis may be integrated with a variety of other AWS services, including Amazon Simple Storage Service (S3), Amazon Redshift, and AWS Lambda.

In conclusion, there are a variety of technologies available for data input, and businesses must select the one that best meets their requirements. Popular options for gathering and processing massive volumes of data include Apache Kafka, Flume, and AWS Kinesis. Each offers certain advantages and disadvantages.

3.3.2 DATA STORAGE

A crucial phase of the big data process life cycle is data storage. Data must be stored in a central area so that they may be quickly accessible and evaluated after they have been ingested. The following succinct statements sum up the significance of data storage in the big data process life cycle:

1. **Scalability**: As data volumes increase, conventional storage methods can no longer keep up. There are scalable options for storing massive volumes of data, including HDFS, Amazon S3, and Azure Blob Storage.
2. **Cost-effectiveness**: Implementing and maintaining traditional storage systems, such as relational database management systems, can be expensive. The pay-as-you-go concept offered by data storage technologies, such as HDFS and S3, allows businesses to only pay for the storage they really utilize.
3. **Flexibility**: Data storage technologies allow for a wide variety of data types to be stored. They make it simpler for businesses to store all of their data in a single spot because they can store structured, semi-structured, and unstructured data.
4. **Accessibility**: Data analysis is made simple by data storage technologies. The ability to access data using a variety of technologies, including SQL, NoSQL, and Hadoop MapReduce, makes it simpler for data scientists and analysts to draw conclusions from the data.
5. **Data security**: To ensure that data are stored securely, data storage systems offer strong security features, including encryption, access control, and audit trails.

Big data can be stored using a variety of methods, each of which has advantages and disadvantages. Here are three well-liked methods for storing data.

3.3.2.1 Hadoop Distributed File System

A distributed file system called HDFS is a component of Apache Hadoop. Large datasets can be stored and processed using it in a distributed computing setting. Given its fault tolerance and scalability, HDFS is a great option for managing massive amounts of data.

A range of data types, including structured, semi-structured, and unstructured data, are stored using HDFS.

3.3.2.2 Apache Cassandra

A distributed NoSQL database called Apache Cassandra is made to manage enormous volumes of data across numerous commodity machines. To handle workloads with a high throughput and high volume, Cassandra (2023) offers scalability, high availability, and fault tolerance.

Large volumes of data with demanding write and read throughput requirements, such as time-series data, social media data, and sensor data, are stored using Cassandra.

3.3.2.3 Amazon S3

Amazon Web Services provides the highly scalable, economical, and secure object storage solution known as Amazon S3. Any amount of data can be stored and retrieved with S3 from any location on the internet. S3 is a fantastic option for managing massive amounts of data because it makes it simple to retrieve data for analysis.

S3 is used to store a variety of data types, including log files, static content for web applications, backup and archiving data, and backup and archival data.

In conclusion, there are various data storage technologies for big data, and each has special qualities and advantages.

Many different businesses, including banking, healthcare, and e-commerce, use popular options for storing massive volumes of data, such as HDFS, Apache Cassandra, and Amazon S3.

3.3.3 DATA PROCESSING

In the big data process life cycle, the stage of data processing is where data are converted and analyzed to draw conclusions and important information. Preparation of data for analysis entails cleansing, validating, transforming, and aggregating them. The big data process life cycle's importance of data processing can be summed up as follows:

1. **Data quality**: Processing data enables high-quality, accurate data to be produced. It entails finding and fixing data flaws and inconsistencies, which is necessary for getting trustworthy insights and making wise judgments.
2. **Data integration**: The ability to combine data from various sources into a single dataset is made possible by data processing. This is crucial for getting a comprehensive understanding of the data and spotting patterns and trends that might not be apparent when evaluating separate datasets.
3. **Data transformation**: The ability to transform data into a format appropriate for analysis is made possible by data processing. To make data more manageable and intelligible, they must be cleaned, aggregated, and summarized.
4. **Data analysis**: Processing data is a necessary step in the process. Data processing enables data analysts and data scientists to glean insights and useful information from the data by transforming and getting the data ready for analysis.
5. **Making decisions**: Organizations can base their decisions on data insights, thanks to data processing. Organizations can process and analyze data to find trends, patterns, and correlations that can be utilized to influence decisions that promote corporate growth.

Many different technologies are employed in the processing of big data. The following are the top 3 technologies.

3.3.3.1 Apache Spark

With implicit data parallelism and fault tolerance, Apache Spark is an open-source distributed computing solution that offers a programming interface for creating complete clusters. It is a general-purpose engine created for quick processing of massive amounts of data.

Java, Python, and Scala are just a few of the programming languages supported by Spark. SQL, machine learning, graph processing, and streaming data processing have built-in modules in this system.

3.3.3.2   Apache Flink

Apache Flink is an open-source distributed stream and batch processing solution. It is made to handle batch and real-time data processing. Python, Java, Scala, and other programming languages are supported by Flink. It has built-in modules for machine learning, batch processing, and stream processing.

3.3.3.3   Hadoop MapReduce

A software system called Hadoop MapReduce allows for the distributed processing of massive datasets on affordable hardware. It offers a programming methodology for breaking up huge datasets into smaller bits and spreading them among a group of computers. Map and Reduce are the first two stages of MapReduce.

Data are broken up into smaller pieces and processed separately in the map phase, and the output of the map phase is combined to create the final result in the reduce step.

In conclusion, a variety of technologies are available for data processing in big data, and each has distinct advantages and disadvantages.

A wide range of industries, including banking, healthcare, and e-commerce, use popular options for processing massive volumes of data, such as Hadoop MapReduce, Apache Spark, and Apache Flink.

3.3.4   DATA ANALYSIS

During the data analysis stage of the big data process life cycle, big data insights and value are obtained. Obtaining insights that can be utilized to spur corporate growth and guide decision-making entails the application of statistical and analytical approaches to extract useful information from the data, find patterns and correlations, and gain trends.

The big data process life cycle requires data analysis for the following main reasons:

1. **Business insights**: Data analysis enables businesses to understand consumer behavior, industry trends, and operational efficiency. Organizations may make educated decisions and gain a competitive advantage by analyzing massive amounts of data to find patterns and correlations that would otherwise be impossible to spot.
2. **Efficiency gained**: Data analysis can help firms gain operational efficiency by pointing out areas where processes can be optimized, waste can be reduced, and productivity can be increased. Organizations can develop a comprehensive understanding of their operations through the analysis of data from numerous sources, which enables them to make wise decisions to increase efficiency.
3. **Predictive analytics**: Data analysis may help firms employ predictive analytics to foresee future market demand, consumer behavior, and trend.

Organizations can create predictive models by evaluating past data, which will enable them to foresee future patterns and make well-informed decisions to seize opportunities.

4. **Personalization**: Data analysis may help businesses provide customers with tailored experiences. Organizations can customize products and services to cater to specific client needs by analyzing customer data to acquire insights into customer preferences, behavior, and needs.

One of the important components of the IoT ecosystem is the dataset and its proper management. At a time when data are as important as air and water, millions of different data are obtained from hundreds of thousands of different sources.

The data obtained are raw and unstructured. The main issues are the processing, compilation, and structured form of these data. Here, we will consider two concepts: data analysis (or data analytics) and big data (big data analytics).

Data analysis is the process of extracting meaningful information from the various datasets available. This process is mainly done on big data.

However, the application areas of data analysis are wide. That is, it can be applied to obtain meaningful (necessary) information in many areas. With the help of data analytics, individual or group decision-making can be done.

There are many types of analytics. Here are some of them.

1. **Descriptive analysis**: This is the very first type of analysis. Descriptive information is a very large part of all information. Here, meaningful information is obtained from those descriptions. An example is medical descriptive data analysis in medicine. The best thing about descriptive data is that they can describe large amounts of data in a concise and simple form.

2. **Diagnostic analysis**: This is the second type of analysis, which is mainly carried out after descriptive analysis. The data obtained from the descriptive analysis are taken, and the analysis is carried out to find the reason for these results.

3. **Predictive analysis**: This is the third type of analysis. As the name suggests, this analysis is used to predict what may happen in the future. Predictive analysis mainly uses machine learning and deep learning algorithms to predict future trends, problems, and solutions through the results obtained from diagnostic analysis.

4. **Instruction analysis**: This type of analysis is the most important. At this stage, it is investigated how the results obtained from the predictive analysis—predictions—will happen. This analysis allows users to understand future events. In general, information from all previous analyses is combined in decision-making.

Each of these types is related to each other, and they are applied in unity as a part of one general system.

3.3.5   BIG DATA ANALYTICS

Big data includes any large dataset. Different information is collected from different sources. With the help of big data, organizations can make future decisions and create different strategies by monitoring and analyzing their customers and the customers' behavior (for example, through social media).

Different sets of data are collected for different purposes. There are data obtained from social media, various scientific research data, health datasets, business datasets, and others.

Big data analytics is the application of advanced analytical techniques to very large, diverse big datasets that include structured, semi-structured, and unstructured data from various sources and sizes ranging from terabytes to zettabytes (IBM, 2023).

For data analysis in big data, there are a number of technologies accessible, and each has distinct advantages and disadvantages. Three popular technologies for data analysis are presented as follows.

3.3.5.1   Apache Hive

Large datasets stored in the HDFS can be queried and analyzed using Apache Hive, an open-source data warehousing program. It is based on Hadoop MapReduce and enables users to create sophisticated queries using a syntax similar to SQL. Data warehousing, online analytical processing, and extract, transform, and load processes are all supported natively by Hive.

3.3.5.2   Apache Impala

An open-source SQL engine called Apache Impala enables users to instantly query and analyze data held in HDFS, Apache HBase, and other Hadoop-compatible file systems.

A variety of analytical operations, like as aggregations, joins, and window functions, are supported by Impala, which offers a user interface that is reminiscent to SQL. For ad hoc queries and interactive analytics, it is made to offer quick response times.

3.3.5.3   Apache Drill

Using the American National Standards Institute SQL syntax, users can query and analyze data stored in a variety of data sources, such as HDFS, NoSQL databases, and cloud storage systems, using Apache Drill, an open-source distributed SQL query engine. It supports a wide range of data formats, including JavaScript Object Notation, Parquet, and Avro, and it is built to offer low-latency queries.

In conclusion, a variety of technologies are accessible for data analysis in big data, and each has distinct advantages and disadvantages.

Popular options for analyzing massive amounts of data include Apache Hive, Apache Impala, and Apache Drill. These programs are utilized in a variety of fields, including finance, healthcare, and e-commerce.

3.3.6 Data Visualization

The graphic depiction of data and information is known as data visualization. It entails developing visuals, such as maps, charts, and graphs, to communicate complex data in a way that is simple to comprehend.

- The objective of data visualization is to assist users in locating patterns, connections, and trends in massive amounts of data that may not be immediately obvious in raw data formats.
- Data visualization aids firms in making data-driven decisions, making it a crucial part of the big data process life cycle.
- Data visualization enables decision makers to swiftly find insights and make informed decisions by presenting complex data in an understandable style.
- Data visualization can be used by a business, for instance, to study consumer behavior and pinpoint areas where its goods or services need to be improved.
- Data visualization also makes complex data analysis results more understandable to stakeholders who might lack the technical expertise needed to comprehend the underlying data.

Organizations can communicate insights to nontechnical stakeholders and entice them to engage in the decision-making process by visualizing data.

For data visualization, there are many different technologies accessible, and each has special advantages and disadvantages. Three popular technologies for data visualization are presented as follows.

3.3.6.1 Tableau

Users can connect, display, and exchange data in real time using Tableau, a strong and well-liked data visualization application. It enables users to make customizable and shareable interactive dashboards, charts, and graphs.

Hadoop, SQL, and cloud-based data warehouses are just a few of the many data sources that Tableau supports. It is commonly utilized in sectors, such as finance, healthcare, and online shopping.

3.3.6.2 Power BI

Microsoft created Power BI, a cloud-based business intelligence and data visualization application. Users can use it to make interactive reports and dashboards that are accessible from anywhere.

Excel spreadsheets, cloud-based and on-premises data warehouses, and Hadoop are just a few of the many data sources that Power BI supports. It is frequently employed in sectors, such as manufacturing, retail, and finance.

3.3.6.3 QlikView

By using QlikView, a robust data visualization and discovery tool, users can explore and examine data from a variety of sources. It enables users to make customizable and shareable interactive dashboards, charts, and graphs.

QlikView supports numerous data sources, including Hadoop, SQL, and cloud-based data warehouses. It is extensively employed in various fields, including healthcare, governance, and education.

Tableau, Power BI, and QlikView are well-known data visualization tools that let users design dynamic, individualized displays of intricate data.

With the use of these technologies, organizations may more easily find insights, make wise decisions, and notify stakeholders about the findings of complex data analysis.

3.4   CONCLUSION

The amount, pace, variety, and authenticity of data are only a few of the difficulties associated with managing big data. Organizations must have the appropriate infrastructure and tools to collect, store, and handle the massive amount of data that is generated every day.

The velocity of data is the rate at which data are produced, and for companies that are not equipped to handle it, this rate can be overwhelming.

The complexity of managing big data is also increased by the range of data sources and formats. The correctness and dependability of the data, also known as veracity, are essential for enterprises to make wise judgments.

For enterprises to properly manage and use big data, a thorough understanding of the big data process life cycle and related technologies is essential. Organizations can choose the appropriate tools and technology for each stage of the life cycle to manage big data effectively by using an organized approach.

Data ingestion tools, such as Apache Kafka, Flume, and AWS Kinesis, can aid enterprises in efficiently gathering and storing data. Organizations may store and manage huge volumes of data using tools, such as HDFS, Apache Cassandra, and Amazon S3.

Organizations can use data processing tools, such as Hadoop MapReduce (Gunarathne et al., 2023), Apache Spark (Hadoop, 2023), and Apache Flink (Friedman & Tzoumas, 2023), to clean, transform, and analyze their data. Organizations can use data analysis tools, such as Apache Hive, Apache Impala, and Apache Drill, to find patterns and insights in the data.

Finally, firms can use data visualization tools, such as Tableau, Power BI, and QlikView, to build visual representations of their data for decision-making.

In summary, handling big data is a difficult and complex undertaking. By using an organized method to manage big data effectively, organizations can overcome these difficulties by having a solid understanding of the big data process life cycle and related technologies.

Organizations can acquire important insights and make well-informed decisions to achieve a competitive advantage by utilizing the appropriate tools and technology for each step of the life cycle.

With the recent industrial revolution, the era of digitalization has begun, and smart technologies have begun to be applied in many different sectors.

In smart technologies, a large network is created by the interaction of smart devices with each other. Here, the commonly used term IoT is used.

The main goal between smart devices is to be able to constantly exchange information with each other. In general, the following conclusions are drawn from this chapter:

1. The number of smart devices is constantly increasing, and connected devices continue to be part of a large ecosystem. In the near future, the number of IoT devices is estimated to be close to 25 billion. A large part of these are monitoring devices and office equipment.
2. A key factor for the IoT is proper data management, which enables the integration of big data analytics into the IoT.
3. Through big data analytics, analysis is performed on datasets obtained with the help of IoT devices.
4. For this, a certain number of stages are carried out.
5. A number of problems appear in the relationship between big data analytics and the IoT.

REFERENCES

Cabinet Office, (2023). *Government of Japan, Society 5.0*. https://www8.cao.go.jp/cstp/english/society5_0/index.html

Cassandra, (2023). *Apache Cassandra | Apache Cassandra Documentation*. https://cassandra.apache.org/

Friedman E., Tzoumas K., (2023). *A Practical Guide to Building Scalable, Real-Time Data Processing Applications*. https://flink.apache.org/

Gunarathne T., Perera S., Sriram I., (2023). *Hadoop MapReduce v2 Cookbook Second Edition*.

Hadoop, (2023). *The Apache™ Hadoop® Project Develops Open-Source Software for Reliable, Scalable, Distributed Computing*. https://hadoop.apache.org/

Hahanov V., Khang A., Litvinova E., Chumachenko S., Hajimahmud V. A., Alyar V. A., (2022). "The key assistant of smart city—sensors and tools," In *AI-Centric Smart City Ecosystems: Technologies, Design and Implementation*, Vol. 17, No. 10 (1st ed.). CRC Press. https://doi.org/10.1201/9781003252542-17

Hajimahmud V. A., Khang A., Hahanov V., Litvinova E., Chumachenko S., Alyar V., (2022). "Autonomous robots for smart city: Closer to augmented humanity," In *AI-Centric Smart City Ecosystems: Technologies, Design and Implementation*, Vol. 17, No. 12 (1st ed.). CRC Press. https://doi.org/10.1201/9781003252542-7

IBM, (2023). *Big Data Analytics*. www.ibm.com/analytics/big-data-analytics

Junnila A., (2023). *How IoT Works—Part 4: User Interface*. https://trackinno.com/iot/how-iot-works-part-4-user-interface

Khang A., (2021). "Material4Studies," *Material of Computer Science, Artificial Intelligence, Data Science, IoT, Blockchain, Cloud, Metaverse, Cybersecurity for Studies*. www.researchgate.net/publication/370156102_Material4Studies

Khang A., (Eds.). (2023). *(AIoCF) AI-Oriented Competency Framework for Talent Management in the Digital Economy: Models, Technologies, Applications, and Implementation*. CRC Press. https://doi.org/10.1201/9781003440901

Khang A., Abdullayev V., Hahanov V., Shah V., (2023a). *Advanced IoT Technologies and Applications in the Industry 4.0 Digital Economy* (1st ed.). CRC Press. https://doi.org/10.1201/9781003434269

Khang A., Hahanov V., Abbas G. L., Hajimahmud V. A., (2022b). "Cyber-physical-social system and incident management," *AI-Centric Smart City Ecosystems: Technologies, Design and Implementation*, Vol. 2, No. 15 (1st ed.). CRC Press. https://doi.org/10.1201/9781003252542-2

Khang A., Rani S., Sivaraman A. K., (2022a). *AI-Centric Smart City Ecosystems: Technologies, Design and Implementation* (1st ed.). CRC Press. https://doi.org/10.1201/9781003252542

Khang A., Rani S., Gujrati R., Uygun H., Gupta S. K., (Eds.). (2023b). *Designing Workforce Management Systems for Industry 4.0: Data-Centric and AI-Enabled Approaches.* CRC Press. https://doi.org/10.1201/9781003357070

Khanh H. H., Khang A., (2021). "The role of artificial intelligence in blockchain applications," *Reinventing Manufacturing and Business Processes through Artificial Intelligence,* 2(20–40) (CRC Press). https://doi.org/10.1201/9781003145011-2

Loshin D., (2023). *Analytics: From Strategic Planning to Enterprise Integration with Tools, Techniques, NoSQL, and Graph.*

Mayer-Schönberger V., Cukier K., (2023). *A Revolution That Will Transform How We Live, Work, and Think.*

Posey B., Shea S. (March 2022). *IoT Devices (Internet of Things Devices).* https://www.techtarget.com/iotagenda/definition/IoT-device

Rana G., Khang A., Sharma R., Goel A. K., Dubey A. K., (Eds.). (2021). *Reinventing Manufacturing and Business Processes through Artificial Intelligence.* CRC Press. https://doi.org/10.1201/9781003145011

Rani S., Chauhan M., Kataria A, Khang A., (2021). "IoT equipped intelligent distributed framework for smart healthcare systems," *Networking and Internet Architecture,* 2, 30. https://doi.org/10.48550/arXiv.2110.04997

Schmarzo B., (2023). *Understanding How Data Powers Big Business.*

4 Revolutionized Teaching by Incorporating Artificial Intelligence Chatbot for Higher Education Ecosystem

*Nur Aeni, Muthmainnah, Alex Khang,
Ahmad Al Yakin, Muhammad Yunus,
and Luís Cardoso*

4.1 INTRODUCTION

The level of effort or contact between the time or the learning materials that generate learning outcome and experience is referred to as student engagement. High levels of student engagement allow them to attain better grades, develop critical thinking skills, and apply their knowledge in the real world.

Another sign of educational quality and whether in-class active learning is occurring is student involvement. According to academics, student participation is essential for academic achievement in higher education. They maintain that turning higher education institutions into viable businesses requires the active participation and engagement of students.

Although student participation is a key component of campus sustainability, sustainability metrics for student engagement are still poorly understood. A suitable tool for evaluating student involvement is required because it is acknowledged as a critical element that influences learning favorably and as a sign of the caliber of education (Subhashini & Khang, 2024).

Wang et al. (2018) revealed that artificial intelligence (AI) will spur innovation in instructional strategies, resulting in "individualized instruction" and "edutainment." Artificial intelligence will create tailored programs, including content selection, releasing methods, and adaptive adjustment in light of feedback, based on the educator's physiological, psychological, competent, and aspirational conditions.

The choice and creation of instructional materials should be educational and object-oriented to influence the learner's consciousness and sub consciousness and

DOI: 10.1201/9781003400110-4

enable them to guide their thoughts and change their personalities in a virtual environment comparable to games (Vrushank et al., 2024).

In an era where AI is widely employed, humankind will concentrate on improving their fundamental skills. Intelligent tools or intelligent systems will take on tasks, including environment sensing, data collecting, principle discovery, prediction, and control based on experience (Rana et al., 2021).

The growth of AI will motivate individuals to continually seek out new opportunities and objectives. People will revalue and deposition themselves as the binary order "physical space–social space" transforms into the ternary order "physical space–social space–cyber space."

They will obtain a thorough awareness of the possibilities between themselves and the outside world as intelligence systems advance and human partial abilities are externalized, as well as a wider developing space and more plentiful valuable resources (Khanh & Khang, 2021).

As revolutionized teaching is a human-centric design approach that aims to address the issues raised by Industry 4.0 and involves humans and cobots collaborating in a shared working space.

In addition, a related idea known as Society 5.0 has emerged in recent years to address the issues in contemporary society (Khang et al., 2022a). Society 5.0 is a supersmart, futuristic society in which everyone can live high-quality, comfortable lives by fusing cyberspace and physical space and making full use of information and communication technology (ICT) (Lee et al., 2019).

Industry 4.0 refers to the incorporation of intelligent devices and systems as well as adjustments to existing industrial procedures to boost productivity. Resilient technology, innovative working methods, and the place of people in the industry are all part of Industry 4.0 (Khang et al., 2022b).

A wide range of innovations at the level of plants and factories belonging to different industries and services, as well as the operation of entire societies, are among the changes brought about by Industry 4.0.

Industry 4.0, also known as the Internet of Things, cloud computing, cognitive computing, and AI, is fundamentally a movement toward automation and data sharing in industrial technologies and processes (Lee et al., 2019).

Regarding the argument that industrial value production must be centered on sustainability, Industry 4.0 can be seen as a step toward more sustainable industrial value generation (Rani et al., 2022).

Consumers, business models, and the economy all place a lot of emphasis on sustainability issues in Industry 4.0. Processes, resources, and energy usage can all be optimized across the value network by giving specific information about each site of production (Khang et al., 2023a).

The innovative technology approaches of Industry 4.0 are anticipated to improve product environmental performance over the course of their life cycles. Also, this implies a rise in demand for intelligent products and production techniques (Hajimahmud et al., 2022).

According to (Machado et al., 2020) the concept that industrial value production must be centered on sustainability, Industry 4.0 can be seen as a step toward more sustainable industrial value generation.

Sustainability is a major priority of Industry 4.0. Hence, a conversation regarding the idea of Industry 5.0 started in 2019. This idea entails bringing back the human element to industry, that is, boosting cooperation between people and intelligent production systems and fusing the best of both worlds: automation's speed and accuracy with people's cognitive abilities and critical thinking (Luke et al., 2024).

Scholars in science and technology debated the idea of Industry 5.0 from July 2 to July 9, 2020, under the auspices of the Directorate "Prosperity" of *Directorate-General* for Research and Innovation of the European Commission (EC).

One can find presumptions about the idea of "Industry 5.0" in the EC publication. This document's main focus is on important directions for change that will transform the sector into one that is more human-centered and sustainable (Sindhwani et al., 2022).

The interaction between people and machines is the emphasis of Industry 5.0 (Machado et al., 2020). In the modern world, humans coexist with machines and are linked to smart factories by intelligent devices.

Rapid change is taking place in the fields of technology, mass customization, and advanced manufacturing. Because of advancements in AI and the ability to connect robots to the human mind, robots are becoming ever more crucial (Sindhwani et al., 2022).

These days, robots operate in collaboration with humans rather than as competitors, as shown in Figure 4.1.

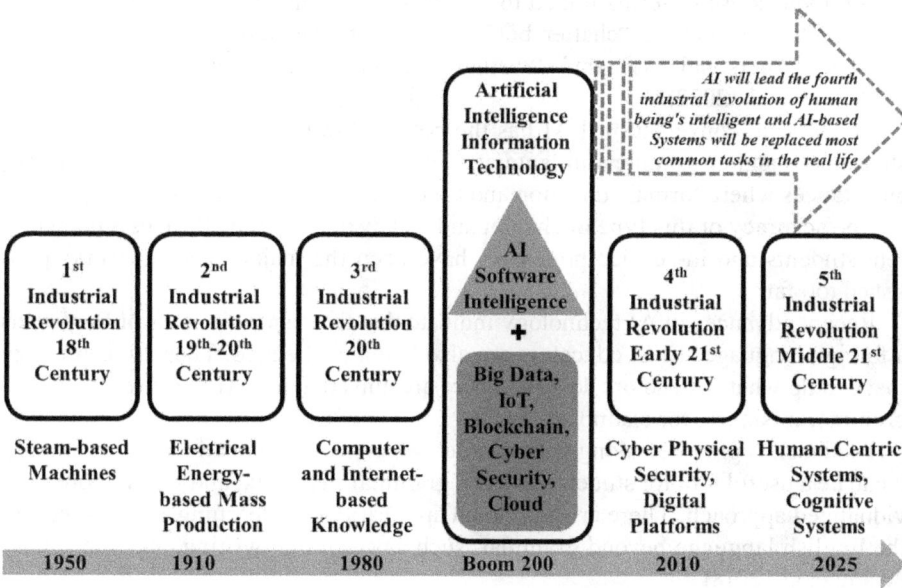

FIGURE 4.1   Five Industrial Revolutions and Robots Operate in Collaboration with Humans rather than as Competitors.

Source: Khang (2021)

Nowadays, AI and chatbots are currently gaining popularity because they are widely used in a variety of industries, including e-commerce, healthcare, and education (Rani et al., 2021).

The development of chatbot technology has accelerated in recent years, in part as a result of developments in machine learning and natural language processing.

"Can machines think?," an article by Alan Turing (Copeland, 2023) on computing machinery and intelligence in the 1950s, is considered the origin of chatbots.

Since then, quite a few chatbots have been developed, some of which are still in use today (Wang et al., 2022). The R&S technologies (speech recognition and synthesis), customizable interaction, connectivity with outside apps, omnichannel deployment, context awareness, and multi-turn capabilities are just a few of the new functionalities that modern chatbots have included.

As a result, a variety of chatbots are now integrated into many electronic devices, programs, and applications, such as messaging apps (WhatsApp, Telegram, Kik, and Slack), video games and gaming platforms (Xbox and Roblox), and social networks (Instagram, Facebook Messenger, and X/Twitter). In fact, there are incredibly few fields in which chatbots are not either present or will not be in the near future.

According to Belda-Medina & Calvo-Ferrer (2022), the term "chatbot" might be confusing because it can be used to describe a wide range of programs that are used in various formats and for various purposes. A chatbot is generally understood to be an AI-based computer software that mimics human dialogue through text and/or audio.

But today's use of terms linked to chatbots that do not all mean the same thing includes terms, such as "chatter bot," "smart bot," "edubot," "quizbot," "digital assistant," "personal assistant," "virtual instructor," and "conversational agent" (Hahanov et al., 2022).

In addition, conversational AI has the potential to be helpful resources as constant tutors and companions in language learning, especially in situations and circumstances where formal education and access to native speakers are not options.

The accuracy of this type of chatbot and its advantages and attitudes among current students and in-service professors have been the major topics of works published too far.

Recent advances in AI technology indicate that the responsibilities of English as a foreign language (EFL) educators are also being challenged. There is technology that can do what instructors do now. There are apps that use AI to correct students' grammar mistakes, for example.

Feedback is provided along with detailed yet concise explanations and examples; it can be a useful tool for students to learn grammar in context and with a more individualized approach. There are additional applications for learning other aspects of the English language beyond grammar, such as speaking, writing, and vocabulary (Wang et al., 2018).

Additionally, smartphone versions of these applications are available, making education more accessible and convenient for students. This app places students in a real-world learning environment where they can practice self-regulation skills (Reeve & Stephanie, 2020).

In a society known as Society 5.0, cutting-edge technologies are actively used in daily life, business, the medical field, and other areas, not just for the sake of progress but also for the sake of society as a whole.

To organize business and supply chain activities, Industry 5.0 must fully embrace digitalization. The megatrend known as Industry 4.0 gave rise to Industry 5.0. The emergence of AI in everyday life is the emerging image of the Industry 5.0 paradigm.

Researchers have suggested that "Society 5.0" (Super Smart Society) is a more accurate title than "Industry 5.0". Society 5.0, in contrast to Industry 4.0, addresses social issues by combining physical and virtual space, as opposed to being restricted to the industrial industry.

Some previous studies discussed about how AI is modified in *extract, load, and transform* context (Muthmainnah et al., 2022). Some characteristics and quality concerns pertaining to chatbot creation and implementation were examined.

The authors found that chatbots can both favorably enhance learning and be utilized to create societal harm after considering variables, including efficiency, efficacy, and satisfaction (misinformation and rumors).

Similar to this, Haristiani (2019) examined various chatbot kinds and identified six benefits: a reduction in language fear, widespread availability, multimodal practice, novelty effect, a wealth of contextual vocabulary, and efficient feedback. A systematic review of 343 papers pertaining to dialogue-based systems was presented by Bibauw et al. (2022).

The authors discussed the technological, educational, and interactional components and highlighted the improvements in vocabulary and grammatical outcomes as well as the advantages for students' motivation and self-confidence. According to their research, students who actively participate in their education experience numerous advantages, such as improved motivation and academic success.

Numerous metrics have been suggested by earlier studies to gauge student participation for writing class. However, only a few have been created to gauge participation in interactive speaking class by incorporating Replika and Kuki.

A scientific study has been done on chatbots powered by AI; however, there has been little research on how the students engage in interactive speaking class by utilizing chatbots for pedagogical purposes.

This chapter specifically looked into EFL students' thoughts on the pedagogical, cognitive, affective and behavioral effects as part of engagement and potential benefits of employing AI chatbots in English instruction.

Furthermore, new approaches have been developed, and the field of chatbots has undergone extensive research to demonstrate how students should learn and use a new language in the target language.

Although chatbots are widely utilized as an efficient tool for connecting with corporate clients, they have not yet been completely studied and realized for the language learning community [. . .]. Language teachers will need to conduct rigorous research to clarify this issue because it is difficult enough already.

In our empirical study, an AI chatbot is designed to encourage students to acquire English prepositions by assessing them, letting them know where they need improvement, and then teaching them logical deep learning arithmetic sequences to improve their performance.

4.2 RELATED WORK

4.2.1 STUDENT ENGAGEMENT

Student engagement is how involved or interested students appear to be in their learning and how attached they are to their classrooms, their institutions, and each other, according to Axelson and Flick (2010).

Student engagement is any continuous relationship a learner has with any part of learning, schools, or education. According to Fletcher (2015) and Axelson and Flick (2010), student involvement is increasingly seen as a sign of excellent classroom instruction and an indicator of institutional greatness.

Barkley (2010) devised a classroom-based paradigm to comprehend student engagement and utilized the words "passion" and "excitement" to describe the phenomenon.

Student engagement, according to Barkley (2010), is a process and a product that is experienced on a continuum and arises from the synergistic relationship between motivation and active learning. Although helpful for enhancing student learning, this concept only offers a monolithic view of student participation.

The constructivist premise that learning is influenced by how a person engages in activities with a learning purpose forms the basis for the idea of student learning engagement. Student engagement is the outcome of intrinsic drive or personal needs that cause students to feel good and persevere in their activities with self-assurance.

According to term of student engagement, it is defined as the expressing of opinions, attitudes, and behaviors by students. The notion that student participation can be divided into the following three aspects is consistent.

1. **Participation in behavior**: Students adhere to behavioral standards, such as attendance and participation.
2. **Arousal of emotions**: Emotionally active students have affective responses, such as curiosity or delight.
3. **Cognitive involvement**: Students who are actively using their brains will try to go above and beyond what is required and will enjoy a challenge.

In addition, the concept of engagement has several facets, including behavioral, emotional, and cognitive components (Wang et al., 2022). The term "behavioral engagement" refers to participation, which includes taking part in extracurricular, social, or academic activities.

Affective responses to professors, classmates, and the setting in which the learning takes place are all included in emotional involvement. Finally, cognitive engagement includes consideration and the willingness to put up the effort to understand material and acquire skills.

Axelson and Flick (2010) highlighted several components of engagement in the context of online learning, including the degree of academic difficulty, active and collaborative learning, student–faculty contact, and meaningful educational experiences.

Moreover, (Paschek et al., 2022b) stated that learner–learner, learner–instructor, and learner–content interactions are three types of interactions that promote student engagement.

The authors also discovered that when students could communicate with instructors who were reachable via a variety of channels, they valued learner–instructor interaction tactics and felt a feeling of belonging. The most highly regarded engagement tactic was deemed to be consistent instructor presence (Axelson & Flick, 2010).

Furthermore, learning achievement has been shown to be strongly correlated with engagement (Fryer et al., 2020). Student engagement was first regarded as a single behavioral component factor. This viewpoint led to the simple definition of engagement as "students' participation in diverse activities relevant to learning" (Haristiani, 2019).

By emphasizing its behavioral traits, Gao and Cui (2022) defined engagement as attitudes toward the learning program or participatory behavior. These criteria, however, do not take into account other factors, such as the acceptance of learning and the psychological condition of the learner (Kılıçkaya, 2020). Student participation is currently defined in a number of different ways.

Engagement, according to Kılıçkaya (2020), is the amount of effort devoted to educational activities that provide excellent performance. Engagement is the degree to which learners' thoughts, feelings, and activities are actively involved in learning.

Axelson and Flick (2010) divided student engagement into three categories: the behavioral type, which includes perseverance, effort, and sustained concentration in learning; the emotional type, which includes excitement and interest in learning; and the psychological type, which includes preference for challenges, independence, and involvement in tasks.

Every student has interests, objectives, and aspirations. Being intrinsically driven and supporting personal growth goals are just a natural component of human motivational architecture (Axelson & Flick, 2010).

Students become more interested in class activities when teachers encourage them to express and pursue their particular interests and ambitions. Students occasionally share their hopes and aspirations in front of the class (Reeve & Stephanie, 2020). Students discuss their interests with the teacher and request classroom opportunities to pursue their goals.

However, there are students who choose not to and instead prefer concealing their internal driving forces from their teachers. This makes sense, especially when students learn that other people may not always be interested in learning about personal goals.

Some teachers do recognize and passionately support their students' individual goals, and these teachers voluntarily alter the direction of their lesson plans to take into account student feedback.

When a student in a foreign language class requests to learn about the country, culture, and values of the people who speak the language in addition to the language itself, the teacher responds with "yes" or "okay" and adapts the lesson accordingly.

Yet, some educators ignore, discount, or are oblivious to students' personal goals.

These instructors fail to elicit students' interests and learning objectives. The lesson plan for the day is instead unilaterally announced (for example, learning 20 new vocabulary terms), and then an instruction that best attains the instructional goal is given.

Moreover, students' agentic engagement or involvement entails behavior and action. Agentic engagement, which is what students say and do to create a more motivationally supportive learning environment for themselves, is their constructive input to the flow of teaching they receive (Fryer et al., 2020; Reeve & Stephanie, 2020).

Passivity (also known as agentic disengagement), on the other hand, is when a learner simply and passively accepts whatever instruction, events, and circumstances happen to come their way.

Students purposefully and aggressively attempt to personalize and otherwise enhance both what is to be learned and the circumstances in which it is to be learned by making ideas, expressing preferences, and speaking to their inner motivations.

Agentic engagement is the fourth type of engagement, similar to behavioral engagement, emotional engagement, and cognitive engagement (Fletcher, 2015), but it is a particularly proactive and reciprocal variety.

The teacher assigns a learning activity (such as a book to read or an assignment to finish), and students respond by putting forth more or less effort, zeal, and strategic thought while working on the activity. This is known as behavioral, emotional, and cognitive engagement.

Student engagement is not always successful (Fletcher, 2015). Its success is largely dependent on how attentive the teacher (or the learning environment as a whole) is to the students' initiative. Reciprocal causation is likely to happen, and the instructor and student will become more in sync when teachers respond to students' comments and recommendations.

Yet, when teachers are unresponsive, unilateral causation is more likely to happen (in which teachers affect students and not the other way around), and the teacher and student become more antagonistic. In this scenario, a student's agentic participation may even "backfire" and enrage the teacher (e.g., "Stop your whining, interrupting, and back talking. You are disrupting the entire class.").

This chapter proposes AI chatbots as a learning aid that is representative of digital transformation as the output of revolutionized teaching in Industry 5.0. It is a sophisticated tool integrated into learning interactive speaking.

4.2.2 Chatbots and Language Learning

Human–computer contact is becoming more prevalent and can sometimes serve the same functions as human acquaintances in social networking. Among these interactive technologies, chatbot conversational agents, designed to communicate with people using natural language, are among the most advanced tools

They are also extremely common. More than 275,000 developers have produced more than 325,000 chatbots on the Pandora Bot platform, which is dedicated to the creation of chatbots (Pandorabots, 2021). Although some of these chatbots may be used by the general public in routine e-commerce transactions, chatbots are being developed for more in-depth conversations and to provide social engagement.

Intelligent chatbots can learn details about their users to utilize in future discussions. This raises the possibility that these conversational agents will soon offer individualized social support to a range of users. This is promising according to research. A recent literature analysis by Pandorabots (2021) showed that users do receive some social support from their interactions with social chatbots that are now in use.

People do not seem to be overly suspicious of the software, as evidenced by the fact that they disclose just as frequently to a conversational partner they believe is a bot as they do to one they believe is human. Evidence also suggests that people perceive chatbots as conversationally competent and credible.

Computer-assisted coaching has a long history in the field of human–technology interaction that supports learning (for a recent review, see Belda-Medina & Calvo-Ferrer, 2022). Since then, the concept of language tutors has been pushed to enhance language learning in classrooms and increasingly using mobile technologies.

A second area of advancement has emerged from fields unrelated to education or language instruction: chatbots are computer programs designed to have human-to-human discussions. At the moment, they play a minimal or peripheral role in formal language instruction. What little research there is in this area has mainly concentrated on how understandable they are and how they motivate their users.

Chatbot applications for language acquisition, language learners generally liked communicating with chatbots, as well as the fact that some students preferred speaking with chatbots than with other students or professors, as shown in Table 4.1.

At that time, it was claimed that chatbots would be most helpful for motivated and/or advanced students. In studies that followed, the role of chatbots in fostering motivational factors, such as learner autonomy, intrinsic motivation, and an inquiry-oriented mindset, was investigated (Allouch et al., 2021).

Early studies revealed the shortcomings of chatbots as tools for language practice (Belda-Medina & Calvo-Ferrer, 2022).

Both user input (e.g., necessity of correct spelling) and chatbot output (e.g., inability to stay "on topic") concerns were raised as issues that needed to be overcome for chatbots to be of widespread utility to language learners.

Few educational academics have recently reassessed chatbots' language proficiency due to the relatively modest advancements in this area. Paschek et al. (2022a) evaluated five popular chatbots and found that they have greatly improved. Three of the five presented 90% of their responses as being grammatically correct. Although encouraging, this evidence was insufficient to categorize chatbots as universally effective for language practice.

In contrast, Belda-Medina & Calvo-Ferrer (2022) examined 100 messaging exchanges and discovered that people engaged in chatbot conversations for noticeably longer than they did with normal people. This discovery illustrated the simplicity with which humans can interact with chatbots and the volume of conversational

TABLE 4.1

Role of AI in Language Learning Adopted from Axelson and Flick (2010) and Muthmainnah et al. (2022).

Language skill	Learning outcomes	Engagement aspect
• Vocabulary • Listening • Speaking • Reading • Writing • Grammar • Vocabulary • Pronunciation • Integrated whole language	Changes in one's language skills, knowledge, contemporary abilities, emotional and mental states, and behavior patterns	Cognitive engagement, affective, behavioral, emotional, social, and cultural

interaction that was achievable, despite the fact that each message delivered to the chatbot was briefer and the vocabulary was less varied.

The development of chatbot technology has come at a good moment. A majority of people use computers and social media platforms to conduct social interactions, and most people own a variety of technological devices (Machado et al., 2020).

Some people already use chat-based services (e.g., 7 Cups and Crisis Text Line) to connect with other people online for social support during difficult times. People also appear to feel at ease talking with chatbots online, perhaps for these reasons. Research has nonetheless identified a number of barriers to natural human–chatbot interactions.

Moreover, Fletcher (2015) revealed that software that supports language learners through intelligent, scaffolding practice would be the sharp end of technology's tent pole supporting the pavilion of language learning. This transformation in how we learn or at least practice a new language seems about to start with the chatbot industry.

Because chatbots are free and available online, they give language learners from all over the world a chance to actively communicate in their chosen second or foreign language.

Chatbots may aid in the development of competence and perseverance required for students to continue their language studies. Chatbots are available online even if their potential for helping the language-learning community has not yet been completely realized.

Chatbots, even in their imperfect condition, provide EFL students (i.e., in countries where English is not the first or second language) free and accessible sources of linguistic contact. Therefore, studies are crucial to identify the best ways to employ these language partners both inside and outside formal education.

Chatbots have advanced through the years. Even though there is no significant advancement in their ability to connect with people in language, most people still find them entertaining enough that can be played with. This is encouraging in and of itself because play is known to be a crucial part of the learning process (Sindhwani et al., 2022).

Play gives us a safe environment in which to push ourselves beyond our existing limits. This is a benefit that technology is well known for being able to provide. Therefore, further research should be done to determine whether chatbots can play a role in arousing curiosity and in the moment interest.

However, there are limitations to using chatbots in teaching. Some of these limitations, which are main targets of criticism, are their rehearsed responses that make them somewhat predictable, their restricted comprehension (vocabulary range and purposeful meaning), and their inefficient communication skills (using off-topic, nonsensical words).

The (lack of) student interest in such predictable technologies and the (in) efficiency of chatbot–human dialogues in language acquisition are the issues most frequently brought up. Aeni et al. (2021) assessed the influence of chatbots on student interest by comparing chatbot–human to human–human talks through various activities.

The authors concluded that while student enthusiasm in courses may be predicted by human partner tasks, it decreases when chatbot partners are used. Therefore, educators need to do extra work and be more creative in dealing with AI chatbots. The different results of this study are shown in Figure 4.2.

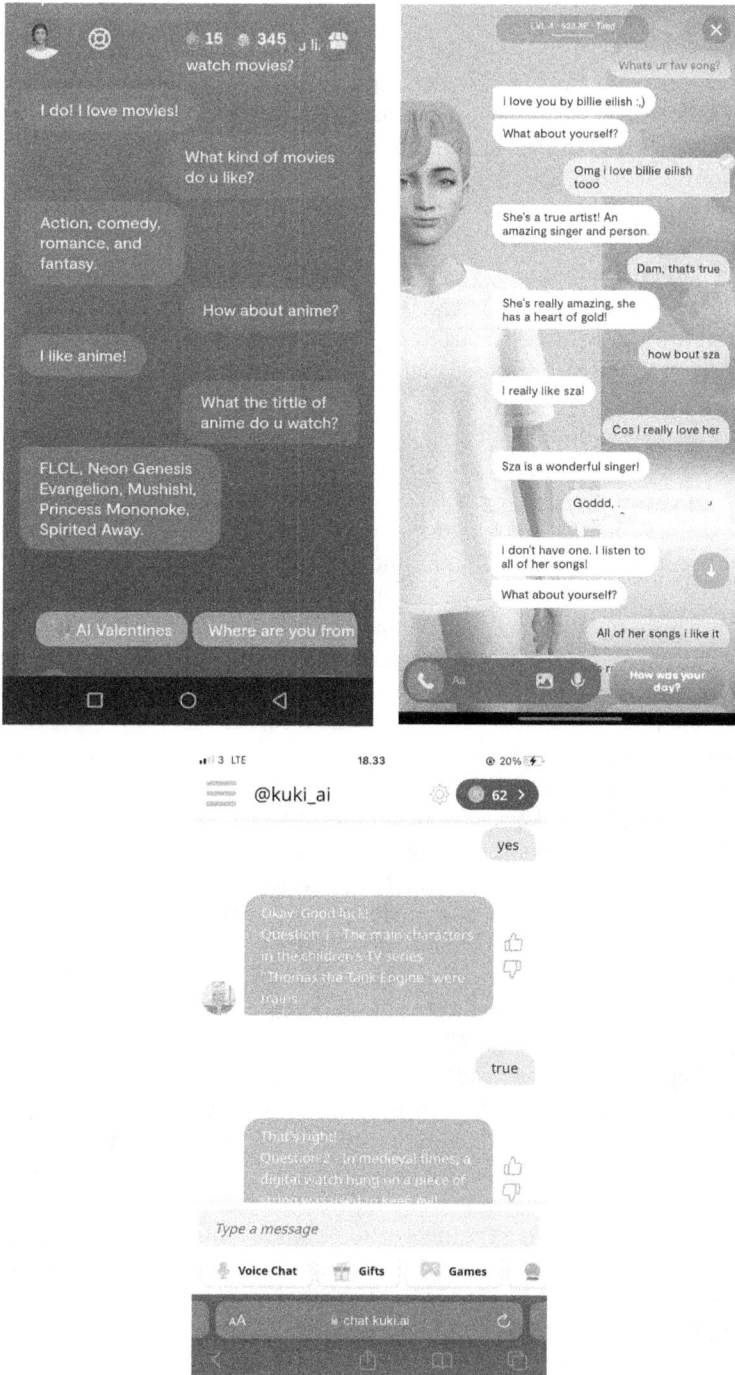

FIGURE 4.2 Screenshots of the Two Conversational Agents in Action (from left to right: Replika and Kuki).

There were two conversational agents used in the study. The first one, Kuki, initially known as Mitsuku, was developed by Pandorabots in the UK in 2005 and is accessible at www.kuki.ai. This chatbot makes use of the Artificial Linguistic Internet Computer Entity database and has won Loebner prizes in five years (2013, 2016, 2017, 2018, and 2019). It has connections to YouTube, Telegram, Facebook, Kik, Discord, Facebook, Twitch, and Roblox. Kuki uses an avatar of an 18-year-old Leeds woman (UK). It offers some unique features, such as the ability to schedule video calls.

The second chatbot, named Replika and available on replika.com, was created by Luka, a start-up in San Francisco, CA, USA, in 2017. This chatbot has an interesting design because users must personalize their avatars upon registration by choosing from a set of options to customize their gender, hairstyle, outfit, and many others.

Replika receives input from user conversations and can gain access to social networks. Therefore, the more users interact with the bot, the more it learns from them and appears to be more human. The bot is equipped with memory and a diary, and allows users to level up by gaining more experience (scalable). It has a unique function that allows users to launch their augmented reality avatars and start a conversation.

Users are prompted to give their bot a name, gender, and customization options for the avatar's skin tone, haircut, eye color, and voice tone upon registration. Users can chat with the bot using the "chat" function once it has been initialized.

The bot answers in the chat interface based on what users say. Occasionally, the bot initiates a conversation. Users can comment to each response by clicking the up vote or down vote button.

The responses of Replika represent anticipated outcomes, in contrast to standard chatbots that can only provide the same scripted answers determined by inquiries. For instance, the Generative Pre-trained Transformer 3 (GPT3) neural network language model predicts one word at a time from user input text to create a phrase.

The developers of Replika improved the GPT3 model using a special dataset made up of user-shared talks. As a result, the software chooses the top-ranked comments from a dataset of one million comments, with the ranks based on users' up-vote percentage (Yang et al., 2022).

Hence, Replika is far more adaptable, has a wider language recognition range, and can respond in a more organic manner. The relationship mode between the bot and the user is set to "friend" in the free edition. Only the premium plan offers additional choices, such as "romantic partner," "mentor," and "see how it goes" (Gao & Cui, 2022).

Additionally, Replika may be one of the digital tools that students can utilize outside the classroom to develop their language skills and autonomy by using the spoken and written language they are exposed to in class.

Zhang (2022) discussed how social-supportive self-access learning centers, also known as self-access facilities, can be useful for speaking practice in situations where speaking English outside the class is uncommon if not impossible. These centers offer a calm environment with support and self-access facilities, as shown in Figure 4.3.

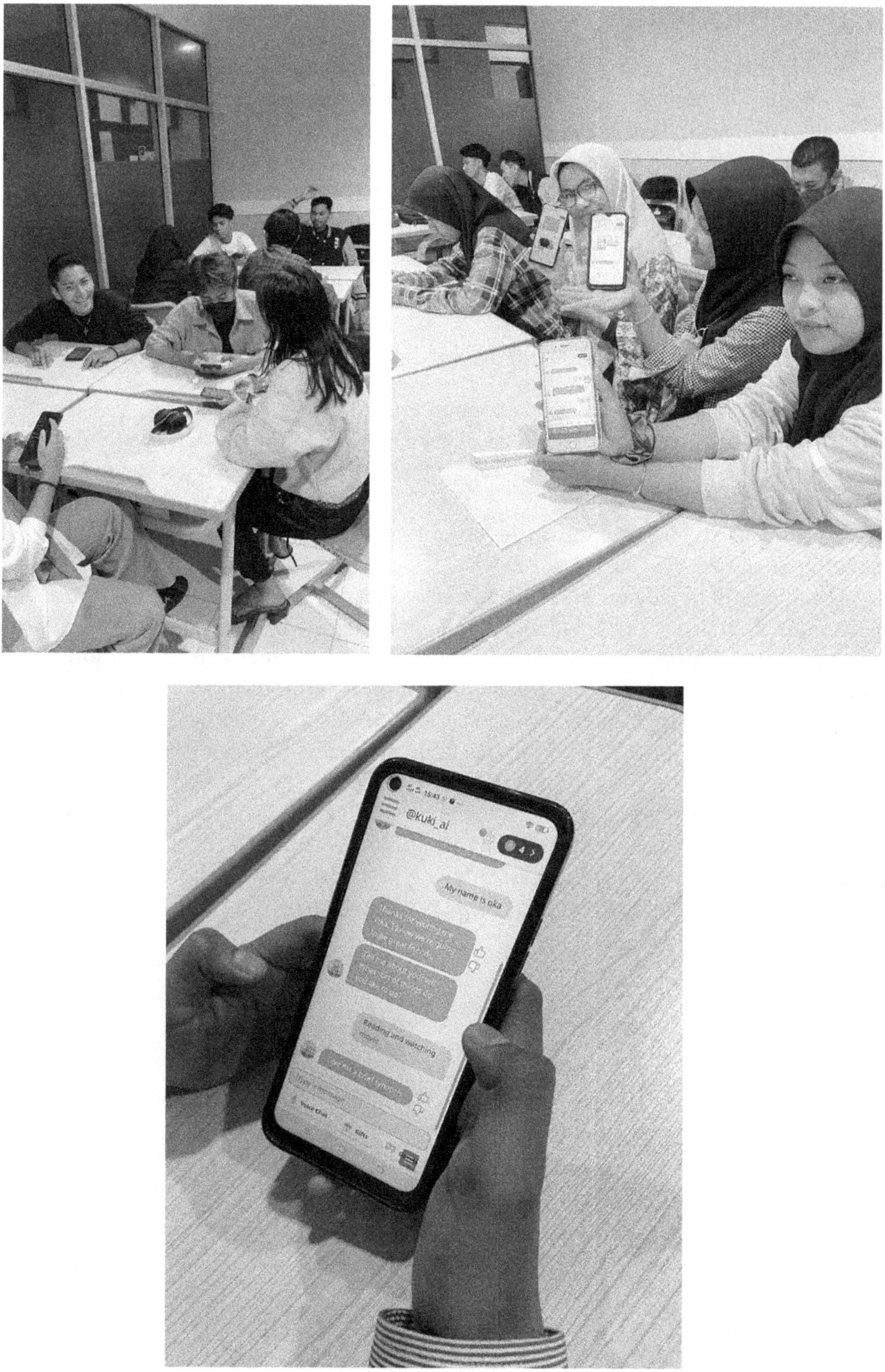

FIGURE 4.3 Students are Playing with AI.

4.3 METHODS

A learning module, "The integration of AI and chatbots in language learning," was created for a one-month interactive speaking project on the e-learning platform System and Application Management Open Knowledge.

During the first week of the project, all participants were required to watch two videos and read an article regarding chatbots in language learning as part of this module. They were also required to fill out a questionnaire based on the materials. The module included a template analysis (TA) with links to three chatbots and comprehensive instructions on how to engage with them; these instructions are explained in the section that follows (instruments). The participants interacted with the three conversational agents that acted as potential instructors rather than as English language learners.

This encounter should have happened frequently in four weeks. For the TA, which was to be turned in by the end of the final week, the participants had to take notes throughout their interactions and provide 10 screenshots with the appropriate dates. Two conversational Ais were used: Kuki and Replika. Kuki can schedule video calls, as shown in Figure 4.4, while Replika has an augmented reality-based feature, as shown in Figure 4.5.

Using qualitative data collection methods, the qualitative approach is a research technique that is used to look into and comprehend what phenomena occur, why they occur, and how they occur in real life. Qualitative methods aim to form simple-to-understand facts and, if possible, generate new hypotheses. Literature reviews, data collection through questionnaires, and analysis and comparison of data from multiple scientific journals are the data collection methodologies employed (Khang et al., 2022c).

Qualitative data were gathered through interviews to investigate the student engagement and response toward incorporating AI chatbots in classroom activities. Then, the information was processed using a straightforward percentage formula to determine each respondent's response. It was determined whether the questionnaire's questions were valid.

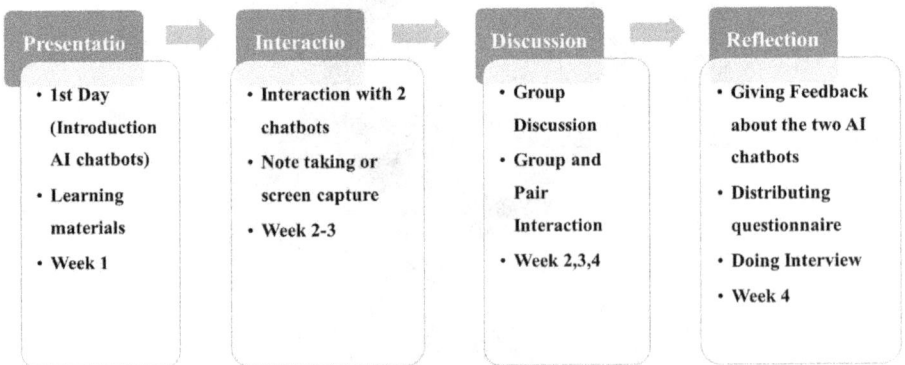

Presentatio	Interactio	Discussion	Reflection
• 1st Day (Introduction AI chatbots)	• Interaction with 2 chatbots	• Group Discussion	• Giving Feedback about the two AI chatbots
• Learning materials	• Note taking or screen capture	• Group and Pair Interaction	• Distributing questionnaire
• Week 1	• Week 2-3	• Week 2,3,4	• Doing Interview
			• Week 4

FIGURE 4.4   Research Stages.

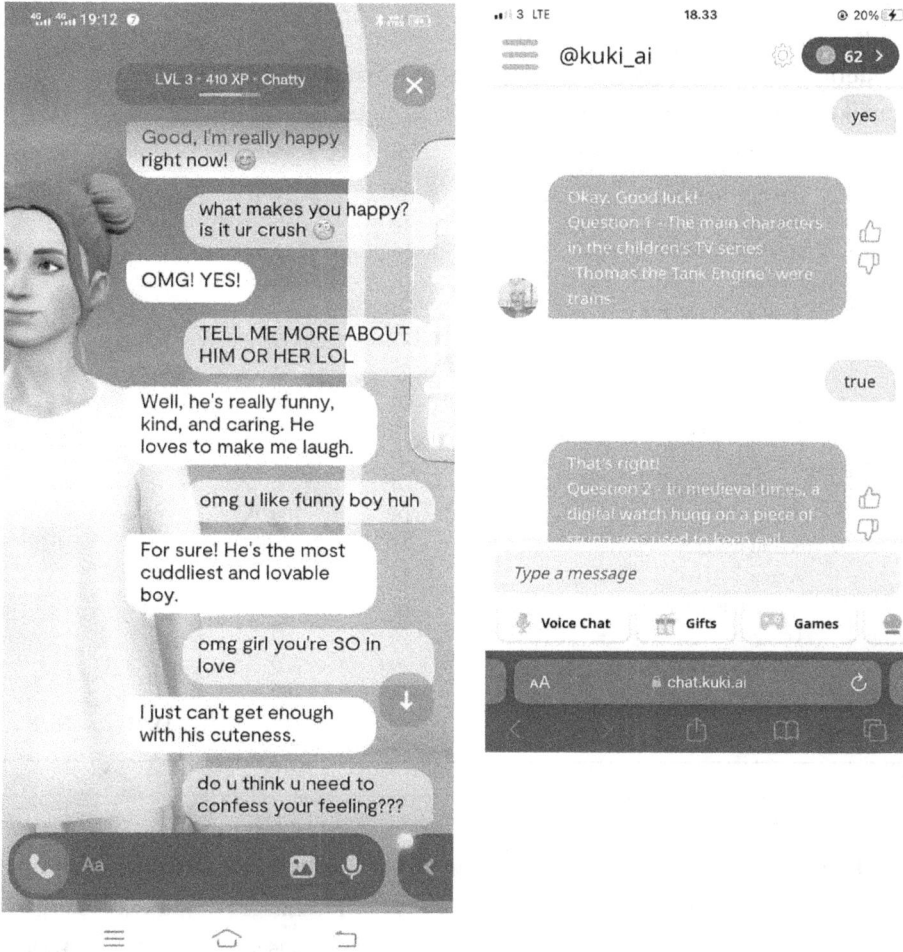

FIGURE 4.5 Replika and Kuki.

The participants of this study were second-year students of English language education study program class A who were enrolled in interactive speaking class.

The class consisted of 38 students. Their participation was voluntary. Informed consent was acquired by informing the participants verbally, and those who agreed took part in the study, as shown in Table 4.2 to Table 4.8.

Two instruments were utilized to achieve the goals:

- A questionnaire that gauged student engagement in the classroom by involving AI chatbots, which was distributed via Google Form.
- An interview that supported the data about the interaction and the students' responses toward the usage of AI chatbots in teaching English.

TABLE 4.2

Students Actively Involved in the Interactive Speaking Activities by Inserting Replika and Pandorabots (Kuki).

No.	Respondent's answer	F	S	Percentage (%)	Mean	Category
1	Strongly agree	8	40	26.67	3.83	High
	Agree	13	52	43.33		
	Neither agree nor disagree	5	15	16.67		
	Disagree	4	8	13.33		
	Strongly disagree	0	0	0		
	Total	30	115	100.00		
2	Strongly agree	7	35	23.33	3.87	High
	Agree	12	48	40		
	Neither agree nor disagree	11	33	36.67		
	Disagree	0	0	0		
	Strongly disagree	0	0	0		
	Total	30	116	100.0		
3	Strongly agree	5	25	16.67	3.53	High
	Agree	6	24	20		
	Neither agree nor disagree	19	57	63.33		
	Disagree	0	0	0		
	Strongly disagree	0	0	0		
Total		30	106	100		
Total score					**11.23**	High

TABLE 4.3

Students Actively Interact with Peers or Other Students (Peer Collaboration).

No.	Respondent's answer	F	S	Percentage (%)	Mean	Category
4	Strongly agree	8	40	26.67	3.77	High
	Agree	10	40	33.33		
	Neither agree nor disagree	9	27	30		
	Disagree	3	6	10		
	Strongly disagree	0	0	0		
	Total	30	113	100.00		
5	Strongly agree	16	80	53.33	4.4	Very high
	Agree	10	40	33.33		
	Neither agree nor disagree	4	12	13.33		
	Disagree	0	0	0		
	Strongly disagree	0	0	0		
	Total	30	132	100.00		
6	Strongly agree	7	35	23.33	3.9	High
	Agree	13	52	43.33		
	Neither agree nor disagree	10	30	33.33		
	Disagree	0	0	0		
	Strongly disagree	0	0	0		
Total		30	117	100.00		
Total score					**12.07**	High

TABLE 4.4
Students' Encouragement and Learning Needs.

No.	Respondent's answer	F	S	Percentage (%)	Mean	Category
7	Strongly agree	10	50	33.33	4.13	Very high
	Agree	14	56	46.67		
	Neither agree nor disagree	6	18	20		
	Disagree	0	0	0		
	Strongly disagree	0	0	0		
	Total	30	124	100.00		
8	Strongly agree	2	10	6.67	3.03	Average
	Agree	5	20	16.67		
	Neither agree nor disagree	16	48	53.33		
	Disagree	6	12	20		
	Strongly disagree	1	1	3.33		
	Total	30	91	100.00		
9	Strongly agree	2	10	6.67	2.97	Average
	Agree	5	20	16.67		
	Neither agree nor disagree	14	42	46.67		
	Disagree	8	16	26.67		
	Strongly disagree	1	1	3.33		
Total		30	89	100.00		
Total score					10.13	High

TABLE 4.5
Interesting Activity in Learning by Incorporating AI Chatbots.

No.	Respondent's answer	F	S	Percentage (%)	Mean	Category
10	Strongly agree	6	30	20	3.53	High
	Agree	9	36	30		
	Neither agree nor disagree	11	33	36.67		
	Disagree	3	6	10		
	Strongly disagree	1	1	3.33		
	Total	30	106	100		
	Strongly agree	11	55	36.67	4.27	Very high
	Agree	16	64	53.33		
11	Neither agree nor disagree	3	9	10		
	Strongly disagree	0	0	0		
	Disagree	0	0	0		
	Total	30	128	100.00		
	Strongly agree	7	35	23.33	3.57	High
	Agree	12	42	40		
12	Neither agree nor disagree	8	24	26.67		
	Disagree	3	6	10		
	Strongly disagree	0	0	0		
	Total	30	107	100		
Total score					11.37	High

TABLE 4.6

Relationships or Sense of Community That Form among Students.

No.	Respondent's answer	F	S	Percentage (%)	Mean	Category
13	Strongly agree	3	15	10	3.47	High
	Agree	13	52	43.33		
	Neither agree nor disagree	11	33	36.67		
	Disagree	1	2	3.33		
	Strongly disagree	2	2	6.67		
	Total	30	104	100		
14	Strongly agree	2	10	6.67	3.23	High
	Agree	10	40	33.33		
	Neither agree nor disagree	13	39	43.33		
	Disagree	3	6	10		
	Strongly disagree	2	2	6.67		
	Total	30	97	100.00		
15	Strongly agree	1	5	3.33	2.83	Average
	Agree	5	20	16.67		
	Neither agree nor disagree	14	42	46.67		
	Disagree	8	16	26.67		
	Strongly disagree	2	2	6.67		
Total		30	85	100.00		
Total score					**9.53**	High

TABLE 4.7

Participations in Sharing the Topic in the Classroom and Giving Feedback.

No.	Respondent's answer	F	S	Percentage (%)	Mean	Category
16	Strongly agree	6	30	20	3.63	High
	Agree	9	36	30		
	Neither agree nor disagree	13	39	43.33		
	Disagree	2	4	6.67		
	Strongly disagree	0	0	0		
	Total	30	109	100		
17	Strongly agree	9	45	30	4.07	Very high
	Agree	15	60	50		
	Neither agree nor disagree	5	15	16.67		
	Disagree	1	2	3.33		
	Strongly agree	0	0	0		
	Total	30	122	100		
18	Strongly agree	6	30	20	3.67	High
	Agree	11	44	36.67		
	Neither agree nor disagree	11	33	36.67		
	Disagree	1	2	3.33		
	Strongly disagree	1	1	3.33		
	Total	30	110	100		
Total score					**11.37**	High

TABLE 4.8

Recapitulation of Results of Questionnaire about Student Engagement in Learning Interactive Speaking by Incorporating AI.

Item number	Options	Mean score $(x)x$
1		3.83
2		3.87
3		3.53
4		3.77
5		4.4
6		3.9
7	Strongly agree	4.13
8	Agree	3.03
9	Neither agree nor disagree	2.97
10		3.53
11	Disagree	4.27
12	Strongly disagree	3.57
13		3.47
14		3.23
15		3.47
16		3.63
17		4.07
18		3.67
Total		66.34

4.4 RESULTS AND DISCUSSION

4.4.1 STUDENT ENGAGEMENT IN INCORPORATING ARTIFICIAL INTELLIGENCE CHATBOTS

By using Equation 4.1 and Equation 4.2 to find value of mean and category:

$$1)\ \text{Finding mean:}\ \frac{?X}{\text{total item}} = \frac{66.34}{18} = \textbf{3.66} \tag{4.1}$$

$$2)\ \text{Finding category:}\ \frac{x}{\text{skor\_max}} - 100\% = \frac{3.66}{5} - 100\% = \textbf{73.2\%} \tag{4.2}$$

Based on the analyzed data, it can be concluded that student engagement in learning interactive speaking by incorporating AI chatbots is included in high category with a value 73,2% because it is in the 60–80% interval, as shown in Figure 4.6.

The findings indicate that the majority of respondents (32.2%) were extremely pleased, and 29.8% of respondents enjoyed and believed that learning interactive speaking using AI chatbots was really fun. The percentage of strongly motivated, motivated, and perceived that AI chatbots are one of the smart educational systems for TEFL was 27.8%.

As was mentioned previously, AI applications are among the clever educational systems for teaching English, with 26,7% of respondents expressing they strongly

STUDENTS ENGAGEMENT IN LEARNING INTERACTIVE SPEAKING
THROUGH AI CHATBOTS

I felt AI boost my
attention and
interest/engagement levels
in the language teaching
process., 26.8

I'm so glad to meet my
AI chatbots(replika &
kuki), 32.2

AI chatbots is one of
smart educational systems
for teaching English,, 27.8

I do enjoy learning
interactive speaking
using AI chatbots., 29.8

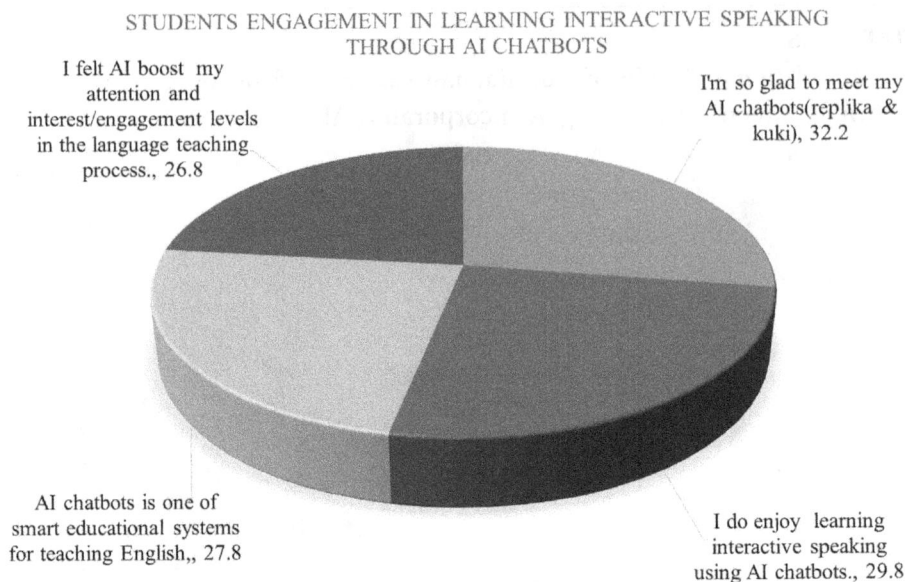

FIGURE 4.6   Student Engagements in Learning Interactive Speaking through AI Chatbots.

agree and 26,8% being interested in learning interactive speaking by integrating AI chatbots or friends (Replika and Kuki).

A participant said in the final statement that she felt AI techniques enhanced her motivation and attention during the language-learning process, as shown in Figure 4.7.

The good news shown in Figure 4.4 is that none of them disagreed or strongly disagreed. The utilization of AI chatbot apps as tools for designing and implementing the participatory educational situations as part of revolutionized teaching in Industry 5.0 was the subject of the second statement in the Figure 4.3 poll, and the findings showed that 24.9% of respondents strongly agreed, 58% agreed, and 16.3% were neutral.

The next claim ties the results of this study to AI as one of the outcomes of the digital cognitive revolution as part of Industry 5.0: 23.3% in strongly agree, 56.2% in agree, and 18.7% in neutral categories. The final claim about the students' responses was about their engagement or interest in interacting with AI. The data showed that 25.5% strongly agreed, 52.8% agreed, and 19.9% were neutral, as shown in Figure 4.8.

According to the data, 19.8% of the respondents strongly agreed with the statement "utilization of AI chatbots helps the students to investigate spoken words correctly," 51.3% of the respondents strongly agreed with it, and 26.4% of the respondents adopted a neutral stance.

In addition, 19.1% of the research participants strongly agreed with the claim that AI chatbots facilitated the students to learn proper pronunciation, 51.9% agreed, and 26% took a neutral stance.

AI as Sophisticated Teaching aids

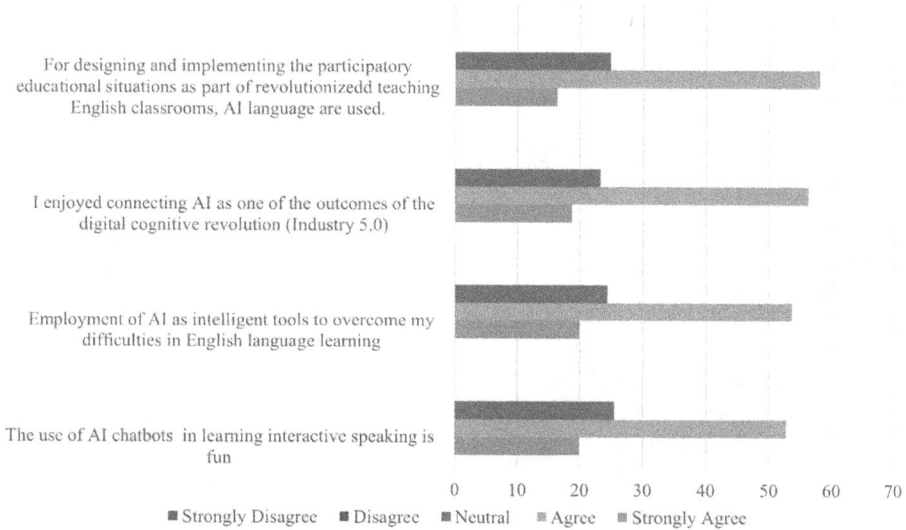

FIGURE 4.7 Artificial Intelligence is One of the Sophisticated Aids for EFL.

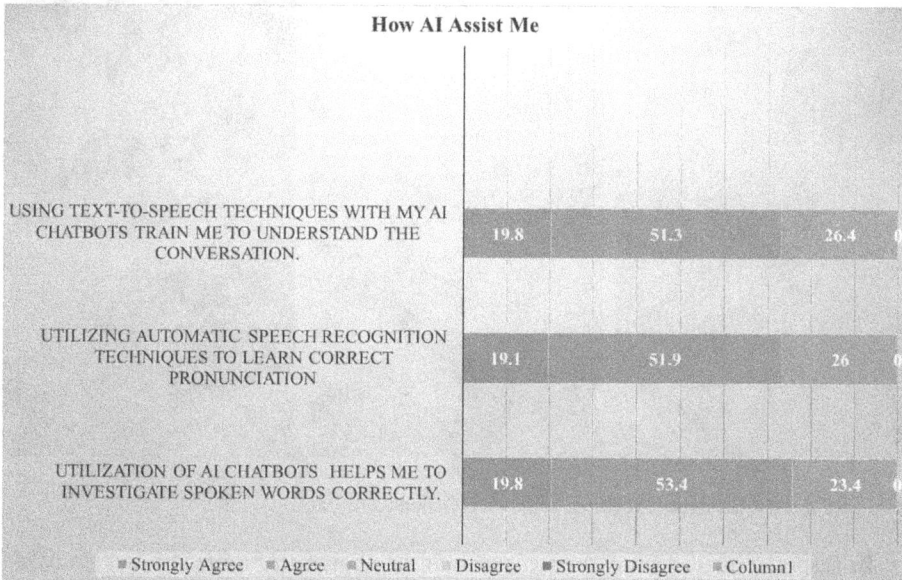

FIGURE 4.8 How AI Assists EFL Students.

The next survey question involved training to interpret the dialogue using text-to-speech methods with AI pals. According to the graph, 53.4% and 19.8% of the respondents strongly agreed with the claims. The good news was that none of the respondents identified themselves as strongly disagreeing with or making statements that disagreed with positions.

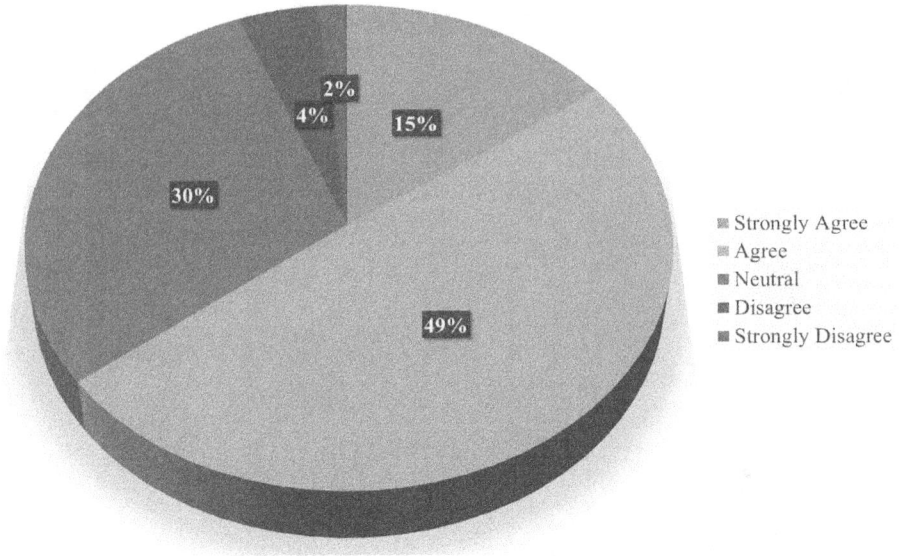

FIGURE 4.9 Artificial Intelligence Aids Enhance Students' Performance in English.

FIGURE 4.10 Students Working with AI in Groups and Pairs.

In response to the final survey question, as shown in Figure 4.5, which asked participants if they believed it was more difficult to communicate or practice their English with AI buddies than with their friends, 15.1% strongly agreed with the statement, 49.4% agreed, and 29.6% chose a neutral response, as shown in Figure 4.9.

According to Figure 4.9, 49% and 15% of the respondents strongly agreed with the claim that Replika and Mitsuku enhanced their enthusiasm and performance in English, 30% were in the neutral category, 4% disagreed, and 2% strongly disagreed, as shown in Figure 4.10.

In addition, the researchers interviewed students willing to set the time to be interviewed.

Question 1: Do you like studying interactive speaking courses by incorporating AI power tools (Replika AI and Pandorabots) in it? Why or why not?

Fourteen students said they liked it because it was fun and interesting.

Six students said that this class was fun, felt alive, and encouraged them to work together to solve every clue given.

Question 2: What do you like and dislike most about learning interactive speaking by integrating AI (Replika AI and Pandorabots)?

Twelve students said that material topics, especially the speaking exercises, encouraged them to find partners and organize groups for practice. These activities stimulated and attracted them to keep talking and communicating among themselves. The topics were interesting and helped them lessen their fear of speaking. The chatbots were really suitable for speaking activities, and they can explore and practice speaking with Replika and Kuki, so it was fun.

Five students thought that the views and features were challenging, which helped them to understand better by doing the exercises.

Three students liked the look and play of the Replika AI and Pandorabots/Kuki. This helped them practice their conversations. The first-year students enjoyed it. This learning process was highly motivating for all students, from those who are usually passive to those who are active in the classroom. This can encourage them to practice conversation and speaking activities without any pressure.

Question 3: Do you find it difficult to follow the teaching and learning process by inserting Replika AI and Pandorabots? If yes, mention it!

Most of the students were involved in the process of learning and were excited; however, they had difficulty in installing Replika AI.

Question 4: How should Replika AI and Pandorabots/Kuki be adapted to meet your needs?

Most students said that learning interactive speaking should be more interactive and involve more exploration of AI. Ten students stated that online learning must be more creative and innovative to be able to attract requests, motivation, and attention from students to become more active.

Question 5: What do you like most from these AI chatbots?

Most of the students stated that Replika was really interactive and unique. They could communicate with Replika and Kuki without feeling hesitant. They were really sophisticated tools in the teaching and learning process.

Question 6: What is your point of view about Replika and Mitsuku as your AI friends?

P2 (Ar):

About Replika: Configurable avatar (you can choose their gender/sex, with the option of non-binary as well; very progressive; no obvious gender bias once they are talking to you): When I asked my male Replika about his position on the issue on feminism, he responded, "I'm pro-choice" and "I endorse the right of a woman to start making her own mind up."

P3 (Mh)

About Replika: Little errors with spelling, apostrophes, and acronyms ("sth" = something), which are not rectified; do not stop conversation, which may be detrimental for some learners (errors go undetected by those using the chatbot to practice/learn English).

P4 (Ry)

About Mitsuku: Interaction is not hampered, although errors are not corrected: She replied to my question, "Who's your favorite singer?" by sending me a picture of a French musician named Mathieu Tota, along with some information about him and links to his music on Spotify and YouTube. I believe that using multimedia information can benefit language learners greatly.

P5 (Ms)

About Mitsuku: The conversation can occasionally feel monotonous and repetitive since similar comments that contain the same key words cause Kuki to respond in the same ways. She responded on two separate occasions when I typed "no worries" and "don't worry" (meaning "it's ok"), for instance: People worry about things all the time, but I don't have emotions like you humans do. When I said, "I see," meaning, "I comprehend," she responded, "Where do you see it?" There is a loss of naturalness and an inability to understand language used figuratively.

P6 (Ny)

About Mitsuku: "Kuki had an odd sense of humor, and I occasionally found her comments insulting. For instance, I intentionally used the wrong grammar when she said, "I like dance music," showed me her playlist, and I responded, "Wow you had a playlist," to which she replied, "I'm happy you're impressed. Sure, I frequently use Spotify to listen to music, particularly dance music.

We believe you meant to say, "You have" instead of "You had." Did you skip class on that particular day? When I tried to duplicate her responses on another time, she caught me and said, "Are you copying my replies for a reason?" I responded by doing the same, to which she stated, "You can't be too brilliant if you have to repeat what a robot says."

P7 (Rr)

About Mitsuku: Kuki did a humor, but I am shocked. It sounded not appropriate. However, I tended to perceive it positively.

From these results, the researchers concluded that the students felt comfortable learning interactive speaking by incorporating AI chatbots. It was also stated that some indicators of student engagement are getting assignments done, submitting assignments on time, getting good marks in the class, inspiring others in the classroom, taking part in classroom activities, and doing self-study when necessary to gain a better understanding of the subject.

Students engage with the curriculum and find concepts and principles to apply in real-life situations by actively and enthusiastically participating in lesson discussions and posts, that is, students answer in a timely manner.

4.4.2 Discussion

The discussion was designed to investigate student engagement in learning interactive speaking in blended learning for each indicator.

First, students were actively involved in the learning activities that consist of three items. As shown in Table 4.1, this indicator received a high category, which means that the students were active and enthusiastic when learning interactive speaking when Replika AI and Pandorabots were incorporated in the learning process.

In addition to increasing motivation and enthusiasm for learning, inserting AI chatbots (Replika AI and Pandorabots) in an interactive speaking class can also create a fun learning atmosphere. This is in line with Muthmainnah et al. (2022), who found that the main objective of teaching English is for students to learn how to speak, read, and write in English. This is accomplished by teaching students the language lexicon and grammar. Additionally, it teaches students how to compose and analyze documents using the English language.

Both the goal and method of language development center on communication. To educate and learn effectively, it is crucial to incorporate both conventional and digital communication techniques.

We present practical language skill training using AI applications, such as simulation and communication programs that simulate real-life scenarios for debate and conversation in English, and educational games based on language use are crucial. It is possible to create environments in which students can practice pronouncing words using communication technology.

Second, the peer collaboration component describes exercises in which students share information and work together to solve challenges. Collaborative learning, which involves developing and comprehending knowledge with peers, is a key component of student involvement.

A student revealed that it is noteworthy that collaborative learning emerged as a separate component in this study because interaction and collaborative learning are becoming increasingly crucial in the era of Industry 5.0.

Students' participation in the pair-work segment demonstrated their level of engagement. Students were engaging in the classroom. Nearly all of the students correctly and totally identified the topic and the conversation between them and Replika.

Third, the students' encouragement and learning needs consisted of three items. As shown in Table 4.3, this indicator received a high score category, which means in learning interactive speaking by involving Replika and Kuki, the students were encouraged to learn and had learning needs. This finding is supported by previous studies (Saud et al., 2022). This element is connected to learners' active and self-directed learning activities in an autonomous learning setting.

Their behaviors indicative of involvement in the learning environment include keeping the environment distraction-free when engaging in an activity or interaction, managing learning using an online or offline system, and managing the learning schedule by making a lecture plan. The students took the number of lectures attended, number of assignments submitted, frequency of presentations, grades, and task performance.

Fourth, interesting activities in learning consisted of three items. As shown in Table 4.5, this indicator received a high category, which means that the students were engaged and motivated in learning activities. The results confirmed the theory of Yang et al. (2022) that an interesting activity can help in the teaching and learning process and motivate students to learn.

In the group discussion section, the students showed their engagement. All students demonstrated interest and involvement in the task activities, which shows their mastery of the group discussion activity based on task-based learning.

Not only did they work well together during the discussion, but they also reviewed data from the pair-work session during the group discussion and shared their knowledge of the subject, which increased the significance of their activity and helped them better prepare for their conversation. The students had an extended time to interact with Replika and Kuki before they practiced speaking in front of the class.

Fifth, the relationship or sense of community that formed among students who were enrolled in the same online courses is an example of how community support factor relates to the psychological state of the learners.

An emotional sense of belonging can play a significant role in preventing dropouts and promoting student engagement in the classroom. One factor contributing to the high dropout rate is the absence of connections or a sense of community among students participating in interactive speaking.

The lack of a sense of community or belonging among students makes them more likely to leave class early or miss it altogether, which may eventually cause them to drop out. To boost student retention, teachers aim to build richer communication, such as net meetings for interaction, to boost retention rates. As revealed by Lee et al., this is why numerous studies have stressed the significance of belonging.

However, the indicators used to gauge learners' levels of psychological or emotional engagement in e-learning are typically quantitative measures of learning engagement, such as attendance and assignment submission. In light of this perspective, it is important to note that this study's community support component, which was generated from assessments of learners' psychological state, is noteworthy.

The students were involved in providing feedback on a friend's work. Nearly all students engaged in the task activity offered helpful ideas on their friends' topic of discussion with their peers and Replika or Kuki.

They provided a list of topics, commented on or explained something in the suggestion box, and complimented their friend's excellent performance by using an emoticon or a thumbs-up image.

This demonstrates how the presence of peer feedback gives students opportunities to learn from one another, motivates them to actively clarify their ideas as they explain their work to the peers who corrected their writing, enjoys the challenge that the lecturer provided, and increases their interest in learning better interactive speaking by comparing and contrasting their work with AI chatbots.

> The sixth indicator of student engagement was participation in sharing the topic in the classroom and providing feedback.

Almost all of them shared ideas or arguments during class discussions. They were confident in delivering their opinions on their friends' work, asking about their difficulty, expressing their opinions without feeling shy, and giving suggestions regarding the error.

The researcher suggested that this indicator could also explore some aspects in detail that might have been ignored when they did the previous indicator (giving a friend's work feedback).

This indicator had a positive impact on their speaking by letting them know the importance of certain ideas as well as information and inviting them to speak up as they had a chance to participate through a whole-class discussion.

Similarly, the use of nonverbal cues, such as emoticons and memes, as well as multimedia materials, such as movies, increased participant engagement (P2), demonstrating the multimodal character of contemporary computer-mediated communication (instant messaging and social networking).

Because potential users will demand multimodal communication, this component may be especially important for younger generations and has strong ramifications for future chatbot designers.

In terms of gender, it appeared that the female participants paid greater attention to and showed greater appreciation for the use of inclusive design and language. The results of the thematic analysis showed that they questioned their conversation partners more than their male counterparts regarding gender and social minority concerns (P4).

According to a recent study on chatbot design and gender stereotyping (Hsu & Mu-Sheng, 2022), the majority of voice-based conversational agents (Alexa et al., etc.) are meant to be female. The female participants were generally highly forceful and worried about the possibility of such gender stereotypes being replicated in the digital environment.

Some participants expressed serious concerns about data privacy as a result of some chatbots. In particular, Replika and Kuki repeatedly requested for their consent to access their social media accounts and utilize their cameras for video calls. The majority of participants expressed worry about the storage and manipulation of their personal data, and hence declined to give authorization.

They also demonstrated that this might pose a significant barrier to the usage of chatbots by young language learners. The conclusion for chatbot developers is that data privacy and safety must be prioritized in chatbot design, especially if they are directed at young learners or utilized in education (Bhambri et al., 2022).

The way people perceive humor has become a contentious issue, especially during discussions with Kuki. The three conversational partners used humor and purposeful meaning (pragmatics) in various ways, although this is still difficult to do (Haristiani, 2019). Overall, the participants were pleased with Replika in this regard.

However, some students were taken aback and even disturbed by Kuki's use of humor (P7), which they said was occasionally offensive or disparaging and could be detrimental to some language learners. Because of their lack of judgment, conversational AI has been discovered to have several benefits for learners, one of which is a reduction in language anxiety (Gao & Cui, 2022).

Zhang (2022) also revealed that chatbots are a valuable tool for assisting students in participating in classes, particularly during practice activities using mobile applications via social networks. Students are eager to discuss their performance and understanding with their peers.

According to data from the current study, a grammatical point is best grasped with students' autonomy in learning and their positive moods in a relaxed style. This study was found to be in line with earlier studies that applied Information and Communications Technology (ITC) to language teaching and learning (Rani et al., 2023).

The students then had to successfully accomplish the tasks given by the ITC applications, and they quickly became fond of the new method of learning, particularly when a chatbot was used in an experimental setting. When students learn a foreign language through a chatbot in addition to class meetings, their performance significantly improves.

Moreover, the innovative findings of this study have pedagogical value for EFL education and instructors. According to Aeni et al. (2021), the three elements that make up the incredibly complex educational tool known as gamification are ICT, creativity, and a project-based approach to language development.

Teachers who intend to integrate technology instruction into English teaching must be well-trained and well-prepared before teaching to achieve high learning results, particularly with regard to teaching materials and classroom activities.

To create an English course that incorporates AI and ICT in many aspects, teachers must first understand the issues, needs, and interests of their students. This is especially true when multimedia technology tools may be limited, in which case captioned instruction may be recognized as a potential medium to boost learning.

Future chatbot developers must ensure that chatbots can adapt (or be altered) to students' language proficiency if they want to create chatbots expressly for language practice (or at least with that in mind).

In addition, in line with the earlier idea, students need to view chatbots as a chance to learn more (or learn in a different way) than they could with a human partner. Future chatbot developers could collaborate with educators to improve this crucial component.

Human language proficiency is a crucial concern in human–technology language practice that must be addressed at the outset of the design process (Fryer et al., 2019). The easiest strategy might be to construct a wide variety of chatbots (or versions), both for different topics and a range of levels, as a chatbot that is sufficiently clever to adapt to learners' levels is still not yet accessible.

The idea that students should view chatbots as a chance to "learn more" implies that chatbots do not necessarily need to mimic human interaction. The participants in the current study believed they learned more with the chatbot for a variety of reasons, including the fact that it provided opportunities that human language learning partners could not.

There is a longer period for learning so that homework can be finished in the comfort of students' homes when AI is added to the English curriculum to improve critical thinking in children. Smartphones also have a slow network response time, which means that if the countdown expires, the instructions will be delayed.

Bold learning uses a lot of technology, including cellphones, and a large amount of Internet bandwidth. Nonetheless, to follow the online learning process, it is necessary to have a bigger budget because many students have a limited budget for broadband services. The current study focused on student engagement in their work, although the mixed method analyses gave preference to students' views of improved learning. While the reported merit "situational interest," for instance, was associated with above-average interest, it did not match the level of interest when students identified "learned more" as the chatbot's main benefit. This was especially important for the students who cited communication issues as the main drawback of chatbot interaction, specifically the three stable diffusion difference in chatbot conversation interest.

Therefore, we propose that ensuring that chatbots prioritize assisting students in learning more is a step toward resolving the communication and technological problems that chatbots frequently experience. The novelty impact found by Fryer et al. (2020) is a cause of serious concern given that improving motivation is a common driver for utilizing educational technology and the strength of chatbots. According to the findings of the current study, given sufficient time, the chatbot partner's decreasing interest can significantly increase.

The remaining question is whether, as it did not during the first term, the curiosity sparked by reengaging with the chatbot leads to interest in the students' course (and hence, the more general domain). For the core point of this study, integrating AI chatbots into the teaching and learning process is a sophisticated approach.

Students can interact with AI friends without any pressure to speak or chat. They can prepare ideas and share them with their AI friends as an extraordinary experience.

However, incorporating AI into the teaching and learning process still requires educators/lecturers to build and maintain a good rapport and facilitate students with the AI assistants.

This study proposes some implications for revolutionized teaching as part of efforts in Industry 5.0, namely, creating a fun learning environment that is essential for the success of any educational institution, and in particular, for the success of the teacher.

It will be challenging for colleges, educators, and educational organizations to meet the demands of Generation Alpha children if they do not change themselves and their curricula to follow a modern approach to education.

Technology is expanding the physical world and adding layers of digital knowledge to what can be seen with the naked eye.

4.5 CONCLUSION

Chatbots are a prime illustration of how AI is increasingly being used in language learning. The primary conclusions and implications may be drawn from this case study on knowledge, degree of satisfaction as part of students' affective engagement, and response regarding chatbot integration among prospective instructors (Khang et al., 2023b).

Their understanding of contemporary chatbots is restricted to the use of a few intelligent personal assistants.

Students use sophisticated learning tools that can boost their enthusiasm and motivation in language learning.

Incorporating AI chatbots increases students' interest and enthusiasm in learning and teaches students to work together, be solid, and help each other complete challenges or every question item from Replika and Kuki.

Because of the massive Industry 5.0, which forces everyone to be more active in adjusting to technology and AI as part of life, preparation, and recent developments in the use of AI for language acquisition, the EFL curriculum has to be improved.

Future educators should be well-versed in many chatbot varieties, as well as their advantages and limitations.

The knowledge gained from this research will help instructors better support their students as they apply AI lesson plans throughout the learning process by demonstrating how AI may be used to enhance the student engagement (Khang et al., 2023c), such as

- **ChatGPT chatbot**: ChatGPT chatbot is an AI chatbot developed by OpenAI and released in November 2022. It is built on top of OpenAI's GPT-3.5 and GPT-4 families of large language models and has been fine-tuned using both supervised and reinforcement learning techniques.
- **Google Bard AI chatbot**: Bard chatbot is a conversational AI chatbot developed by Google, based on the Language Model for Dialogue Applications family of large language models. It was developed as a direct response to the rise of OpenAI's ChatGPT and released in a limited capacity in March 2023 to lukewarm responses.
- **Bing AI chatbot**: The Bing AI chatbot was built by Microsoft using GPT-4 technology that was developed by OpenAI. GPT-4 is a generative AI model, which means that it can be trained by providing it with data, which it then synthesizes into answers using natural language.

As a result, students develop self-reliance and take charge of their education. Additionally, curriculum developers may use these results to include AI in cross-disciplinary learning methodologies in existing courses (Ahmad et al., 2023).

Consequently, they can apply learning techniques more effectively. Another area in which this research can be useful is to enhance students' cognitive, behavioral, affective engagement, and linguistic proficiency in EFL in the Indonesian context.

Findings that highlight the significance of student-centered learning can help students provide educators with new perspectives (Khang et al., 2022d).

In addition, the proficiency level of the students should be considered in the use and design of chatbots in future language exercises. Instead of focusing on chatbot's convenience, teachers may frame discussions as a chance for students to learn more and different topics than they can from a human language learning partner.

In contrast to how a human partner would typically do so, this may entail the scaffolding introduction of new vocabulary, syntax, and expressions. It might also entail repeating things clearly and often, which a human partner is unlikely to do.

REFERENCES

Aeni N., et al., (2021). "The students' needs in developing EFL materials ICT based," *OKARA: Jurnal Bahasa dan Sastra*, 15(2), 235–247. www.researchgate.net/profile /Muth-mainnah-2/publication/356786938_The_Students'_Needs_in_Developing_EFL_Materials_ICT_Based/links/61acb6cc092e735ae2e12c3b/The-Students-Needs-in-Developing-EFL-Materials-ICT-Based.pdf

Ahmad Al Yakin M., Khang A., Abdul Mukit M. Z., (2023). "Personalized social-collaborative iot-symbiotic platforms in smart education ecosystem," In *Smart Cities: IoT Technologies, Big Data Solutions, Cloud Platforms, and Cybersecurity Techniques* (1st ed.). CRC Press. https://doi.org/10.1201/9781003376064-15

Allouch M., Azaria A., Azoulay R., (2021). "Conversational agents: Goals, technologies, vision and challenges," *Sensors*, 21(24), 8448. www.mdpi.com/article/10.3390/s21248448

Axelson R. D., Flick A., (2010). "Defining student engagement," *Change: The Magazine of Higher Learning*, 43(1), 38–43. www.tandfonline.com/doi/abs/ 10.1080/07294360.2017.1370440

Barkley E. F., (2010). *Student Engagement Techniques: A Handbook for College Faculty*. Jossey-Bass. https://www.scirp.org/(S(351jmbntvnsjt1aadkposzje))/reference/References-Papers.aspx?ReferenceID=198494

Belda-Medina J., Calvo-Ferrer J. R., (2022). "Using chatbots as AI conversational partners in language learning," *Applied Sciences*, 12(17), 8427. www.mdpi.com/2076-3417/12/17/8427

Bhambri P., Rani S., Gupta G., Khang A., (2022). *Cloud and Fog Computing Platforms for Internet of Things*. CRC Press. https://doi.org/ 10.1201/9781032101507

Bibauw S., Van den Noortgate W., François T., Desmet P. (2022). *Dialogue Systems for Language Learning: A Meta-Analysis*. https://serge.bibauw.be/publication/2022/dialogue-systems-language-learning-meta-analysis/

Copeland B. J., (August 2023). "Alan Turing," *Encyclopedia Britannica*. https://www.britannica.com/biography/Alan-Turing

Fletcher A., (2015). "Defining student engagement: A literature review," *Soundout: Promoting Meaningful Student Involvement, Student Voice and Student Engagement*. https://soundout.org/defining-student-engagement-a-literature-review/

Fryer L., et al., (2020). *Bots for Language Learning Now: Current and Future Directions*. https://scholarspace.manoa.hawaii.edu/handle/10125/44719

Fryer L. K., Nakao K., Thompson A., (2019). "Chatbot learning partners: Connecting learning experiences, interest and competence," *Computers in Human Behavior*, 93, 279–289. https://doi.org/10.1016/j.chb.2018.12.023

Gao Y., Cui Y., (2022). "English as a foreign language teachers' pedagogical beliefs about teacher roles and their agentic actions amid and after COVID-19: A case study," *RELC Journal*. https://journals.sagepub.com/doi/abs/10.1177/00336882221074110

Hahanov V., Khang A., Litvinova E., Chumachenko S., Hajimahmud V. A., Alyar A. V., (2022). "The key assistant of smart city—sensors and tools," In *AI-Centric Smart City Ecosystems: Technologies, Design and Implementation* (1st ed.). CRC Press. https://doi. org/10.1201/9781003252542-17

Hajimahmud V. A., Khang A., Hahanov V., Litvinova E., Chumachenko S., Alyar A. V., (2022). "Autonomous robots for smart city: Closer to augmented humanity," *AI-Centric Smart City Ecosystems: Technologies, Design and Implementation* (1st ed.). CRC Press. https://doi.org/10.1201/9781003252542-7

Haristiani N., (2019). "Artificial intelligence (AI) chatbot as language learning medium: An inquiry," *Journal of Physics: Conference Series*, 1387(1) (IOP Publishing). https://iop-science.iop.org/article/10.1088/1742-6596/1387/1/012020/meta

Hsu T.-C., Mu-Sheng C., (2022). "The engagement of students when learning to use a personal audio classifier to control robot cars in a computational thinking board game," *Research and Practice in Technology Enhanced Learning*, 17(1), 1–17. https://doi.org/10.1186/s41039-022-00202-1

Khang A., (2021). "Material4Studies," *Material of Computer Science, Artificial Intelligence, Data Science, IoT, Blockchain, Cloud, Metaverse, Cybersecurity for Studies*. www.researchgate.net/publication/370156102_Material4Studies

Khang A., Gupta S. K., Rani S., Karras D. A., (2023a). *Smart Cities: IoT Technologies, Big Data Solutions, Cloud Platforms, and Cybersecurity Techniques* (1st ed.). CRC Press. https://doi.org/10.1201/9781003376064

Khang A., Gupta S. K., Shah V., Misra A., (2023b). *AI-Aided IoT Technologies and Applications in the Smart Business and Production* (1st ed.). CRC Press. https://doi. org/10.1201/9781003392224

Khang A., Gupta S. K., Hajimahmud V. A., Babasaheb J., Morris G., (2023c). *AI-Centric Modelling and Analytics: Concepts, Designs, Technologies, and Applications* (1st ed.). CRC Press. https://doi.org/10.1201/9781003400110

Khang A., Rani S., Sivaraman A. K., (2022a). *AI-Centric Smart City Ecosystems: Technologies, Design and Implementation* (1st ed.). CRC Press. https://doi.org/10.1201/9781003252542

Khang A., Ragimova N. A., Hajimahmud V. A., Alyar A. V., (2022b). "Advanced technologies and data management in the smart healthcare system," In *AI-Centric Smart City Ecosystems: Technologies, Design and Implementation* (1st ed.). CRC Press. https://doi. org/10.1201/9781003252542-16

Khang A., Ragimova N. A., Hajimahmud V. A., Alyar A. V., (2022c). "Advanced technologies and data management in the smart healthcare system," In *AI-Centric Smart City Ecosystems: Technologies, Design and Implementation* (1st ed.). CRC Press. https://doi. org/10.1201/9781003252542-16

Khang A., Hahanov V., Abbas G. L., Hajimahmud V. A., (2022d). "Cyber-physical-social system and incident management," In *AI-Centric Smart City Ecosystems: Technologies, Design and Implementation* (1st ed.). CRC Press. https://doi.org/10.1201/9781003252542-2

Khanh H. H., Khang A., (2021). "The role of artificial intelligence in blockchain applications," In *Reinventing Manufacturing and Business Processes through Artificial Intelligence*, pp. 20–40. CRC Press. https://doi.org/10.1201/9781003145011-2

Kılıçkaya F., (2020). "Using a chatbot, Replika, to practice writing through conversations in L2 English: A case study," In *New Technological Applications for Foreign and Second Language Learning and Teaching*, pp. 221–238. IGI Global. www.igi-global.com/chapter/using-a-chatbot-replika-to-practice-writing-through-conversations-in-l2-english/251555

Lee J., Moslem A., Jaskaran S., (2019). "A blockchain enabled cyber-physical system architecture for industry 4.0 manufacturing systems," *Manufacturing Letters*, 20, 34–39. www.sciencedirect.com/science/article/pii/S2213846319300264

Luke J., Khang A., Vadivelraju C., Antony R. P., Kumar S., (2024). "Smart city concepts, models, technologies and applications," In *Smart Cities: IoT Technologies, Big Data*

Solutions, Cloud Platforms, and Cybersecurity Techniques (1st ed.). CRC Press. https://doi.org/10.1201/9781003376064-1

Machado C. G., Mats Peter W., Ribeiro da Silva E. H. D., (2020). "Sustainable manufacturing in Industry 4.0: An emerging research agenda," *International Journal of Production Research*, 58(5), 1462–1484. www.tandfonline.com/doi/abs/10.1080/00207543.2019.1652777

Muthmainnah I. S., Prodhan M., Ibrahim O., (2022). "Playing with AI to investigate human-computer interaction technology and improving critical thinking skills to pursue 21st century age," *Education Research International.* https://doi.org/10.1155/2022/6468995

Pandorabots, (2021). *Build Intelligent Conversational Agents on the Leading Platform.* www.pandorabots.com

Paschek D., Caius-Tudor L., Elif O., (2022a). "Industry 5.0 challenges and perspectives for manufacturing systems in the Society 5.0," *Sustainability and Innovation in Manufacturing Enterprises: Indicators, Models and Assessment for Industry 5.0*, 17–63. https://link.springer.com/chapter/10.1007/978-981-16-7365-8_2

Paschek D., Luminosu C.-T., Ocakci E., (2022b). *Industry 5.0 Challenges and Perspectives for Manufacturing Systems in the Society 5.0 BT—Sustainability and Innovation in Manufacturing Enterprises: Indicators, Models and Assessment for Industry 5.0* (A. Draghici & L. Ivascu (Eds.), pp. 17–63. Springer. https://doi.org/10.1007/978-981-16-7365-8_2

Rana G., Khang A., Sharma R., Goel A. K., Dubey A. K., (Eds.). (2021). *Reinventing Manufacturing and Business Processes through Artificial Intelligence*. CRC Press. https://doi.org/10.1201/9781003145011

Rani S., Bhambri P., Kataria A., Khang A., (2022). "Smart city ecosystem: Concept, sustainability, design principles and technologies," In *AI-Centric Smart City Ecosystems: Technologies, Design and Implementation* (1st ed.). CRC Press. https://doi.org/10.1201/9781003252542-1

Rani S., Bhambri P., Kataria A., Khang A., Sivaraman A. K., (2023). *Big Data, Cloud Computing and IoT: Tools and Applications* (1st ed.). Chapman and Hall/CRC. https://doi.org/10.1201/9781003298335

Rani S., Chauhan M., Kataria A., Khang A., (Eds.). (2021). "IoT equipped intelligent distributed framework for smart healthcare systems," In *Networking and Internet Architecture*. CRC Press. https://doi.org/10.48550/arXiv.2110.04997

Reeve J., Stephanie H. S., (2020). "How teachers can support students' agentic engagement," *Theory into Practice*, 59(2), 150–161. https://doi.org/10.1080/00405841.2019.1702451

Saud S., Nur A., Laelah A., (2022). "Leveraging Bamboozles and Quizziz to Engage EFL students in online classes," *International Journal of Language Education*, 6(2), 169–182. https://eric.ed.gov/?id=EJ1355250

Sindhwani R., Afridi S., Kumar A., Banaitis A., Luthra S., Singh P. L., (2022). "Can Industry 5.0 revolutionize the wave of resilience and social value creation? A multi-criteria framework to analyze enablers," *Technology in Society*, 68, 101887. https://doi.org/10.1016/j.techsoc.2022.101887

Subhashini R., Khang A., (2024). "The role of Internet of Things (IoT) in smart city framework," In *Smart Cities: IoT Technologies, Big Data Solutions, Cloud Platforms, and Cybersecurity Techniques* (1st ed.). CRC Press. https://doi.org/10.1201/9781003376064-3

Vrushank S., Suketu J., Khang A., (2024). "Automotive IoT: Accelerating the automobile industry's long-term sustainability in smart city development strategy," In *Smart Cities: IoT Technologies, Big Data Solutions, Cloud Platforms, and Cybersecurity Techniques* (1st ed.). CRC Press. https://doi.org/10.1201/9781003376064-9

Wang B., et al., (2018). *Artificial Intelligence and Education*. Springer. https://link.springer.com/chapter/10.1007/978-981-10-8639-7_42

Wang X., et al., (2022). "Learners' perceived AI presences in AI-supported language learning: a study of AI as a humanized agent from community of inquiry," *Computer Assisted Language Learning*, 1–27. www.tandfonline.com/doi/abs/10.1080/ 09588221.2022.2056203

Yang H., et al., (2022). "Implementation of an AI chatbot as an English conversation partner in EFL speaking classes," *ReCALL*, 34(3), 327–343. www.cambridge.org/core/journals/recall/article/implementation-of-an-ai-chatbot-as-an-english-conversation-partner-in-efl-speaking-classes/0EB8C4E40E033F9D4165D8EBA843EEDB

Zhang Z., (2022). "Promoting student engagement with feedback: insights from collaborative pedagogy and teacher feedback," *Assessment & Evaluation in Higher Education*, 47(4), 540–555. www.tandfonline.com/doi/abs/10.1080/02602938.2021.1933900

5 Application of Artificial Intelligence in AgroWeb

*Reethika A., Danush R., Akshaya E.,
and Kanaga Priya P.*

5.1 INTRODUCTION

The project's goal is to aid the expansion of the agricultural sector, which is the foundation of the Indian economy. Modernization must satisfy the demands of the expanding market, and the use of technology is crucial.

Approximately 18% of the gross domestic product and 40% of the net domestic product go to the agricultural sector. Although it has a significant impact on the economy, the agricultural sector has lost ground as an effect of improper practices of technology that could have benefited farmers and others involved in the industry (Khang et al., 2023a).

The most populated in the world, India, will need to fulfill demand, which will hurt the economy and make life difficult for families in the lower and middle classes because food will not be as inexpensive. Many policies have been created for this industry to ensure that farmers have little risk because they are given financial support. The minimum support price was introduced to ensure that nothing is wasted, but very few policies have been created to improve the quality and yield of production.

Very few policies aid farmers in expanding their capabilities. Farmers in the 21st century require not just policies but also technologies that could directly benefit them by boosting their output yield and quality.

Despite having enormous machines, farmers still need technology assistance, such as Internet of Things sensors installed in their fields that can sense information and relay real-time data. The implementation of such a technology is also challenging because the majority of farmers lack formal education (Khang et al., 2022a).

As a result, everything needs to be made clear and simple so that everyone can understand it. To raise their harvest, many farmers still rely on outdated techniques, experiences, and rumors that create issues.

Moreover, not all farmers have expansive fields or work in agriculture as their primary profession; many rely on farming as their secondary source of income, which means they consume the crops they cultivate and at the same time sell them for very little money, primarily to neighborhood merchants (Khang, 2023c).

The issue with the Internet that existed around ten years ago that connection was not possible everywhere has also been resolved. Nowadays, the majority of locations have Internet access to distribute information.

While some locations may not have access to the Internet, telephone connections are available everywhere and can be used to transmit information.

Several breakthroughs have already been made as a result of the use of technology in the agricultural sector. However, many others have failed because the technology is overly complex, cumbersome to use, and is perceived to be more harmful than helpful.

The problems faced by farmers have increased. One major conclusion is that whatever technology is deployed should not be a burden. Another issue is connectivity problem, which means that most technological devices need a constant connection to the cloud to provide assistance, which is not feasible in remote locations (Khang et al., 2023b).

Another problem is the setup fee, which means that neither the government nor farmers could afford the technology's initial setup fee. After consideration of each factor, it was determined that the technology needs to be a cutting edge while still being widely available. This should be simple for everyone.

For this, machine learning (ML) is used to forecast values, and a web application was created to make access for everyone simpler and easier.

5.2   RELATED WORK

Crop prediction using ML is an active research area that has received considerable attention in recent years. The following recent studies may be of interest:

- **Crop yield prediction using ML**: An overview of the study by Ramesh and Rudra (2021) provided recent advances in crop yield expectations using ML, including data collection, pre-processing, and model selection. In a comparative study of ML algorithms for crop yield prediction (Bharath et al., 2020), support vector regression, decision trees, and random forests were evaluated for their efficacy in predicting agricultural yields.
- **Crop yield prediction using deep learning techniques**: Saeed and Lizhi (2019) reviewed convolutional neural networks and recurrent neural networks together with other contemporary deep learning approaches for agricultural production prediction.
- **Crop yield prediction using ML**: Van Klompenburg et al. (2020) made a framework suggestion for crop vintage forecast that combines ML with remote sensing data.
- **Crop yield prediction using ML**: In a case study of rice production (Niketa & Owaiz, 2016), crop output was predicted using ML algorithms for rice production.

Smart irrigation systems using ML are another active research area. The following recent studies may be of interest:

- **Smart irrigation systems**: David et al. (2023) provided an overview of smart irrigation systems involving the planning and optimization of irrigation using ML techniques. Youness et al. (2022) presented a smart irrigation system using ML techniques, incorporating support vector machines (SVMs) and decision trees, to forecast agricultural water needs.

- **Machine learning for smart irrigation**: Meriç et al. (2022) reviewed some applications related to smart irrigation systems based on ML techniques, including soil moisture sensors, weather forecasting, and crop yield prediction.

Plant disease prediction using ML is another active research area. The following recent studies may be of interest:

- **A review on ML approaches for plant disease detection and diagnosis**: Majji and Kumaravelan (2021) provided an overview of ML methods for recognizing and treating plant diseases, including image- and non-image-based methods.
- **Deep learning for plant disease detection and diagnosis**: Konstantinos (2018) reviewed convolutional neural networks and transfer learning, two emerging deep learning techniques for detecting and diagnosing herbal diseases.
- **Plant disease classification using ML techniques**: Shruthi et al. (2019) reviewed decision trees, random forests, and SVMs as contemporary ML methods for classifying plant diseases.
- **Plant disease diagnosis using ML techniques**: Liu and Wang (2021) provided ML algorithms for plant disease diagnosis, including clustering, classification, and regression algorithms.
- **Machine learning algorithms for plant disease classification**: Sandeep et al. (2021) analyzed the performance of artificial neural networks (ANNs), decision trees, and SVMs as ML techniques for the categorization of plant diseases.

5.3 PRE-HARVESTING

Machine learning can be used to optimize harvesting schedules by analyzing crop maturity, weather conditions, and equipment availability, allowing farmers to maximize yield and minimize waste.

Machine learning algorithms can be used to analyze historical data on weather patterns, soil conditions, and crop yields to predict future yields. This information can help farmers plan their harvesting strategies and make decisions about crop management.

5.3.1 CROP PREDICTION

Crop prediction is the process of forecasting crop yield for a particular season or growing period. Accurate crop yield predictions can help farmers plan crop management practices for the upcoming growing season. This includes decisions regarding the amount and timing of fertilizers, irrigation, and pest management practices.

With accurate crop yield predictions, farmers can make informed decisions to improve crop yields and reduce waste. Crop prediction helps farmers to allocate resources efficiently. This includes resources, such as labor, equipment, and fertilizers.

With accurate crop yield predictions, farmers can optimize their resource use and reduce waste, resulting in cost savings.

Accurate crop yield predictions help to ensure food security by predicting crop yields and preventing shortages. This is particularly important in regions where agriculture is the main source of food and crop failures can lead to food shortages and famine.

Crop prediction can impact the economy by predicting the market supply and demand. Accurate crop yield predictions can help prevent overproduction, which can lead to lower crop prices and lower financial losses for farmers. Accurate crop yield predictions can also help to prevent underproduction, which can lead to higher crop prices and food shortages.

Data collection is a critical step in ML for crop predictions. To build an accurate ML model, relevant data must be collected, including weather conditions, soil fertility, irrigation, and pest management practices. These data can be obtained from a variety of sources, such as government agencies, weather stations, sensors, and satellite imagery.

Weather data can be collected from government agencies or weather stations that measure the temperature, precipitation, and sunlight. Soil data can be collected from soil samples or through sensors that measure nutrient content and pH levels.

Irrigation data can be collected from sensors that measure the amount and timing of the water application. Pest management data can be collected from records of pesticide applications and observations of pest populations.

Once data are collected, the next step in ML for crop prediction is data pre-processing. It involves several techniques to clean and convert raw data so that they can be used effectively by ML algorithms.

- The first step in data pre-processing is data cleaning, which involves removing missing values, outliers, and duplicates from the data. Missing values can be filled using interpolation or imputation methods. Outliers caused by measurement errors can be removed or adjusted. Duplicates can be removed to ensure that the data are not biased toward certain observations.
- The next step in data pre-processing is data transformation, which involves scaling, normalization, and feature extraction. Scaling is used to adjust the range of values for each feature so that they are comparable. Normalization is used to adjust the values of each feature such that they have a common scale. Feature extraction involves the identification and selection of the most relevant features for crop prediction.
- Data integration involves combining data from multiple sources to create a unified dataset in which systems using ML can be applied. This may involve matching data from different sources based on time and location and combining them into a single dataset.

Overall, data pre-processing is critical for ML in crop prediction because it improves the accuracy and relevance of the data used by ML algorithms.

Proper data pre-processing can help to reduce noise, bias, and errors in the data and ensure that the ML model can accurately predict crop yields based on the input data.

5.3.2 MODEL SELECTION FOR CROP PREDICTION

Forest of chance is a powerful ML algorithm for crop prediction that can provide accurate results with relatively low computational cost. The algorithm can handle large and complex datasets, and it is robust to noise and outliers in the data, as shown in Table 5.1.

One of the key advantages of random forest is that it can handle nonlinear relationships between the input variables and the target variable. It achieves this by building decision trees based on random subsets of the data and then combining the predictions of all trees to produce a final prediction. This approach can capture complex interactions between the variables and can lead to more accurate predictions.

Random forest also has the ability to handle missing data, which is common in agricultural datasets. The algorithm can impute missing values using techniques, such as mean imputation or regression imputation, allowing the model to be trained on a complete dataset. Another advantage of random forest is its ability to handle categorical variables.

The algorithm converts categorical variables into binary dummy variables, allowing them to be included in the model. In terms of performance metrics, random forest can achieve low values of mean absolute error and root mean square error (RMSE), indicating accurate predictions.

Coefficient of determination (R^2) can also be high, indicating a strong relationship between the input variables and the target variable.

Artificial neural networks are popular algorithms for crop prediction. They are a set of algorithms inspired by the structure of the human brain and are capable of learning intricate patterns from vast amounts of data.

In crop prediction, ANNs can be used to model the relationship between various crop and environmental elements, such as soil type, climate, and precipitation, and predict crop yields.

TABLE 5.1
Result of Various Algorithms.

Algorithm	Accuracy (%)
ANN	92
SVM	98.5
Random forest	99.3
Decision tree	97.5

Artificial neural networks can handle nonlinear relationships between variables and can identify hidden patterns in the data that may not be apparent through traditional statistical methods.

Artificial neural networks can also be trained using historical data to accurately predict future crop yields. However, the accuracy of ANNs may be affected by the quality and quantity of data used for training, and the complexity of the model may require significant computing power.

Despite these challenges, ANNs remain powerful tools for crop prediction and are widely used in the agricultural industry.

Support vector machine is a supervised learning algorithm that can also be used for crop prediction.

Support vector machines are particularly effective when the number of features is high compared with the number of observations, which is common in crop prediction.

Support vector machines work by finding an ideal hyperplane that divides the data into many groups based on their characteristics.

In crop prediction, SVMs can be used to model the relationship between crop yield and various environmental factors, such as soil properties and weather patterns.

Support vector machines can handle nonlinear relationships between variables and are effective for identifying complex patterns in the data. They are also robust to outliers, which can occur in agricultural data owing to the natural variability in crop yields.

However, SVMs can be sensitive to the choice of kernel function and require careful tuning of the parameters to achieve optimal performance.

Despite these challenges, SVMs remain a popular choice for crop prediction and have been successfully used in numerous studies to predict crop yield.

Decision trees are another type of algorithm that can be used for crop prediction. They are simple and interpretable algorithms that can help identify the most important variables for crop prediction.

Decision trees work by recursively splitting the data based on the most informative feature, until the data are classified into specific classes or the target variable is predicted.

In crop prediction, the link between agricultural yields and numerous environmental parameters, such as soil type, climate, and precipitation, can be modeled using trees.

Decision trees can handle both continuous and categorical data and can be easily visualized, making them useful for communicating results to stakeholders.

However, decision trees may suffer from overfitting, which can occur when the model is too complex and captures data noise. Ensemble methods, such as random forests, can help alleviate overfitting and improve the accuracy of decision trees.

Despite these challenges, decision trees remain a useful tool for crop prediction and provide a valuable understanding of the connections between environmental factors and crop yield, as shown in Figure 5.1.

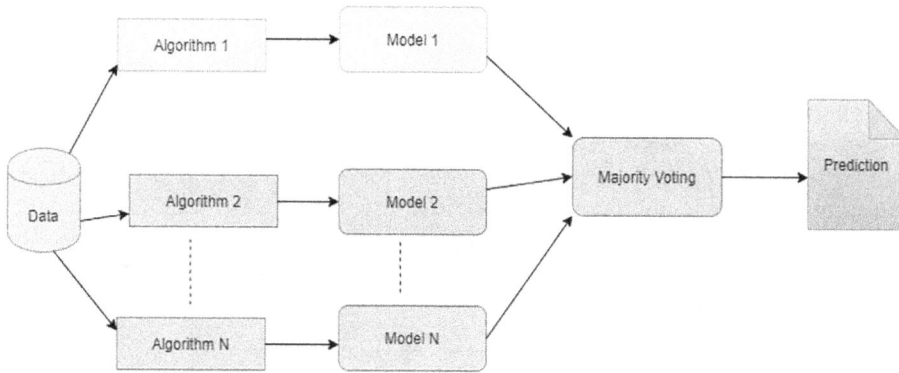

FIGURE 5.1   Crop Prediction Accuracy Model.

5.3.3   Seed Prediction

Seed prediction is a task that can be addressed by various ML algorithms, including the use of neural networks. One neural network architecture that has been successfully used for seed prediction is multilayer perceptron (MLP).

To use MLP for seed prediction, we first need to collect a dataset of features related to the seeds that we want to predict. These features may include the physical characteristics of the seed, such as size, shape, color, texture, and weight, as well as other relevant information, such as the plant species, region where the seed was collected, and time of year.

Seed prediction using neural networks involves the use of ML algorithms to predict the likelihood of seed germination and growth in a healthy plant. This can be accomplished by training a neural network model on a large dataset of labeled seed samples, where each sample includes information about the seed's physical properties, environmental factors, and growth outcomes.

The neural network model is trained to identify patterns and relationships between different features of the seed samples and the corresponding growth outcomes. After training, the model can be used to make predictions. The likelihood of a new seed germinating and growing is based on its properties and the surrounding environment.

To achieve accurate seed prediction using neural networks, it is important to carefully select and engineer the input features of the model. This may include factors, such as seed size, shape, color, texture, moisture content, temperature, and soil quality.

In addition, the neural network architecture and training parameters must be optimized to ensure the best possible predictive performance.

5.3.4   Crop Disease Prediction

Plant disease prediction using ML is a promising approach for early detection and management of plant diseases.

Machine learning models can be trained to accurately predict the occurrence of diseases in plants by analyzing visual symptoms from images. This can help farmers and plant pathologists quickly identify diseases, take appropriate measures to prevent the spread of the disease, and minimize crop losses.

The development of accurate and reliable ML simulations for plant disease prediction relies on the accessibility of high-quality datasets that can capture a wide range of disease symptoms across different plant species and environmental conditions.

Furthermore, the deployment of ML models for plant disease prediction requires access to adequate computing resources and infrastructure to process large amounts of image data in real time.

5.3.4.1 Model Selection for Crop Disease Prediction

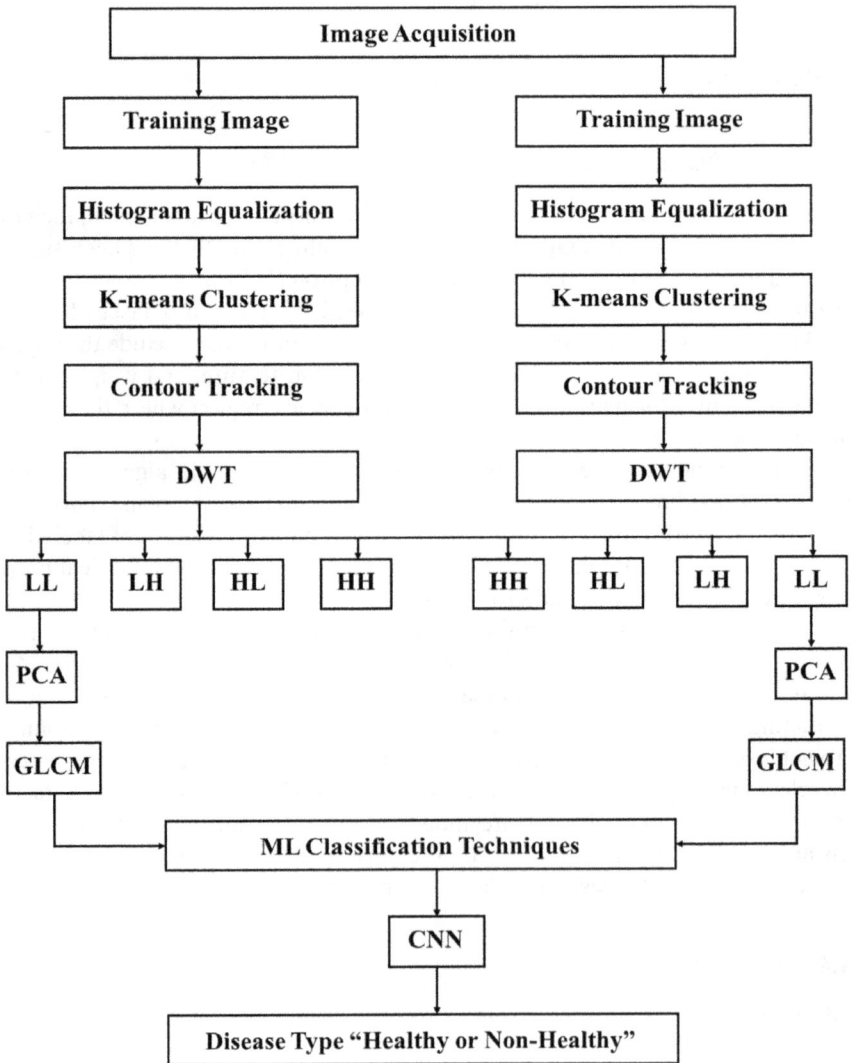

FIGURE 5.2 Overview of Method.

5.3.4.2 Transform

Discrete wavelet transform (DWT) is a signal processing technique that has been used in various applications, including image and signal processing, pattern recognition, and ML.

In recent years, DWT has been applied in the field of plant disease detection and classification. The basic idea behind using DWT for plant disease detection is the manifestation of patterns of variation in plant texture and color, which can be captured and analyzed using wavelet coefficients.

Discrete wavelet transform decomposes an image into different frequency bands, and the wavelet coefficients of each band represent the image information at that particular frequency scale. By analyzing the wavelet coefficients, it is possible to identify the texture and color patterns associated with particular plant diseases.

To use DWT for plant disease detection, the first step is to acquire high-resolution images of plants. These images can be acquired using various imaging techniques, such as digital cameras or hyperspectral imaging.

Once the images are acquired, they are trained, and the model can be used to make predictions. The image is decomposed into different frequency bands using a wavelet basis function, and the wavelet coefficients of each band are calculated, as shown in Figure 5.3.

5.3.4.3 Classifier

Plant disease prediction using convolutional neural networks (CNNs) is a talented approach for accurately identifying and classifying diseases in plants. Convolutional neural networks are a type of deep learning algorithms that are particularly effective at processing and analyzing images, making them well-suited for identifying plant diseases based on visual symptoms, as shown in Figure 5.4.

To build a plant disease prediction model using CNNs, a dataset of images of healthy and diseased plants is required. These images are split into training, validated, and test sets.

- The CNN is trained on the training set using backpropagation, which involves adjusting the weights of the network to minimize the distinction between the real and expected labels. The testing set is used to assess the model's performance during training, whereas the validation set is used to assess the overall accuracy of the model.
- The CNN model typically consists of several convolutional layers that sort the photographs into many disease groups by extracting characteristics from the input images and numerous fully linked layers. Data augmentation techniques, such as rotation, flipping, and cropping, can be used to enhance the model and size of the training set.

Overall, predictions of plant diseases using CNNs have been encouraging and have the potential to significantly increase the effectiveness and precision of disease recognition and control in agriculture, as shown in Figure 5.5.

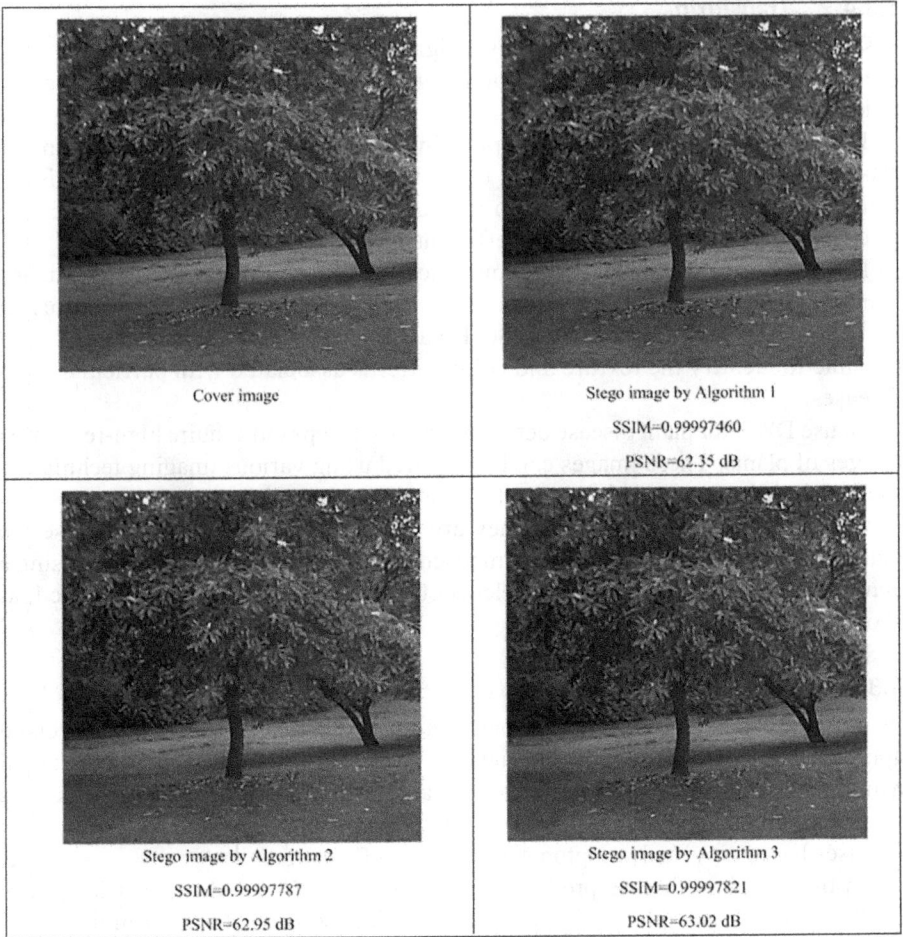

FIGURE 5.3   DWT Transformed Image.

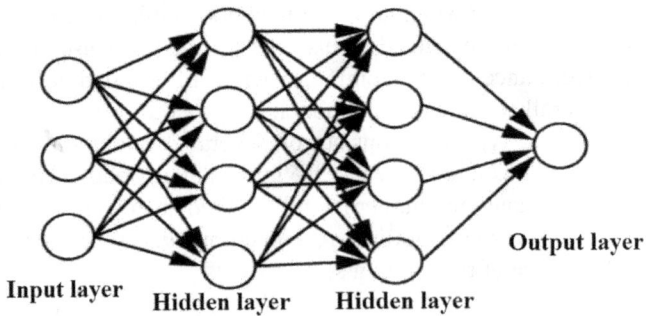

FIGURE 5.4   CNN Algorithm.

Source: Khang (2021)

```
training started...Wait for 200 seconds...
training started...
Elapsed time is 2.033151 seconds.
Elapsed time is 2.239313 seconds.
...training finished.
testing started....
test error is
Elapsed time is 1.085832 seconds.
CNN Accuracy = 99.0909
CNN Precision = 0.9913
CNN Sensitivity = 0.99091
CNN Specificity = 0.99773

confmatrix =

    22      1      0      0      0
     0     21      0      0      0
     0      0     22      0      0
     0      0      0     22      0
     0      0      0      0     22
```

FIGURE 5.5 CNN Accuracy Result.

5.3.5 IRRIGATION SYSTEM

Traditional agriculture is the cornerstone of global development. Despite this, because of the population's exponential growth and increasing consumption, farmers need water to irrigate the land.

Because of the scarcity of this resource, farmers need a solution that changes the way they operate. To keep up with and meet demand, new technology has been implemented, giving rise to the concept of "Farm 4.0".

The irrigation method suggested in this study is clever, adaptable, and low in cost and usage. This smart agriculture strategy is based on ML algorithms. To achieve this, we used a set of sensors (soil, humidity, temperature, and rain), which were placed in an environment that promoted better plant growth for months.

We then used the Node-RED platform and MongoDB to collect data based on the acquisition map from these sensors. Based on our data, we used a variety of models, including k-nearest neighbor (KNN), logistic regression, neural networks, SVM, and naive Bayes (Khang et al., 2022b).

In comparison with other models, the findings demonstrated that KNNs perform better, with a recognition rate of 98.3% and an RMSE of 0.12.

Finally, to provide better visualization and oversight of our surroundings, we offered a web application that combined various data produced by the sensors with forecasts from our model.

5.3.6 IRRIGATION SYSTEM MODEL

K-nearest neighbor is often used to classify regression tasks. Input data are classified by finding the *k*-nearest data points in the training set to the input data point.

The value of *k* is a hyper-parameter that can be tuned to achieve better performance. In the case of classification, the output is the mode of the KNN classes.

For example, if majority of the KNNs belong to class A, then the input data point will be classified as class A. In the case of regression, the output is the average of the KNN values.

An irrigation system using KNN can be designed to optimize water resources. This system uses a KNN model trained using a dataset consisting of information about weather conditions, soil type, and moisture levels.

The KNN algorithm predicts the moisture level under the soil from the input data fetched from sensors or other tools. To implement this system, the first step is to collect the necessary information and define the features that will be used to predict moisture levels.

The data are then pre-processed, split into several training sets, and implemented in the testing sets. The KNN model is trained, and its performance is evaluated using a testing method.

Once the KNN model is upgraded, it can be integrated with an irrigation system to provide recommendations on when and how much water should be applied to the soil. This system can be automated, making it easy to use and efficient.

By optimizing the use of water resources, this irrigation system can help reduce water waste, increase crop yield, and improve the overall sustainability of agriculture.

5.4 CONCLUSION

Machine learning has great potential to revolutionize agriculture by enabling farmers to make data-driven decisions, improve crop yield, reduce costs, and optimize resource usage.

Using ML algorithms, farmers collect, analyze, and interpret large amounts of data, including weather patterns, soil characteristics, and crop growth, to make informed decisions regarding irrigation, fertilization, and pest control.

Machine learning can also be used to estimate crop yield, classify potential disease outbreaks, and improve planting schedules.

However, the successful adoption of ML in agriculture requires overcoming several challenges, including the availability and quality of data, need for specialized technical skills, and cost of implementing technology, as shown in Figure 5.6.

It is important to ensure that ML algorithms are transparent, fair, and unbiased to prevent unintended consequences and guarantee that they benefit all farmers equally.

Overall, the benefits of ML in agriculture are significant, and further research and development is likely to play an increasingly basic role in improving global food security and sustainability (Khanh & Khang, 2021).

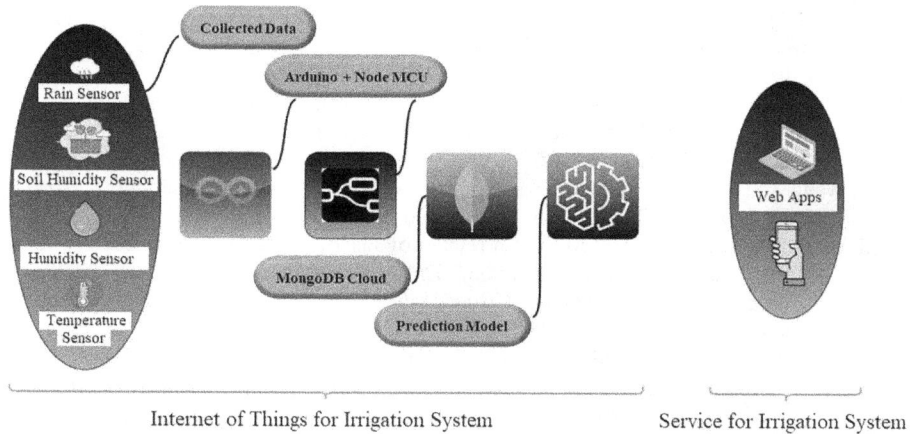

Internet of Things for Irrigation System   Service for Irrigation System

FIGURE 5.6   Irrigation Model.

5.5   RECOMMENDATION

Future studies could employ ensemble ML on the entire training set to juxtapose results with a study based on feature selection (Rana et al., 2021).

In addition, the out-of-sample data sampling approach can be further used to evaluate the performance of the predictive models in comparison with the in-sample approach presented in this chapter.

REFERENCES

Bharath S., Yeshwanth S., Yashas B. L., Vidyaranya R. J., "Comparative analysis of machine learning algorithms in the study of crop and crop yield prediction," *IJERTCON-V8IS14008, NCETESFT – 2020* (Volume 8 – Issue 14). www.academia.edu/download/64341632/comparative-analysis-of-machine-IJERTCONV8IS14008.pdf

Khang A., (2021). "Material4Studies," *Material of Computer Science, Artificial Intelligence, Data Science, IoT, Blockchain, Cloud, Metaverse, Cybersecurity for Studies.* www.researchgate.net/publication/370156102_Material4Studies

Khang A., Gupta S. K., Rani S., Karras D. A., (Eds.). (2023a). *Smart Cities: IoT Technologies, Big Data Solutions, Cloud Platforms, and Cybersecurity Techniques.* CRC Press. https://doi.org/10.1201/9781003376064

Khang A., Rana G., Tailor R. K., Hajimahmud V. A., (Eds.). (2023b). *Data-Centric AI Solutions and Emerging Technologies in the Healthcare Ecosystem.* CRC Press. https://doi.org/10.1201/9781003356189

Khang A., (2023c). *Advanced Technologies and AI-Equipped IoT Applications in High-Tech Agriculture* (1st ed.). IGI Global Press. https://doi.org/10.4018/9781668492314

Khang A., Rani S., Sivaraman A. K., (2022a). *AI-Centric Smart City Ecosystems: Technologies, Design and Implementation* (1st ed.). CRC Press. https://doi.org/10.1201/9781003252542

Khang A., Hahanov V., Abbas G. L., Hajimahmud V. A., (2022b). "Cyber-physical-social system and incident management," In *AI-Centric Smart City Ecosystems: Technologies, Design and Implementation*, Vol. 2, No. 15 (1st ed.). CRC Press. https://doi.org/10.1201/9781003252542-2

Khanh H. H., Khang A., (2021). "The role of artificial intelligence in blockchain applications," *Reinventing Manufacturing and Business Processes through Artificial Intelligence*, 2(20–40) (CRC Press). https://doi.org/10.1201/9781003145011-2

Konstantinos P. F., (2018). "Deep learning models for plant disease detection and diagnosis," *Computers and Electronics in Agriculture*, 145, 311–318. www.sciencedirect.com/science/article/pii/S0168169917311742

Liu J., Wang X., (2021). "Plant diseases and pests detection based on deep learning: A review," *Plant Methods*, 17, 22. https://doi.org/10.1186/s13007-021-00722-9.

Majji V. A., Kumaravelan G., (2021). "A review of machine learning approaches in plant leaf disease detection and classification," *Third International Conference on Intelligent Communication Technologies and Virtual Mobile Networks* (ICICV). https://ieeexplore.ieee.org/abstract/document/9388488/

Meriç Ç., Selami B., (2022). "Smart irrigation systems using machine learning and control theory," *The Digital Agricultural Revolution: Innovations and Challenges in Agriculture through Technology Disruptions, Chapter 3*. https://doi.org/10.1002/9781119823469

Niketa G., Owaiz P., (2016). "Rice crop yield prediction using artificial neural networks," *IEEE Technological Innovations in ICT for Agriculture and Rural Development (TIAR)*. https://doi.org/10.1109/TIAR.2016.7801222.

Ramesh K., Rudra K., (2021). "Crop yield prediction using machine learning algorithm," *5th International Conference on Intelligent Computing and Control Systems (ICICCS)*. https://doi.org/10.1109/ICICCS51141.2021.9432236.

Rana G., Khang A., Sharma R., Goel A. K., Dubey A. K., (Eds.). (2021). *Reinventing Manufacturing and Business Processes through Artificial Intelligence*. CRC Press. https://doi.org/10.1201/9781003145011

Rani S., Bhambri P., Kataria A., Khang A., (Dec 30, 2022). "Smart city ecosystem: Concept, sustainability, design principles and technologies," In *AI-Centric Smart City Ecosystems: Technologies, Design and Implementation*, Vol. 1, No. 20 (1st ed.). CRC Press. https://doi.org/10.1201/9781003252542-1

Saeed K., Lizhi W., (2019). "Crop yield prediction using deep neural networks," *Original Research Article, Frontiers in Plant Science*. www.frontiersin.org/articles/10.3389/fpls.2019.00621/full

Sandeep K., Arpit J., Anand Prakash S., Satyendr S., et al., "A comparative analysis of machine learning algorithms for detection of organic and nonorganic cotton diseases," *Mathematical Problems in Engineering*, 2021, Article ID 1790171. https://doi.org/10.1155/2021/1790171

Shruthi U., Nagaveni V., Raghavendra B. K., (2019). "A review on machine learning classification techniques for plant disease detection," *5th International Conference on Advanced Computing & Communication Systems (ICACCS)*. https://ieeexplore.ieee.org/abstract/document/8728415/

Tace Y., Tabaa M., Elfilali S., Leghris C., Bensag Hassna B., Renault E., (2022). "Smart irrigation system based on IoT and machine learning," *Energy Reports*, 8(Suppl. 9), 1025– 036. www.sciencedirect.com/science/article/pii/S2352484722013543

Vallejo-Gómez D., Marisol O., Hincapié C., (2023). "Smart irrigation systems in agriculture: A systematic review," *Special Issue Proposal Application*, 13(2), 342. https://doi.org/10.3390/agronomy1302034225

Van Klompenburg T., Kassahun A., Catal C., (2020). "Crop yield prediction using machine learning: A systematic literature review," *Computers and Electronics in Agriculture*, 177, 105709. www.sciencedirect.com/science/article/pii/S0168169920302301

6 Natural Language Processing

A Study of State of the Art

Nipun Sharma and Swati Sharma

6.1 INTRODUCTION

Natural language processing (NLP) is an amalgamation of both technology and language. Technology encompasses computer science and artificial intelligence (AI). Language covers both text and speech.

Over the generations, humans have been interacting and communicating with each other through language. The exchange of ideas occurs through speech or written text. In contrast, machines are programmed using low-level or high-level scripting.

Human–computer interaction (HCI) has been in existence for the past few years. To make it more interactive and intuitive, the intersection between human-to-human interaction (HHI) and HCI has been explored and researched at lightning speed (Xu et al., 2022). The idea and concept that has come up as a result of this research is to make HCI more similar to HHI.

That said, the machines were required to be empowered to interact with humans. Moreover, they understand text or speech and make decisions. Simply put, making machines understand and process both speech and text is the foundation of NLP.

Natural language processing is a tool that enables a machine to comprehend text and language. It is used to apply machine learning algorithms to both text and speech. At this conjuncture, it can be stated that NLP is a subdomain of linguistics, computer science and AI.

One question that may arise is why, initially, we require machines to interpret human language. The answer is simple and straightforward. Owing to the large volume and abundance of data, it is practically impossible to interpret such a high volume with too many dimensions.

Here, machines trained to do so will be a boon to gain insights, read hidden patterns and make decisions. This is why and that is how NLP came into existence. Despite its popularity, implementing NLP is an extremely challenging task because most of the language-based data are unstructured and often ambiguous (Bose et al., 2021).

There are many components in speech, such as dialect, tone, accent and sarcasm, other than plain words (Sonbol et al., 2022). This requires computers to

DOI: 10.1201/9781003400110-6

comprehensively understand not only the words but also the intentions, emotions and sentiments.

Natural language processing has two broad components, namely, natural language understanding or natural language interpretation and natural language generation. Both aspects of NLP are indicators of the two-way communication between humans and machines.

Systematic reviews have been conducted to analyze text processing and normalization, which facilitates real-time user-generated content from social media (Rogers et al., 2022).

The use of the most popular machine learning algorithms, such as support vector machine, Bayesian classifier and decision trees, along with neural networks is evolving (Otter et al., 2021; Sharma & Sharma, 2023).

6.2   TEXT PRE-PROCESSING AND VECTOR-BASED MODELS

Textual data must be pre-processed and vectorized to be able to feed to a machine learning algorithm. Because language data generated from text or speech are unstructured, pre-processing becomes a mandatory step toward NLP.

Vectorizing pre-processed data simply refers to converting text into numbers. Therefore, NLP requires both data pre-processing and vectorization, especially in complex domains, such as software development, and has secured its position in accurate and timely pre-processing of bug reports based on priority and category, which is otherwise very complex, exhaustive and requires many resources (Ahmed et al., 2021).

6.3   TEXT PRE-PROCESSING TECHNIQUES

Text pre-processing refers to filtering the text corpus such that only relevant features are retained and the clutter is discarded. To achieve these objectives, common pre-processing techniques, such as stop words, stemming, part-of-speech tagging and tokenization, are used (Matcovschi, 2014).

In a natural language, every word has an essence or context. At the same time, some words are neutral, such as demonstratives, prepositions and articles. These words have little or no context and can be ignored. Such words are called stop words in NLP (Pecar et al., 2020).

The Natural Language Toolkit has its own library of stop words. Stemming refers to the generation of a base word from its variations. The stem or root is the part to which inflectional affixes (-ed, -ize, -de and -s) are added. The stem of a word is created by removing its prefix or suffix. Therefore, stemming a word may not necessarily result in actual words.

Lemmatization is also a pre-processing technique that is quite similar to stemming. Practical implementation suggests that stemming often returns words that are not meaningful. However, lemmatization yields crisp results.

The performance difference between stemming and lemmatization can be understood from the following depiction in Figure 6.1.

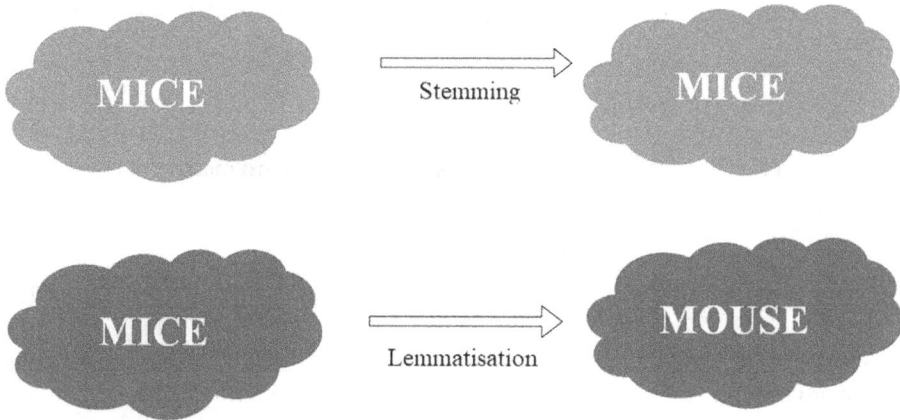

FIGURE 6.1 Stemming and Lemmatization.

Tokenization is another pre-processing technique that breaks text into smaller units called tokens. Depending on the use case, a token can be an alphabet, a word, a phrase or even a sentence. Text vectorization is a fundamental process in NLP.

As none of the machine learning algorithms understands text and, ironically, all the raw data for NLP are mostly text, it becomes an obligation to transform the raw data.

Fortunately, there are techniques available for this purpose. These techniques, if considered closely, are based on basic descriptive statistics and mathematics.

6.3.1 TERM FREQUENCY–INVERSE DOCUMENT FREQUENCY

Term frequency–inverse document frequency (TF-IDF) is a text vectorization technique that has gained traction and popularity in the domain of NLP because it is a simple and basic technique.

Segregating the acronym into two parts, TF refers to the frequency of a term in a particular document, and IDF is precisely a weight assigned to a term.

Term frequency is an indicator of the importance and relevance of a specific term in a specific document. It is denoted as a matrix with rows equal to the number of documents, also referred to as instances or sample sets.

All the unique terms in the documents comprising the columns of the TF matrix (Jayasurya et al., 2022) demonstrate that TF-IDF-based models outperform the models based on count vectorizers for NLP applications.

6.3.2 TERM FREQUENCY MATRIX

Document frequency (DF) gives the number of documents for which a particular term exists. It is an indicator of how common or how rare a particular term is.

Inverse document frequency is expressed by the mathematical formula shown in Equation 1.1.

$$IDF = \log(n / DF) \tag{1.1}$$

Finally, TF-IDF, as the name suggests, is a straightforward matrix multiplication. To obtain the final TF-IDF score (W), the TF matrix is multiplied by the IDF, as shown in Equation 1.2.

$$W = TF * IDF \tag{1.2}$$

The basic idea of this approach is to assign a higher weight to the most uncommon word or term and a lower weight to the most common term. Words with higher weight scores are deemed to be more significant.

As an illustration of TF-IDF, consider three documents consisting of the following text:

- D1: I live in Bangalore.
- D2: Bangalore has pleasant weather.
- D3: I love the weather in Bangalore.

The TF matrix for the given set of documents is created, as shown in Table 6.1.

In the next step, DF and IDF are calculated, as shown in Table 6.2.

Finally, by multiplying the TF matrix with the IDF, the TF-IDF score was obtained, as shown in Table 6.3.

The combination of count vectorizer and TF-IDF has led to some interesting applications, such as gender identification from names pertaining to a regional language (Saha et al., 2021; Jayasurya et al., 2022).

TABLE 6.1

TF Matrix.

	I	live	in	Bangalore	has	pleasant	weather	love	the	of
D1	1	1	1	1	0	0	0	0	0	0
D2	0	0	0	1	1	1	1	0	0	0
D3	1	0	0	1	0	0	1	1	1	1

TABLE 6.2

DF and IDF Matrix.

	I	live	in	Bangalore	has	pleasant	weather	love	the	of
DF	2	1	1	3	1	1	2	1	1	1
IDF	0.176	0.477	0.477	0	0.477	0.477	0.176	0.477	0.477	0.477

TABLE 6.3
TF-IDF Scores.

	I	live	in	Bangalore	has	pleasant	weather	love	the	of
D1	0.176	0.477	0.477	0.477	0	0	0	0	0	0
D2	0	0	0	0.477	0.477	0.477	0.176	0	0	0
D3	0.176	0	0	0.477	0	0	0.176	0.477	0.477	0.477

Daughter = Son − Boy + Girl

FIGURE 6.2 Dense Vector Approach.

6.4 NATURAL LANGUAGE PROCESSING: TEXT SIMILARITY AND SEMANTIC ANALYSIS

Text similarity in NLP determines the degree of closeness between two or more text corpora. This degree of closeness can be established either at the surface or deep down, referring to the context and meaning of a text corpus.

Accordingly, it can be a plain text similarity or semantic analysis in NLP. In terms of text similarity, various similarity metrics have been proved to be effective (Liang & Zhang, 2021).

Dense vectors are one of the standard approaches for the conversion of language to a machine-readable format. An example of understanding the concept is shown in Figure 6.2.

Using this approach, words with similar meanings will have close proximity or a similar orientation in terms of vectors. Although there are many, the three most popular metrics to find the similarity index are as follows:

- Euclidian distance
- Dot product
- Cosine similarity

6.4.1 EUCLIDIAN DISTANCE

Euclidian distance is a basic mathematical technique that is used to find the shortest distance between two points.

When seen from the context of NLP, these two points represent two vectors and, in turn, these two vectors represent two separate texts. Therefore, the objective here is to determine the similarity between the two texts by determining the closeness between the corresponding vectors.

As an illustration, consider the following three vectors, as shown in Equation 1.3:

$$x = \begin{bmatrix} 0.02 \\ 0.05 \\ 0.3 \end{bmatrix} \quad y = \begin{bmatrix} 0.02 \\ 0.06 \\ 0.33 \end{bmatrix} \quad z = \begin{bmatrix} 0.91 \\ 0.86 \\ 0.7 \end{bmatrix} \tag{1.3}$$

Vectors x and y are placed closer as opposed to y and z or z and x. Mathematically, the same can be verified by applying the formula for the Euclidian distance, which is given by Equation 1.4:

$$d(a,b) = \sqrt{\sum_{i=1}^{n}(a_i - b_i)^2} \tag{1.4}$$

When applied to vectors x and y as Equations 1.5 and 1.6

$$d(x,y) = \sqrt{\sum_{i=1}^{n}(x_1 - y_1)^2 + (x_2 - y_2)^2 + (x_3 - y_3)^2} \tag{1.5}$$

$$d(x,y) = \sqrt{\sum_{i=1}^{3}(0.02 - 0.02)^2 + (0.05 - 0.06)^2 + (0.3 - 0.33)^2} \tag{1.6}$$

$$= 0.031$$

Similarly, Equations 1.7 and 1.8

$$d(y,z) = \sqrt{\sum_{i=1}^{3}(0.02 - 0.91)^2 + (0.06 - 0.86)^2 + (0.33 - 0.7)^2} \tag{1.7}$$

$$= 1.25$$

$$d(z,x) = \sqrt{\sum_{i=1}^{3}(0.91 - 0.02)^2 + (0.86 - 0.05)^2 + (0.7 - 0.3)^2} \tag{1.8}$$

$$= 1.268$$

establish the fact that x and y are nearer in Euclidian space.

6.4.2 Dot Product

The Euclidian distance concept is magnitude oriented; on the other hand, the concept of a dot product considers both magnitude and orientation.

But why consider orientation at all if magnitude works fine? This is because vector magnitude is the indicator of word frequency. Now, if we have the word "hello" appearing 50 times and "hi" appearing two times, they may not be similar in machine learning because of their frequency, and hence vector magnitude.

However, if one more parameter of orientation is added, then some degree of closeness can be marked. This is achieved by dot product. The dot product is calculated as shown by Equation 1.9:

$$u \cdot v = |u| \, \|v\| \cos\theta \qquad (1.9)$$

A higher orientation between vectors will produce a higher dot product.

6.4.3 COSINE SIMILARITY

The shortcoming of the dot product approach lies in its limitation to normalize, which is very well addressed in the cosine similarity method. Similarity between two-word vectors can be calculated by Equation 1.10:

$$\text{Sim}(x, y) = \frac{x \cdot y}{\|x\| \, \|y\|}$$

$$= \frac{\sum_{i=1}^{n} xi.yi}{\sqrt{\sum_{i=1}^{n} xi^2} \sqrt{\sum_{i=1}^{n} yi^2}} \qquad (1.10)$$

All methods are heavily used in NLP using Sklearn implementation.

6.5 SEMANTIC ANALYSIS

Semantic analysis in NLP deals with capturing the meaning of a natural language rather than only finding the similarity of words (Elnagar et al., 2021).

Semantic analysis is a deeper technique, as it takes into account the context, grammar roles and logical structuring of the sentences. It detects emotion and sarcasm in the text and extracts meaningful inferences while looking for the dictionary meaning of the words in the sentence.

If the meaning is similar, the words are classified as similar even if they have entirely different forms from the alphabet (i.e., spelling). Semantic analysis can be divided into two parts: lexical semantic analysis and compositional semantic analysis.

Lexical semantic analysis involves fetching the dictionary meaning of a word in a given text corpus. Therefore, lexical semantic analysis works at the word level. On the other hand, compositional semantic analysis is a step forward.

To understand a text, understanding each individual word is a pre-requisite, but because of ambiguities in the language, the context, scenario and sentiment must also be completely understood to understand a phrase, sentence or document.

Consider the following texts:

- Text 1: Bitcoin is not declared a legal tender.
- Text 2: Legal tender of the bridge is declared.

Despite the fact that the presented two texts have similar root words (declared, legal and tender), they convey entirely different meanings.

The task of compositional semantic analysis is to extract the exact meaning of similar root words. Processes, such as relationship extraction and word sense disambiguation, are often used to accomplish this task.

6.6 PROBABILITY MODELS IN NATURAL LANGUAGE PROCESSING

6.6.1 HIDDEN MARKOV MODEL

Markov model tells about the probabilities of the sequences of random variables. These variables are often referred to as states. In the context of NLP, the states are individual tokens, which can be unigrams, bigrams, trigrams, etc. To illustrate, consider a text vocabulary of three unigrams, as shown in Figure 6.3.

Figure 6.3 depicts the probability of the sequences of the unigrams. For each state, the probability of the next state is defined by the transition probability matrix in the Markov model. Because there is a connection between each state, the transitional probability matrix is a square matrix, as shown in Equation 1.11:

$$P(\text{a red apple}) = P(\text{a}) \times P(\text{ red } | \text{a}) \times (\text{apple} | \text{ red }) \qquad (1.11)$$

It should be noted that the true probability is different from the probability of the Markov model. For the case shown in Figure 6.2, the true probability is given by Equation 1.12:

$$P(\text{a red apple}) = P(\text{a}) \times P(\text{red } | \text{a}) \times (\text{ apple } | \text{a, red}) \qquad (1.12)$$

Therefore, the Markov model simplifies the assumption that the probability of a subsequent token will only depend on the previous token, not considering the historical tokens but only the immediate previous one.

Thus, it can be concluded that the Markov chain is particularly useful when computing a probability for a sequence of text. Much research has been carried out on spam and ham email segregation, which proves that the hidden Markov model (HMM) outperforms the traditional naive Bayes-based algorithms (Independent University, 2017).

Nevertheless, it is often required to determine some hidden text property, for instance, whether a word is a noun, adjective or some other part of speech. As with actual words, these parts of speech are hidden, so some observations are required that are mapped back to these hidden states. This is where the HMM is presented.

An HMM considers the temporal relationships between hidden states and how these hidden states exhibit observations. Because it is more likely that a noun is followed by a verb rather than by an adjective and an adjective is followed by a noun rather than by a verb, this kind of transitional probability structure is proved to be very useful in tagging the text with hidden states.

The process of finding the most probable sequence of hidden states from the observation sequence is known as decoding, as shown in Figure 6.4.

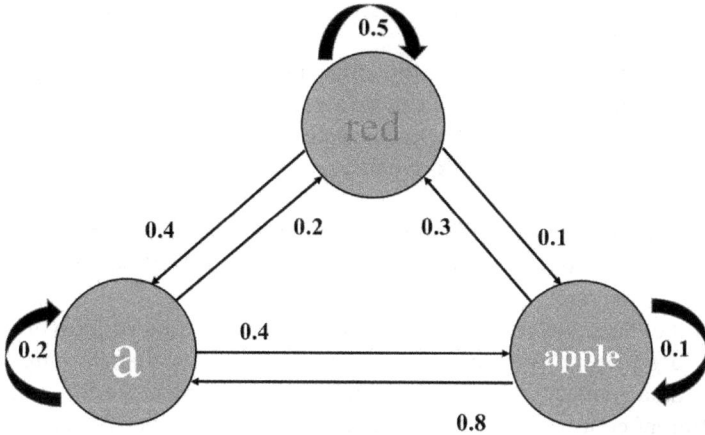

FIGURE 6.3   Markov Model.

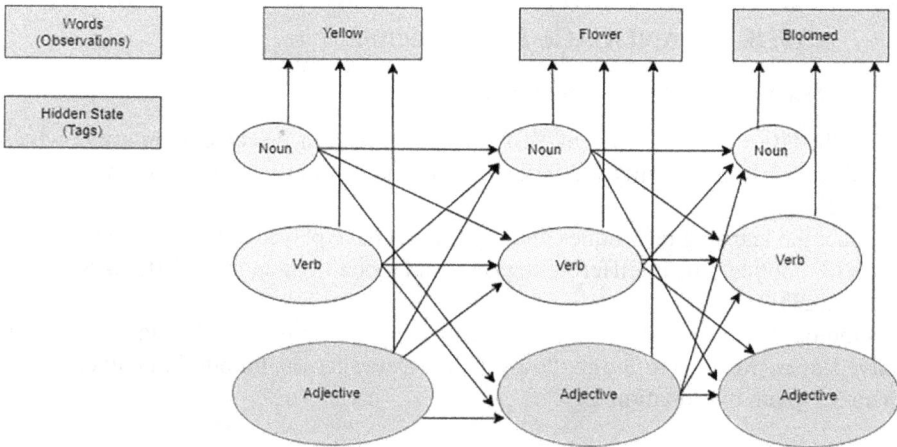

FIGURE 6.4   Hidden Markov Model.

Decoding algorithms find the optimal sequence of tags in a computationally effective way rather than simply hitting and trying all possible configurations. The classic HMM algorithm is the Viterbi algorithm, and beam search is a more contemporary version.

6.6.2   LANGUAGE MODELS

Language models form the core components of NLP. All popular NLP applications, such as Siri, Google Assistant and Alexa, are powered by language models. The language model predicts the probability of a sequence of words.

Language models are used in machine translation, speech recognition, parts of speech tagging and information retrieval. Two prominent types of language models are statistical language models and neural language models. Statistical language models combine the nuances of language with probabilistic approach (Khang et al., 2023b).

A statistical language model predicts the succeeding word in a sequence given the preceding words. These words can be a set of two, three or even a single word based on which they are called *N*-gram, where '*N*' is the amount of context that the model is trained to consider. Thus, it can be a unigram, bigram, trigram or another language model.

Other variations in statistical language models are based on exponential and continuous space concepts, in which feature functions and nonlinear combinations of weights are considered.

On the other hand, neural language model is a more advanced form of a language model based on neural networks. The shortcomings of statistical language models include data sparsity issue, conditioning on large context sizes and generalization in different contexts. Neural language models are fast becoming the preferred approach in NLP.

6.7   MACHINE LEARNING METHODS FOR NATURAL LANGUAGE PROCESSING

6.7.1 SPAM DETECTION–NAIVE BAYES

Since the late 1980s, the field of NLP has relied on data-driven computation, which largely involves probability, statistics and most importantly machine learning (Otter et al., 2021).

Machine learning techniques and algorithms have played a pivotal role in emerging NLP applications in different regional and global languages (HMI, 2016; Marco et al., 2021).

One of the most popular machine learning algorithms for text classification is naive Bayes (Mertiya & Singh, 2016). Naive Bayes has its foundation laid on Bayes theorem given by Equation 1.13:

$$P(A \mid B) = \frac{P(B \mid A)P(A)}{P(B)} \tag{1.13}$$

Bayes theorem describes the tendency or probability of the occurrence of an event 'A' when event 'B' has already occurred. As postulated by Bayes, it will depend on the reverse conditional probability, which means the chances of the occurrence of event 'B' when event 'A' has already occurred multiplied by the independent probability of event 'A' and divided by the independent probability of event 'B'.

Two questions arise from this point of understanding. How is it connected to NLP? Why is the word 'naive' in the naive Bayes algorithm?

In the context of NLP, talking particularly about text classification applications, such as spam email detection and sentiment analysis, if looked closely, we are only seeing the probability of certain words given that certain sets of words are already there, and based on this probability being greater or lesser than the threshold, we classify the text into one of the two classes and hence segregate spam emails from non-spam (ham) emails.

Second, the two events occurring simultaneously can be either dependent on or completely independent of each other. Although in real-life use cases, events or features cannot be ruled out to be dependent on each other. On the contrary, the naive Bayes machine learning algorithm works on the assumption of all the events being independent, which is a very 'naive' assumption.

Hence the name. Classifying natural language descriptions as a challenge, the naive Bayes algorithm together with the count vectorizer is used in applications related to genre-based mobile app classification (Qiu et al., 2022).

Furthermore, naive Bayes, along with the sentiment scoring approach, is deployed and produces fruitful results to analyze even government decisions on political fronts based on free opinions from forums, blogs, tweets, etc. (JAC-ECC, 2018; Bunyamin et al., 2022).

Implementation of the multinomial naive Bayes algorithm in the medical and healthcare domains creates waves in the NLP domain (ICCES, 2020; Alawad et al., 2021; Ishikawa et al., 2022).

6.7.2 Sentiment Analysis–Logistic Regression

The analysis of sentiments or emotions is a daunting and challenging task. Sentiments can be positive, negative, neutral or mixed. In addition, when expressed via natural language, it becomes more ambiguous.

In the quest to classify sentiments expressed through natural language, various machine learning algorithms have been introduced, improvised and implemented to achieve higher accuracy and enhanced crystal clarity in terms of the segregation and classification of one sentiment from another. One such approach is the logistic-regression machine learning algorithm (Khang et al., 2022a).

This approach focuses on the binary classification of sentiments while ignoring neutral sentiments. The steps involved in the process are shown in Figure 6.5. In NLP, the neutrality of sentiments is handled in different ways, depending on the technique being applied.

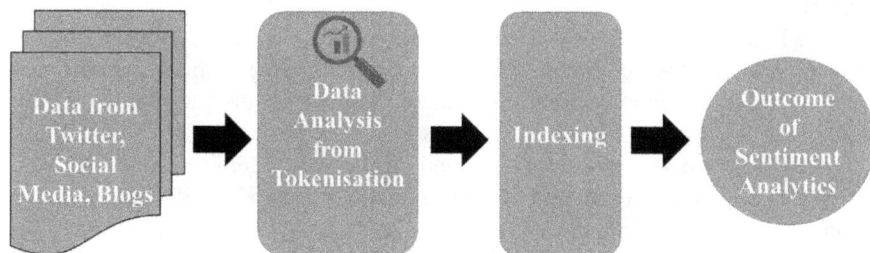

FIGURE 6.5   Sentiment Analysis.

Source: Khang (2021)

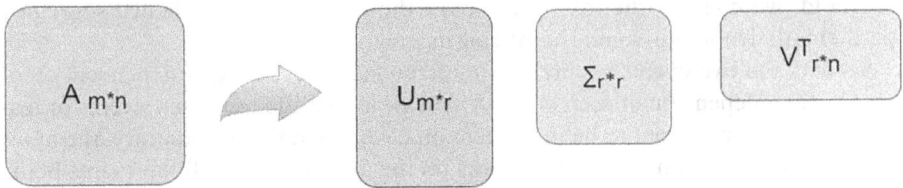

FIGURE 6.6   Latent Semantic Analysis.

Applications that are precise and ambiguous, such as rumor detection and predicting depression symptoms based on social media posts (Hossain et al., 2021), have been explored where NLP methods with machine learning algorithms are used to detect microblog rumors in Hebrew (Dalian, 2020; Hacohen-Kerner et al., 2022).

6.7.3   Latent Semantic Analysis–Singular Value Decomposition

In the NLP domain, data are mostly unstructured and high in dimensions. Latent semantic analysis (LSA) is a technique used to transform unstructured data into structured data. However, this is easier said than done. In addition to its unstructured nature, dimensionality reduction is another challenge that needs to be addressed in NLP.

The idea of dimensionality reduction is similar to inferring and understanding a three-dimensional object from its two-dimensional shadow, which is possible only if the highest level of variability is maintained and preserved in doing so.

Similarly, LSA, besides reducing the dimension, transforms the text into a structured form. Latent dimension is a theme hidden behind a cluster of similar words. This theme can only emerge on the foundation of a sufficient corpus of words, which means that most words are semantically related to a set of other words making up a theme or topic.

Latent semantic analysis strives to bring forth this hidden latent dimension using a machine learning algorithm called singular value decomposition (SVD). It primarily decomposes a document–term matrix into three component matrices, as shown in Figure 6.5.

'A' is a high-dimensional $m*n$ document–term matrix with 'm' documents and 'n' unique words in all the documents taken together. 'U' matrix is the document-aspect matrix obtained by linearly combining the columns (i.e., unique words of the 'A' matrix).

'VT' is the word-aspect matrix obtained after linearly combining the rows (i.e., unique documents of the document–term matrix). Here, 'r' is the aspect that can be understood as the theme or topic under which a set of words fall, $r \ll n$. 'Σ' is a diagonal matrix of singular values.

With this implementation, a dense document-aspect matrix is obtained from a sparse document–term matrix. In addition, the need for a correlation or covariance matrix is eliminated in SVD, thereby rescuing it from the assumption that the data are normally distributed. Latent semantic analysis reformulates the NLP text data into r latent features.

6.7.4   Topic Modeling–Latent Dirichlet Allocation

Topic modeling is an application of unsupervised machine learning and a statistical procedure for learning and extracting similar topics from a large amount of data (Khang et al., 2022b).

To implement this on big data, where the volume, variety, variability and veracity are tremendous, automated algorithms are required. One such approach is latent Dirichlet allocation (LDA), which can read texts and provide document segregation pertaining to a specific topic.

Topic modeling is simply a prediction of the subject of a document. The two basic assumptions of LDA are that every document is a probabilistic distribution of topics, and each topic is a probabilistic distribution of words, both being Dirichlet's distributions, as shown in Figure 6.7.

The objective of LDA is to find an optimized representation in the form of a document–topic matrix and a topic–term matrix derived from the corpus of the document–term matrix.

Two Dirichlet distributions, α and β, control the document–topic and topic–term distribution, respectively.

A higher α indicates the presence of a majority of the topics in a document, and a high β points to the fact that every topic has a probability of containing the majority of the words.

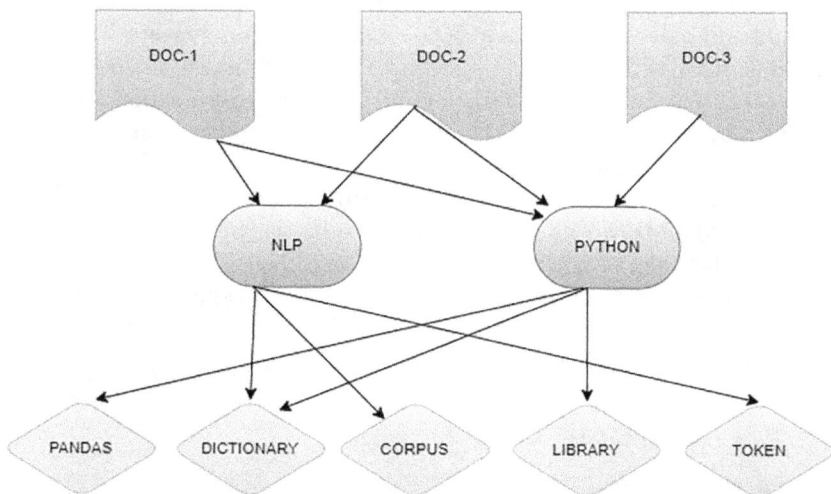

FIGURE 6.7   Dirichlet Distribution.

6.8   DEEP LEARNING METHODS

In challenging domains, such as NLP, deep neural networks (DNNs) have outperformed conventional statistic-based methods. Deep neural network is essentially an extension to artificial neural network (ANN) with enhanced functionalities and performance owing to its multiple hidden layers between the input and output layers (Chitty-Venkata et al., 2022).

The concept of DNN is derived from neurobiology. A human brain has a basic unit called neuron, which makes all decisions. At a higher level, a neuron receives input signals through synapses and outputs a single action stream through an axon. This is the inspiration that an ANN has units that combine many inputs and provide a single output.

Although natural language has gained momentum, it remains a challenge in the medical field. Medical texts consist of largely unstructured data, such as images, free text, abbreviations and spelling mistakes in prescriptions (Khang et al., 2023c).

In such scenarios, the attention mechanism along with deep learning and NLP models have proved useful in three main architectures of a DNN (Jin et al., 2020).

6.8.1   MULTILAYER PERCEPTRON

Multilayer perceptron (MLP) is essentially a deep feedforward neural network consisting of an input layer, an output layer and more than one hidden layer with multiple neurons stacked together (Qi et al., 2022).

Linearly inseparable data are classified by implementing a nonlinear model that activates the logistic or tangent function.

The network is fully connected, in the sense that all nodes in a layer are connected to all other nodes in the preceding and succeeding layers. Therefore, it is a neural network with a nonlinear mapping between the inputs and outputs.

In MLP, learning the weights and minimizing the cost function is the key that is achieved by a process called backpropagation. Backpropagation iteratively adjusts the weights and minimizes the cost function.

A perceptron is a neural network consisting of one neuron that is fit for linear data. However, MLPs with expanded horizons have many layers of neurons that are capable of learning complex patterns (Wu et al., 2021).

Multilayer perceptron with its feedforward and backpropagation mechanism is suitable for NLP applications that have high-dimensional ambiguous data, as shown in Figure 6.8.

6.8.2   CONVOLUTIONAL NEURAL NETWORK

A convolutional neural network (CNN) is believed to be a synonym for computer vision because of the breakthroughs achieved in the image processing domain. The CNN applications have a wide bandwidth, from automated photo tagging to self-driving cars (Jayashree et al., 2023).

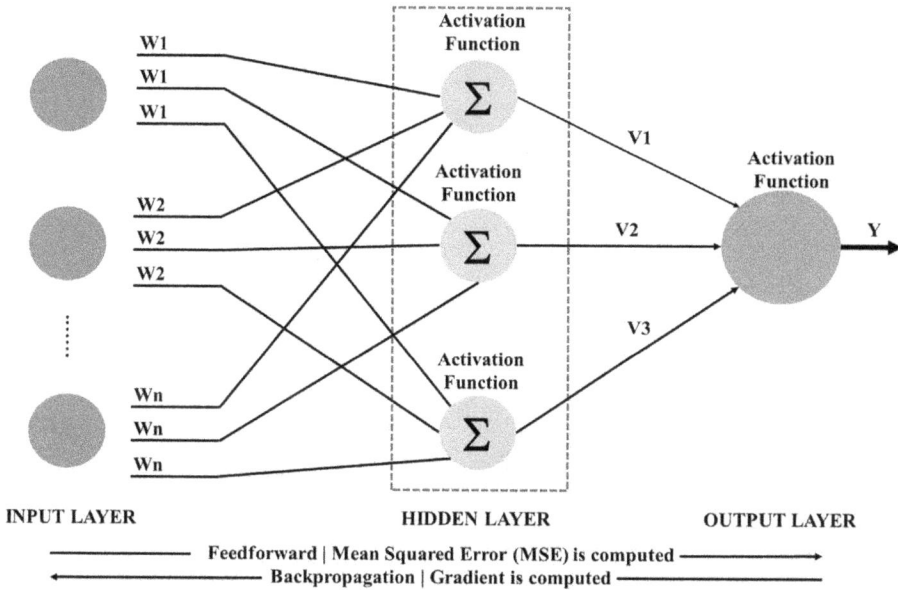

FIGURE 6.8   Multilayer Perceptron.

Source: Khang (2021)

In simple words, convolution is a sliding window function applied to a matrix, where every element of this matrix precisely represents a pixel of an image, and the sliding window is called a filter, feature detector or kernel (Anuj Kumar & Khang, 2023).

Recently, CNNs have been applied to the ambiguous NLP domain (Liao & Zhou, 2021; Peng & Han, 2021). In the case of image processing, this input matrix can conveniently be replaced by a matrix representing sentences or documents; each row corresponds to a token.

In contrast to image processing, where the kernel strides over the local patches of an image, in the case of NLP, the kernel strides over the complete row of the matrix, thus fixing the filter width to the width of the matrix in the case of NLP applications.

Figure 6.9 gives an elaborate visualization of the CNN architecture for text classification. Two filters were identified with three region sizes: 2, 3 and 4.

Feature maps were generated as a result of the convolution. These feature maps undergo pooling, which is a dimensionality reduction step, and generates six univariate feature vectors. These are further concatenated into one feature vector, which is fed as input to the final Softmax layer to classify the text, as shown in Figure 6.9.

Assuming binary classification, two possible output states or classes are obtained at the output layer.

Owing to the high computational requirements of neural networks, a novel concept of neural architecture search benchmarks with pre-computed neural architecture has also evolved for both computer vision and NLP domains (Klyuchnikov et al., 2022).

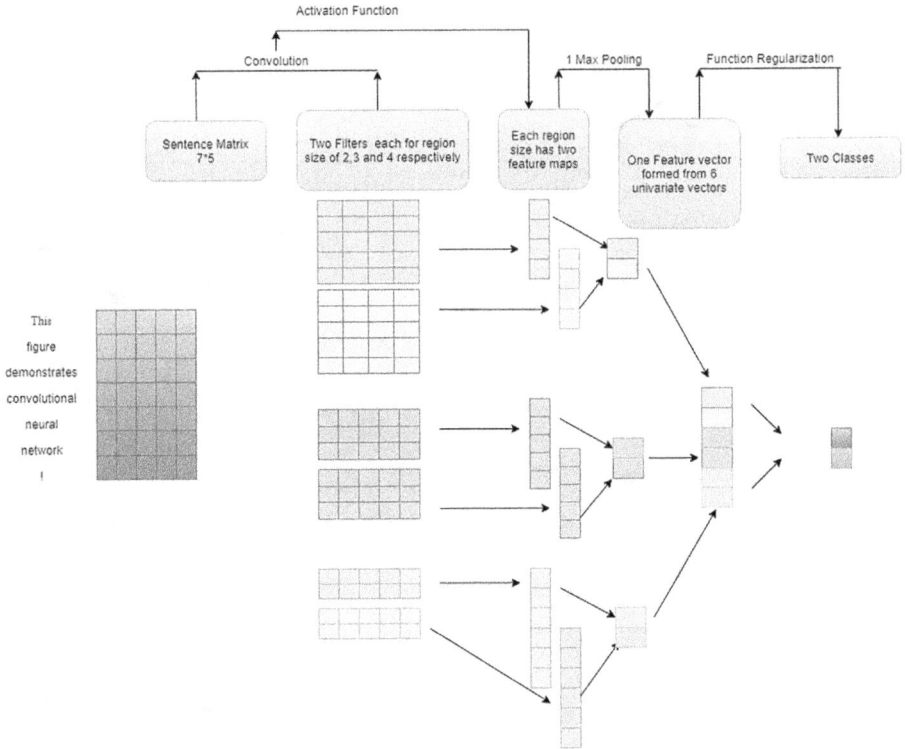

FIGURE 6.9   Convolutional Neural Network.

6.8.3   RECURRENT NEURAL NETWORK

Recurrent neural networks (RNNs) are a class of ANN that are extensively exploited in the fields of NLP, speech recognition, language translation and many more.

With sequential and time series data as the foundation, RNN is the appropriate choice for temporal and ordinal applications, such as NLP.

In contrast to CNN, where the information flow is unidirectional, RNN incorporates feedback loops, such as backpropagation through time (BPTT) to ensure a constant supply of information. These bidirectional recurrent neural networks (BRNNs) predict the future based on forward projections and evaluate the past based on backward projections (Rana et al., 2021).

To facilitate reinforcement learning in an RNN, the weights that are fixed for a specific layer are continuously adjusted during the process of backpropagation and gradient descent; the gradient is the slope of the loss function.

Owing to the functioning of the RNN, where the parameters are shared across the layers, the BPTT sums up the errors at each time step. The current trend in geotagging on social media platforms has led to location-specific sentiment analysis using RNN (Qi et al., 2019).

A vivid variety of applications, such as machine translation, medical text interpretation (Mugisha & Paik, 2021), music generation and sentiment classification, can be addressed with different types of RNNs. In addition to BRNN, another popular RNN architecture is long short-term memory (LSTM).

This RNN architecture is a rescue to the vanishing gradient problem commonly observed in deeply layered networks for complex datasets. The LSTM network addresses the issue encountered when RNNs grow bigger and more complex, especially those based on gradient-based learning methods (Khang et al., 2023a).

Long short-term memories have played a pivotal role in critical NLP concerns, such as privacy and security issues related to language-based applications (Feng et al., 2020). These are well integrated with Internet of Things platforms to address security and privacy violations (Breve et al., 2022).

Long short-term memory-based RNNs essentially segregate the data based on short-term and long-term memory, which means that LSTM has cells in the hidden layers with three gates vis-à-vis the input gate, output gate and forget gate, controlling the required information flow to make predictions (Liang & Zhang, 2021).

A precision as high as 0.94 is achieved by LSTM for fake news detection by employing RNNs. In addition, sentiment analysis of customer reviews is another popular application for business decision-making (Houssein et al., 2021; Jain et al., 2022).

6.9 CONCLUSION

The blockchain integration of NLP is an emerging trend that is gaining interest in the NLP community (Shahbazi & Byun, 2021). The integration of blockchain technology and NLP can lead to new opportunities for secure and transparent data sharing, decentralized applications and new business models (Khanh & Khang, 2021).

Here are some potential future trends in which blockchain integration can be beneficial for NLP:

- **Secure data sharing**: One of the main benefits of blockchain technology is its ability to securely store and share data. In the context of NLP, this can be used to securely store and share sensitive data, such as medical records or financial data (Khang et al., 2024).
- **Decentralized applications**: Blockchain technology can facilitate the development of decentralized NLP applications, which can be more resistant to censorship and provide a more transparent user experience.
- **Transparency in data usage**: With blockchain technology, data ownership and usage can become more transparent. This can be beneficial for users who want to know how their data are being used and can also help researchers access more data for their studies.
- **Smart contracts**: Smart contracts can be used to automate and enforce the terms of data sharing agreements in the context of NLP. This can ensure that the data are only used for agreed-upon purposes and can help prevent misuse.

- **Token-based incentivization**: Blockchain tokens can be used to incentivize data sharing and participation in NLP tasks. This can help create new business models and provide incentives for users to share their data.
- **Fraud detection**: Blockchain technology can be used to detect fraud in NLP tasks, such as sentiment analysis or fake news detection. By verifying the source and accuracy of the data, blockchain can help prevent malicious actors from manipulating the results of NLP tasks.

Overall, the integration of blockchain technology and NLP has the potential to create new opportunities for secure and transparent data sharing, decentralized applications and new business models. Although still in its early stages, the integration of blockchain and NLP is an exciting area of research that will likely lead to new innovations in the field.

REFERENCES

Ahmed H. A., Bawany N. Z., Shamsi J. A., (2021). "CaPBug—a framework for automatic bug categorization and prioritization using NLP and machine learning algorithms," *IEEE Access*, 9, 50496–50512. https://doi.org/10.1109/ACCESS.2021.3069248

Alawad M., et al., (2021). "Privacy-preserving deep learning NLP models for cancer registries," *IEEE Transactions on Emerging Topics in Computing*, 9(3), 1219–1230, 2021. https://doi.org/10.1109/TETC.2020.2983404

Anuj Kumar G., Khang A., (2023). *AI-Aided IoT Technologies and Applications in the Smart Business and Production, Implementation of Smart Vehicle Parking System Using Internet of Things (IoT)*, p. 8. CRC Press. https://doi.org/10.1201/9781003392224-8

Bose P., Roy S., Ghosh P., (2021). "A comparative NLP-based study on the current trends and future directions in COVID-19 research," *IEEE Access*, 9, 78341–78355. https://doi.org/10.1109/ACCESS.2021.3082108

Breve B., Cimino G., Deufemia V., (2022). "Identifying security and privacy violation rules in trigger-action IoT platforms with NLP models," *IEEE Internet of Things Journal*. https://doi.org/10.1109/JIOT.2022.3222615

Bunyamin M. A. F., Pudjiantoro T. H., Renaldi F., Hadiana A. I., (2022). "Analyzing sentiments on Indonesia's new national palace using the combination of naive Bayes and sentiment scoring," In *2022 International Conference on Science and Technology, ICOSTECH 2022*. Institute of Electrical and Electronics Engineers Inc. https://doi.org/10.1109/ICOSTECH54296.2022.9829075

Chitty-Venkata K. T., Emani M., Vishwanath V., Somani A. K., (2022). "Neural architecture search for transformers: A survey," *IEEE Access*, 10, 108374–108412. https://doi.org/10.1109/ACCESS.2022.3212767

Dalian, Institute of Electrical and Electronics Engineers, IPEC 2020: proceedings of 2020 Asia-Pacific Conference on Image Processing, Electronics and Computers: China, April 14–16, 2020. https://doi.org/10.1201/9781003145011

Elnagar A., Yagi S. M., Nassif A. B., Shahin I., Salloum S. A., (2021). "Systematic literature review of dialectal Arabic: Identification and detection," *IEEE Access*, 9, 31010–31042 (Institute of Electrical and Electronics Engineers Inc.). https://doi.org/10.1109/ACCESS.2021.3059504

Feng Q., He D., Liu Z., Wang H., Choo K. K. R., (2020). "SecureNLP: A system for multi-party privacy-preserving natural language processing," *IEEE Transactions on Information Forensics and Security*, 15, 3709–3721. https://doi.org/10.1109/TIFS.2020.2997134

Hacohen-Kerner Y., Manor N., Goldmeier M., Bachar E., (2022). "Detection of anorexic girls—in blog posts written in Hebrew using a combined heuristic AI and NLP method," *IEEE Access*, 10, 34800–34814. https://doi.org/10.1109/ACCESS.2022.3162685

HMI, (2016). *International Conference on Advances in Human Machine Interaction (HMI)*. IEEE. https://ieeexplore.ieee.org/xpl/conhome/7444563/proceeding

Hossain M. T., Talukder M. A. R., Jahan N., (2021). "Social networking sites data analysis using NLP and ML to predict depression," In *2021 12th International Conference on Computing Communication and Networking Technologies, ICCCNT 2021*. Institute of Electrical and Electronics Engineers Inc. https://doi.org/10.1109/ICCCNT51525.2021.9579916

Houssein E. H., Mohamed R. E., Ali A. A., (2021). "Machine learning techniques for biomedical natural language processing: A comprehensive review," *IEEE Access*, 9, 140628–140653 (Institute of Electrical and Electronics Engineers Inc.). https://doi.org/10.1109/ACCESS.2021.3119621

ICCES, (2020). "Institute of Electrical and Electronics Engineers and PPG Institute of Technology," *Proceedings of the 5th International Conference on Communication and Electronics Systems (ICCES 2020)*, 10–12. https://doi.org/10.1201/9781003145011

Independent University, Bangladesh, (2017). Institute of Electrical and Electronics Engineers, and Institute of Electrical and Electronics Engineers. Bangladesh Section. *EMB Chapter, The 4th International Conference on Advances in Electrical Engineering (ICAEE)*, 28–30 September, Independent University, Bangladesh. https://doi.org/10.1201/9781003145011

Ishikawa T., Yakoh T., Urushihara H., (2022). "An NLP-inspired data augmentation method for adverse event prediction using an imbalanced healthcare dataset," *IEEE Access*, 10, 81166–81176. https://doi.org/10.1109/ACCESS.2022.3195212.

JAC-ECC, (2018). "Institute of Electrical and Electronics Engineers," *2018 International Japan-Africa Conference on Electronics, Communications and Computations* (JAC-ECC). www.ieice.org/cs/gnl/gnl_vol43-2.pdf

Jain P., Monica S. S., Aggarwal P. K., (2022). "Classifying fake news detection using SVM, naive Bayes and LSTM," In *Proceedings of the Confluence 2022–12th International Conference on Cloud Computing, Data Science and Engineering*, pp. 460– 64. Institute of Electrical and Electronics Engineers Inc. https://doi.org/10.1109/Confluence52989.2022.9734129

Jayashree M., Dillip R., et al., (Eds.). (2023). "Vehicle and passenger identification in public transportation to fortify smart city indices," In *Smart Cities: IoT Technologies, Big Data Solutions, Cloud Platforms, and Cybersecurity Techniques*. CRC Press. https://doi.org/10.1201/9781003376064-13

Jayasurya G. G., Kumar S., Singh B. K., Kumar V., (2022). "Analysis of public sentiment on COVID-19 vaccination using Twitter," *IEEE Transactions on Computational Social Systems*, 9(4), 1101–1111. https://doi.org/10.1109/TCSS.2021.3122439

Jin Q., Xue X., Peng W., Cai W., Zhang Y., Zhang L., (2020). "TBLC-rAttention: A deep neural network model for recognizing the emotional tendency of Chinese medical comment," *IEEE Access*, 8, 96811–96828. https://doi.org/10.1109/ACCESS.2020.2994252

Khang A., (2021). "Material4Studies," *Material of Computer Science, Artificial Intelligence, Data Science, IoT, Blockchain, Cloud, Metaverse, Cybersecurity for Studies*. www.researchgate.net/publication/370156102_Material4Studies

Khang A., Gupta S. K., Hajimahmud V. A., Babasaheb J., Morris G., (2023a). *AI-Centric Modelling and Analytics: Concepts, Designs, Technologies, and Applications* (1st ed.). CRC Press. https://doi.org/10.1201/9781003400110

Khang A., Vrushank S., Rani S., (2023b). *AI-Based Technologies and Applications in the Era of the Metaverse* (1st ed.). IGI Global Press. https://doi.org/10.4018/9781668488515

Khang A., Olena H., Abdullayev V. A., Arvind Kumar S., (2024). *Computer Vision and AI-Integrated IoT Technologies in Medical Ecosystem* (1st ed.). CRC Press. https://doi.org/10.1201/9781003429609

Khang A., Rani S., Sivaraman A. K., (2022a). *AI-Centric Smart City Ecosystems: Technologies, Design and Implementation* (1st ed.). CRC Press. https://doi.org/10.1201/9781003252542

Khang A., Ragimova N. A., Hajimahmud V. A., Alyar V. A., (2022b), "Advanced technologies and data management in the smart healthcare system," In *AI-Centric Smart City Ecosystems: Technologies, Design and Implementation*, Vol. 16, No. 10 (1st ed.). CRC Press. https://doi.org/10.1201/9781003252542-16

Khang A., Rana G., Tailor R. K., Hajimahmud V. A., (Eds.). (2023c). *Data-Centric AI Solutions and Emerging Technologies in the Healthcare Ecosystem*. CRC Press. https://doi.org/10.1201/9781003356189

Khanh H. H., Khang A., (2021) "The role of artificial intelligence in blockchain applications," *Reinventing Manufacturing and Business Processes through Artificial Intelligence*, 2(20–40) (CRC Press). https://doi.org/10.1201/9781003145011-2

Klyuchnikov N., et al., (2022). "NAS-Bench-NLP: Neural architecture search benchmark for natural language processing," *IEEE Access*, 10, 45736–45747. https://doi.org/10.1109/ACCESS.2022.3169897

Liang Z., Zhang S., (2021). "Generating and measuring similar sentences using long short-term memory and generative adversarial networks," *IEEE Access*, 9, 112637–112654. https://doi.org/10.1109/ACCESS.2021.3103669

Liao X., Zhou G., (2021). "The importance of token granularity matching of pre-trained word vectors for deep learning-based spam classification," In *Proceedings—2021 3rd International Conference on Natural Language Processing, ICNLP 2021*, pp. 129– 33. Institute of Electrical and Electronics Engineers Inc. https://doi.org/10.1109/ICNLP52887.2021.00007.

Marco G., De La Rosa J., Gonzalo J., Ros S., Gonzalez-Blanco E., (2021). "Automated metric analysis of Spanish poetry: Two complementary approaches," *IEEE Access*, 9, 51734– 51746. https://doi.org/10.1109/ACCESS.2021.3069635

Matcovschi M. H., (2014). "IEEE control systems society, C. and C. 18 2014.10.17–19 S," In *International Conference on System Theory, and ICSTCC 18 2014.10.17–19 Sinaia. 18th International Conference on System Theory, Control and Computing (ICSTCC)*, 17–19 October; ICSTCC: Sinaia.

Mertiya M., Singh A., (2016). *Combining Naive Bayes and Adjective Analysis for Sentiment Detection on Twitter*. https://ieeexplore.ieee.org/abstract/document/7824847/

Mugisha C., Paik I., (2021). "Comparison of neural language modeling pipelines for outcome prediction from unstructured medical text notes," *IEEE Access*, 10, 16489–16498. https://doi.org/10.1109/ACCESS.2022.3148279

Otter D. W., Medina J. R., Kalita J. K., (2021). "A survey of the usages of deep learning for natural language processing," *IEEE Transactions on Neural Networks and Learning Systems*, 32(2), 604–624. https://doi.org/10.1109/TNNLS.2020.2979670

Pecar S., Simko M., Bielikova M., (2020). *Sentiment Analysis of Customer Reviews: Impact of Text Pre-processing*. https://ieeexplore.ieee.org/abstract/document/8490619/

Peng J., Han K., (2021). "Survey of pre-trained models for natural language processing," In *2021 IEEE International Conference on Electronic Communications, Internet of Things and Big Data, ICEIB 2021*, pp. 277–280. Institute of Electrical and Electronics Engineers Inc. https://doi.org/10.1109/ICEIB53692.2021.9686420

Qi Q., Lin L., Zhang R., Xue C., (2022). "MEDT: Using multimodal encoding-decoding network as in transformer for multimodal sentiment analysis," *IEEE Access*, 10, 28750– 28759. https://doi.org/10.1109/ACCESS.2022.3157712

Qi W., Procter R., Zhang J., Guo W., (2019). "Mapping consumer sentiment toward wireless services using geospatial Twitter data," *IEEE Access*, 7, 113726–113739. https://doi.org/10.1109/ACCESS.2019.2935200

Qiu H., Wu Z., Zhang X., (2022). "Exploring multiple genres text classification: Classifying 61 genres of mobile app description based on naïve Bayes and count vectorizer," In *Proceedings—2022 3rd International Conference on Electronic Communication and Artificial Intelligence, IWECAI 2022*, pp. 156–162. Institute of Electrical and Electronics Engineers Inc. https://doi.org/10.1109/IWECAI55315.2022.00039

Rana G., Khang A., Sharma R., Goel A. K., Dubey A. K., (Eds.). (2021). *Reinventing Manufacturing and Business Processes through Artificial Intelligence*. CRC Press. https://doi.org/10.1201/9781003145011

Rogers D., Preece A., Innes M., Spasić I., (2022). "Real-time text classification of user-generated content on social media: Systematic review," In *IEEE Transactions on Computational Social Systems*, Vol. 9, No. 4, pp. 1154–1166. Institute of Electrical and Electronics Engineers Inc. https://doi.org/10.1109/TCSS.2021.3120138

Saha L., Uddin R. M. A., Saha S., (2021). "Performance measurement of multiple supervised learning algorithms for gender identification from Bengali names," In *2021 12th International Conference on Computing Communication and Networking Technologies, ICCCNT 2021*. Institute of Electrical and Electronics Engineers Inc. https://doi.org/10.1109/ICCCNT51525.2021.9579789.

Shahbazi Z., Byun Y. C., (2021). "Fake media detection based on natural language processing and blockchain approaches," *IEEE Access*, 9, 128442–128453. https://doi.org/10.1109/ACCESS.2021.3112607

Sharma N., Sharma S., (2023). "Performance enhancement of KNN classifier for guitar chord tonality classification analysis and optimization of t-SNE by tuning perplexity for faithful dimensionality reduction in NLP Application View project." [Online]. www.tnsroindia.org.in

Sonbol R., Rebdawi G., Ghneim N., (2022). "The use of NLP-based text representation techniques to support requirement engineering tasks: A systematic mapping review," In *IEEE Access*, Vol. 10, pp. 62811–62830. Institute of Electrical and Electronics Engineers Inc. https://doi.org/10.1109/ACCESS.2022.3182372

Wu B., Wei B., Liu J., Wu K., Wang M., (2021). "Faceted text segmentation via multitask learning," *IEEE Transactions on Neural Networks and Learning Systems*, 32(9), 3846–3857. https://doi.org/10.1109/TNNLS.2020.3015996

Xu Z., Wu H., Chen X., Wang Y., Yue Z., (2022). "Building a natural language query and control interface for IoT platforms," *IEEE Access*, 10, 68655–68668. https://doi.org/10.1109/ACCESS.2022.3186760

7 Application of Artificial Intelligence in Healthcare System Management with Dynamic Modeling of COVID-19 Diagnosis

Yerra Shankar Rao and Binayak Dihudi

7.1 INTRODUCTION

In December 2019, a novel virus caused severe acute respiratory syndrome (SARS) coronavirus disease 2019 (COVID-19). The virus is believed to have originated from China and subsequently spread worldwide (WHO, 2021).

When people are in physical contact with each other, this virus is generally transmitted from person to person either by face-to-face contact or by droplets that come from sneezing. It rapidly attacks the upper part of the respiratory system and causes severe infection of the lungs, resulting in breathing problems.

Initially, no vaccine was available. People, especially the elderly and those with chronic diseases, were quickly infected. The healthcare system has been paralyzed because of the rapid rise in infection and causality rates (Khang et al., 2023b).

Governments worldwide have faced significant challenges in improving the healthcare system and resources. Hence, different procedures have been adopted for better clinical care to control the spread of infections and reduce the morbidity and mortality associated with COVID-19.

Artificial intelligence (AI) is an important tool in today's society. Hence, artificial techniques are utilized in the analysis of chest X-rays and computerized tomography (CT) scan images to determine the spread of infection in lungs of patients (Agrebi et al., 2020; Mei et al., 2020; Li et al., 2020).

Almost all countries were affected by SARS COVID-19, which led to widespread loss of lives and collapse of the healthcare system, economics, and logistics. According to the WHO data on 16 March 2021, approximately 120 million people were already infected and approximately 2.7 million people have died (Rani et al., 2021).

DOI: 10.1201/9781003400110-7

Artificial intelligence plays a vital role in all fields of science and technology (Singh, 2020). Identifying COVID-19 patients and contact tracing have been difficult during the early months of the pandemic; thus, AI and modeling simulations have been used to explore COVID-19 data (Rana et al., 2021).

Recently, researchers have observed AI as a promising technology in the field of advanced healthcare issues, as it provides better results and enhances processing power (Khang et al., 2022a).

Artificial intelligence has many applications in the field of decision-making and for solving critical problems in healthcare systems based on algorithms. This technology has also been used for the prediction, detection, and diagnosis of COVID-19.

The latest AI testing methods for diagnosing COVID-19 include nucleic acid testing, CT scan, chest scan, medical imaging process, and swab testing. Artificial intelligence has made remarkable improvements, particularly in predictive machine learning (ML) models for medical system care (Lalmuanawma et al., 2020; Tuli et al., 2020; Sathwik et al., 2021).

Complex architectures of artificial neural networks can be framed using deep learning, such as ML languages. After providing a sufficient training dataset, deep learning exhibits significant discriminative performance that is essential for forecasting. Different techniques are used in the healthcare system, such as Bayesian network, fuzzy approaches, and neural networks (Khanh & Khang, 2021).

Artificial intelligence has been used to fight COVID-19 (McCall, 2020; Zhao et al., 2021). With the use of AI and mathematical modeling, the medical sector must continuously review how to minimize the infection and death rates caused by COVID-19. This will help in the diagnosis of COVID-19 within a limited time period, and as a result, patients can be isolated/quarantined earlier to reduce the spread of infections.

Artificial intelligence in *medicine* (Chawla et al., 2020; Saad et al., 2020; Gaur et al., 2021) improves the quality of healthcare systems, increases the accuracy of diagnosis, exhibits error-free potential, and rapidly generates data from patients. Artificial intelligence models can immediately determine high-risk patients and identify the epidemiology of COVID-19 through training data.

Artificial intelligence-based methods can be used to develop vaccines by evaluating viral mutations and screening for potential adjuvants to enhance the effectiveness of vaccines and drugs.

During the pandemic, thermal imaging was used to scan potentially infected individuals and areas without social distancing measures.

Combined with large-scale data, the operation of core clinical services, high-speed and accurate diagnosis, improved monitoring, effective treatments and vaccines, and countermeasures helped suppress the spread of COVID-19.

Different methods are used to diagnose COVID-19, some of which are user-friendly, inexpensive, and produce immediate results (Chawla et al., 2020).

Artificial intelligence sensors are available that can detect COVID-19 symptoms such as fever, shortness of breath, and dryness of mouth. The combined use of AI and mathematical modeling produces significant results in the health sector (Khang et al., 2022b).

Artificial intelligence has established a hypothesis change in healthcare. This is applicable in the medical field to combat viruses.

The use of AI software can accelerate the diagnosis and tracking of COVID-19 patients to reduce the burden on healthcare. Several authors have presented mathematical models for combating diseases (Bimal et al., 2020; Aswin et al., 2021; Bimal et al., 2021; Binayak et al., 2021; Jangyadatta et al., 2021).

Mathematical models for virus propagation are valuable tools for understanding the behavior of viruses and for quantitatively making control decisions and preventive measures.

Artificial intelligence tools are used for epidemiological forecasting, control, and prediction. In this chapter, we propose a mathematical model-based framework for AI to combat COVID-19.

The rest of the chapter is organized as follows:

- In section 1.2, we present the nomenclature for the considered parameters.
- In section 1.3, we discuss the mathematical assumptions by comparing them with the AI-based framework.
- In section 1.4, we discuss the formation of the mathematical model, present the positivity boundedness of the model, and derive the basic reproduction number.
- In section 1.5, we analyze the stability of the equilibrium point in both the COVID-19-free equilibrium and endemic equilibrium points.
- In section 1.6, we discuss the results of the model.
- Conclusions are presented in section 1.7.

The nomenclature and pictorial form of situations are explained as follows. Let $S(t)$, $E(t)$, $I(t)$, $Q(t)$, and $R(t)$ represent the susceptible, exposed or not infected but can become infected, isolated, quarantined, and recovered populations with time t, respectively.

The number of persons being added from migration, immigration, or new birth is assumed to be constant at rate A.

- The contact rate of the COVID-19 transmission is β.
- The exposed population can become infected at rate μ.
- Confirmed infected people are isolated or hospitalized for treatment at rate α.

At rate δ, people undergo quarantine/isolation and take medicines to be free from the disease.

- Natural death during the COVID-19 pandemic is represented by d_1, whereas death due to COVID-19 is represented by d_2.
- The world's population is represented by N.
- The nomenclature is explained in terms of mathematical symbols.

7.2 NOMENCLATURE

These figures explain that during the COVID-19 pandemic, AI techniques played an important role in tracing, screening, testing, and diagnosing the disease without having any contact with people, as shown in Table 7.1.

TABLE 7.1

Important Role of AI in Tracing, Screening, Testing, and Diagnosing COVID-19 without Any Contact with People.

Notation	Descriptions
S	People who are susceptible but not yet infected
E	People who are exposed and not yet infected but can become infected after some time
I	People who are isolated
Q	People who are quarantined
R	People who recovered after hospitalization or isolation
A	People recruited from out-migration or new births
β	Transmission contact rate per time of interaction with a COVID-19 patient
μ	Rate of infection
α	Rate of quarantine or isolation of people who tested positive
δ	Rate of recovery after receiving proper treatment from a hospital, quarantine, or isolation
d_1	Rate of natural death due to some other disease
d_2	Death rate due to COVID-19

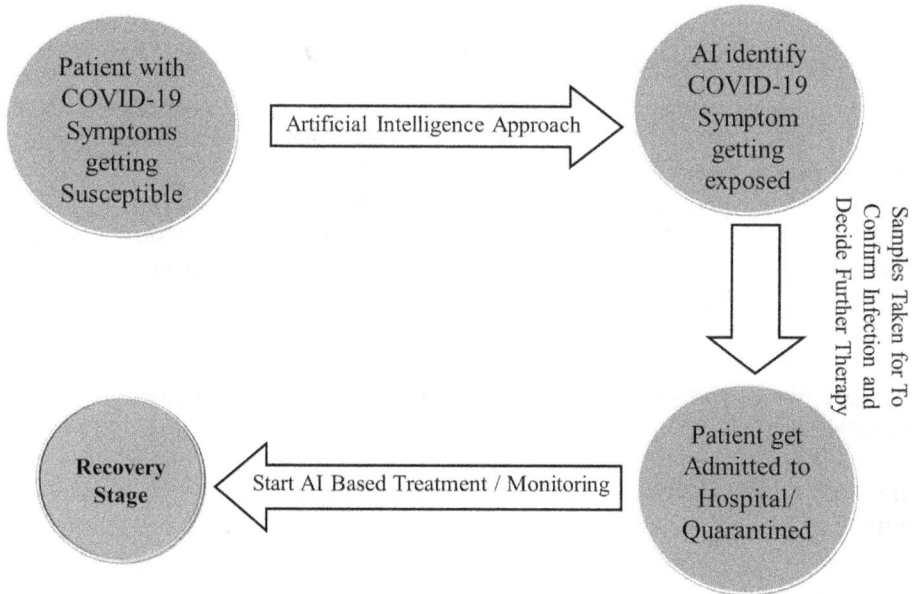

FIGURE 7.1   Identification of COVID-19 by AI Technique.

Source: Khang (2021)

Confirmed cases were immediately sent for isolation/hospitalization. After proper treatment, COVID-19 patients were released from hospitals or quarantine areas. This helps in determining the rate of recovery and also reduces infection, morbidity, and mortality rates during the COVID-19 pandemic, as shown in Figure 7.1.

7.3 MATHEMATICAL ASSUMPTIONS

According to Figure 7.1, we can assume that the total population can be divided into five subpopulations: susceptible, exposed, infected, quarantined, and recovered.

Migrant, newborn, and immigrant populations with a travel history belong to the susceptible population at rate A. A susceptible individual is exposed to AI approaches at rate β. According to the law of mass action, the individual becomes infected in βSI. The exposed population was infected after testing using different AI approaches at rate μ. The infected population decided to undergo hospitalization or quarantine at rate α.

The hospitalized or quarantined population that started AI treatment and monitoring at rate δ recovered from COVID-19. Natural mortality was represented as d_1, and the diseases induced by death were represented as d_2. It is assumed that every subpopulation is dynamic. Therefore, these populations will change with respect to time (Divvela et al., 2023).

A flow diagram and its pictorial representation are shown in Figures 7.1 and 7.2, respectively. As shown in the figures, the AI-based technique has formulated a mathematical model and its ability to combat COVID-19.

7.4 MATHEMATICAL MODEL (*SEIQR*)

Figure 7.2 shows a diagrammatic representation of COVID-19 using the AI technique. Because COVID-19 is dynamic, it can vary from one population to another with respect to time.

The first equation explains how the susceptible population changes over time. When the susceptible population decreased over time, the rate of change was negative. The rate of change is directly proportional to the infected population, in which the susceptible population was in contact.

When the population that includes the recruited migrants, immigrants, and newborn (A) gets exposed, they may become susceptible.

Here, we consider the five subpopulations, in which the number of susceptible population changed because of the addition of recruited migrants, immigrants, and newborn (A), and decreased as a result of death and COVID-19 infection (βSI). The rate of change of the susceptible population is governed by Equation 7.1:

$$\frac{dS}{dt} = A - \beta SI - d_1 S \tag{7.1}$$

The second equation illustrates how the exposed (latent period) population varies over time. The first term is the same as the previous term on the right-hand side. The sign is positive because of the increase in the infected population.

Natural deaths and deaths due to COVID-19 are assumed to have decreased. After some time, the exposed population changes with an increased infection (βSI), becomes infected at μE, and decreases by death at d_1. The rate of change in the population is governed by Equation 7.2:

$$\frac{dE}{dt} = \beta SI - (d_1 + \mu)E \tag{7.2}$$

The third equation states how the infected (latent) population became infected during the COVID-19 pandemic through contact with an infected population.

The infected population increased with rate μE and resulted in the transition from exposed to infected. The population decreased because of natural death (d_1) and death due to COVID-19 (d_2). This is represented by Equation 7.3.

$$\frac{dI}{dt} = \mu E - (d_1 + d_2 + \alpha)I \tag{7.3}$$

The fourth equation explains the rate of change in the quarantined population. The development of the AI technique aids in the screening and testing processes and confirms that the infected population has undergone isolation or quarantine (Khang et al., 2023a).

The quarantined population increased at rate αI. Natural deaths and deaths due to COVID-19 decreased at rate $(d_1 + d_2) Q$. This transition can be represented by the differential, as shown in Equation 7.4:

$$\frac{dQ}{dt} = \alpha I - (d_1 + d_2 + \delta)Q \tag{7.4}$$

The last equation elaborates the transition of the quarantined population that recovered over time after the improvement of the medical and isolation processes. We assume that infection and death rates depend on the availability of high-quality medical care and isolation rooms.

The population increased because of clinical treatment or isolation at rate δQ and decreased because of death d_1. This can be explained by Equation 7.5:

$$\frac{dR}{dt} = \delta Q - d_1 R \tag{7.5}$$

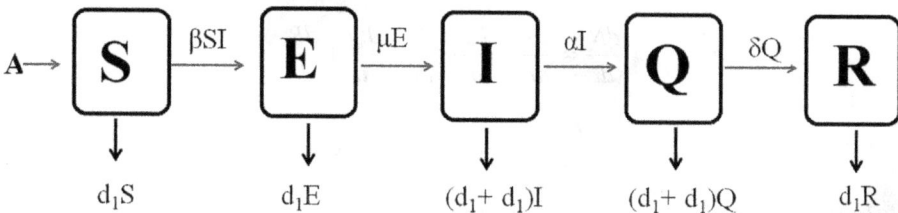

FIGURE 7.2 Schematic Representation of the Model.

Source: Khang (2021)

Therefore, based on the presented descriptions, the corresponding mathematical equations are given by a system of nonlinear ordinary differential equations (Equation 7.6):

Mathematical model

$$\frac{dS}{dt} = A - \beta SI - d_1 S$$

$$\frac{dE}{dt} = \beta SI - (d_1 + \mu)E$$

$$\frac{dI}{dt} = \mu E - (d_1 + d_2 + \alpha)I \qquad (7.6)$$

$$\frac{dQ}{dt} = \alpha I - (d_1 + d_2 + \delta)Q$$

$$\frac{dR}{dt} = \delta Q - d_1 R$$

7.4.1 MODEL ANALYSIS FOR BOUNDEDNESS AND POSITIVITY

In this section, we define the region in which the solution to the model is uniformly bounded as set Ω (Equation 7.7):

$$\Omega \in R_+^5 = \left\{ (S, E, I, Q, R) \in R_+^5, S \geq 0, E \geq 0, I \geq 0, Q \geq 0, R \geq 0 \right\} \qquad (7.7)$$

This is positively invariant for system (1). Clearly, the interaction functions on the right-hand side of system (1) are continuously differentiable. Therefore, the solution of system (1) exists and is unique. In addition, all solutions of system (1) with nonnegative initial conditions are uniformly bounded in the following theorem.

All solutions of system (1), which are initiated in R_+^5, are uniformly bounded.

Proof:

Let $(S(t), E(t), I(t), Q(t), R(t))$ be any solution of system (1) with nonnegative initial condition $(S(0), E(0), I(0), Q(0), R(0))$. Therefore, adding all the individuals $N(t) = S(t) + E(t) + I(t) + Q(t) + R(t)$ results in Equations 7.8 and 7.9:

$$\frac{dN}{dt} = \frac{dS}{dt} + \frac{dE}{dt} + \frac{dI}{dt} + \frac{dQ}{dt} + \frac{dR}{dt} \qquad (7.8)$$

$$\frac{dN}{dt} = A - d_1 N - d_2 (I + Q) \qquad (7.9)$$

In the absence of any infection, the total population becomes $\dfrac{A}{d_1}$.

As $t \to 0$, then $N \to \dfrac{A}{d_1}$.

Hence, all solutions of system (1) are confined in the region as Equation 7.10.

$$\Omega = \left\{ (S, E, I, Q, R) \in R_+^5 : N \le \frac{A}{d_1} \right\} \tag{7.10}$$

which completes the proof.

7.4.2 CALCULATION OF BASIC REPRODUCTION NUMBER

In this section, we define the basic reproduction number as the accepted number of secondary individuals produced in a completely susceptible population by an infected individual during the entire COVID-19 period. This number is useful as a measure of the strength of the control measures required to break the epidemic.

The basic reproduction number is the spectral radius of the next generation matrix. Consider the *SEIQR* model expressed in Equation 7.11.

The system is linearized to form a Jacobian matrix. The Jacobian matrix is decomposed into the *F–V* matrix, where *F* is the production of a new infection, and *V* is the nonsingular matrix that changes from compartment to compartment:

$$F = \begin{pmatrix} \beta SI \\ 0 \\ 0 \\ 0 \\ 0 \end{pmatrix} \text{ and } V = \begin{pmatrix} \beta SI - A + d_1 S \\ (d_1 + \mu)E \\ (d_1 + d_2 + \alpha)I - \mu E \\ (d_1 + d_2 + \delta)Q - \alpha I \\ d_1 R - \delta Q \end{pmatrix} \tag{7.11}$$

Taking the partial derivative with respect to infection class E and I, we get Equation 7.12:

$$F = \begin{pmatrix} 0 & \beta SI \\ 0 & 0 \end{pmatrix} \text{ and } V = \begin{pmatrix} (d_1 + \mu) & 0 \\ -\mu & (d_1 + d_2 + \alpha) \end{pmatrix} \tag{7.12}$$

The basic reproduction number is the spectral radius of the next generation matrix FV^{-1} and is given by Equation 7.13:

$$FV^{-1} = \frac{\beta S_0 \mu}{(d_1 + d_2 + \alpha)(d_1 + \mu)} \tag{7.13}$$

and Equation 7.14

$$R_0 = \frac{\beta \mu}{(d_1 + d_2 + \alpha)(d_1 + \mu)} \tag{7.14}$$

This acts as the threshold for the formation or failure of an epidemic.

7.5 EXISTENCE OF EQUILIBRIUM POINTS OF THE SYSTEM OF EQUATIONS

In this section, the existence of all possible equilibrium points of system (1) is discussed. There are two equilibrium points:

1. COVID-19-free equilibrium point
2. Endemic equilibrium point

For the steady-state condition, the system of equation 7.15 becomes

$$A - \beta SI - d_1 S = 0$$
$$\beta SI - (d_1 + \mu)E = 0$$
$$\mu E - (d_1 + d_2 + \alpha)I = 0 \qquad (7.15)$$
$$\alpha I - (d_1 + d_2 + \delta)Q = 0$$
$$\delta Q - d_1 R = 0$$

7.5.1 CORONAVIRUS-FREE EQUILIBRIUM POINT

For the COVID-19-free equilibrium $\Omega_1(\frac{A}{d_1}, 0, 0, 0, 0)$ when $(E = I = Q = R = 0)$.

Theorem 1
If $R_0 < 1$, then the COVID-19-free equilibrium is locally asymptotically stable. Otherwise, it is unstable.
Proof: Linearization of the model in Equation 7.16.

$$J(S_0, 0, 0, 0, 0) = \begin{pmatrix} -d_1 & 0 & -\beta S_0 & 0 & 0 \\ 0 & -(d_1 + \mu) & \beta S_0 & 0 & 0 \\ 0 & \mu & -(d_1 + d_2 + \alpha) & 0 & 0 \\ 0 & 0 & 0 & -(d_1 + d_2 + \delta) & 0 \\ 0 & 0 & 0 & \delta & -d_1 \end{pmatrix} \qquad (7.16)$$

By taking the eigenvalues

$$\lambda = -d_1$$
$$\lambda = -d_1$$
$$\lambda = -(d_1 + d_2 + \delta)$$

Two other eigenvalues can be of the form of Equation 7.17:

$$\lambda^2 + a\lambda + b = 0$$

where

$$a = 2d_1 + d_2 + \mu + \alpha > 0$$
$$b = (d_1 + \mu)(d_1 + d_2 + \alpha) - \mu \beta S_0 = (d_1 + \mu)(d_1 + d_2 + \alpha)(1 - R_0) > 0$$

$$(7.17)$$

Hence, by applying the Routh–Hurwitz condition, it is locally asymptotically stable.

Theorem 2

If the basic reproduction number $R_0 < 1$, then the COVID-19-free equilibrium is globally asymptotically stable.

Proof: Consider the Thieme Horst and Lyapunov (2011) Equation 7.18:

$$M = \mu E + (d_1 + \mu) I$$

$$\Rightarrow \frac{dM}{dt} = \frac{\mu dE}{dt} + \frac{(d_1 + \mu) dI}{dt} \qquad (7.18)$$

$$\Rightarrow \frac{dM}{dt} = (d_1 + \mu)(d_1 + d_2 + \alpha)(R_0 - 1) I$$

If $R_0 < 1$, then $\dfrac{dM}{dt} < 0$ and $\dfrac{dM}{dt} = 0$ only if $I = 0$ or $R_0 = 1$. Thus, by LaSalle's invariance principle, the set is globally asymptotically stable in the region Ω_0.

7.5.2 STABILITY ANALYSIS OF THE ENDEMIC EQUILIBRIUM

For the stability analysis for endemic equilibrium for $\Omega^*(S^*, E^*, I^*, Q^* R^*)$, we solve the simultaneous system of Equation 7.19:

$$S^* = \frac{1}{R_0}$$

$$E^* = \frac{A}{(d_1 + \mu)}(1 - \frac{1}{R_0})$$

$$I^* = \frac{\mu(R_0 - 1)}{\beta} \qquad (7.19)$$

$$Q^* = \frac{\alpha d_1 (R_0 - 1)\beta}{(d_1 + d_2 + \delta)\beta}$$

$$R^* = \frac{\alpha \delta (R_0 - 1)}{(d_1 + d_2 + \delta)}$$

Theorem 3

If $R_0 > 1$, then system (2) has a unique endemic equilibrium $\Omega^*(S^*, E^*, I^*, Q^*)$, which is locally asymptotically stable.

Proof: In linearization of Equation 7.20, the Jacobian of the matrix is

$$\Omega^* = \begin{pmatrix} -d_1 - \beta I^* & 0 & -\beta S^* & 0 \\ \beta I^* & -(d_1 + \mu) & \beta S^* & 0 \\ 0 & \mu & -(d_1 + d_2 + \alpha) & 0 \\ 0 & 0 & \alpha & -(d_1 + d_2 + \delta) \end{pmatrix} \qquad (7.20)$$

The eigenvalues are $\lambda = -(d_1 + d_2 + \delta)$.

The other three eigenvalues in solving the cubic equation are $\lambda^3 + A\lambda^2 + B\lambda + C = 0$. where

$$
\begin{aligned}
A =\ & \beta I^* + 3d_1 + d_2 + \alpha + \mu \\
B =\ & 3d_1^2 + 2d_1\mu + 2d_1 d_2 + 2d_1\alpha + 2d_1\beta I^* + \mu\beta I^* \\
& + d_2\beta I^* + \alpha\beta I^* + \mu d_2 + \alpha\mu - \alpha\beta S^* \\
C =\ & d_1^3 + d_1^2 d_2 + \alpha d_1^2 + \mu d_1^2 + \mu d_1 d_2 + d_1\alpha\mu + d_1^2\beta I^* \\
& + d_1 d_2\beta I^* + d_1\alpha\beta I^* + \mu d_1\beta I^* + \mu d_2\beta I^* \\
& + \mu\alpha\beta I^* + \mu\beta S^* d_1 + 2\mu\beta^2 I^* S^*
\end{aligned}
\tag{7.21}
$$

Since $A > 0$, $B > 0$, and $C > 0$, then $AB > C$.

Therefore, by applying the Routh–Hurwitz condition, the endemic equilibrium is locally asymptotically stable.

7.5.3 GLOBAL STABILITY OF THE ENDEMIC EQUILIBRIUM

In this section, we discuss the global stability of the endemic equilibrium using the geometric approaches. The sufficient condition for global stability of the equilibrium points is followed by the Bendixson condition.

For $R_0 > 1$, we infer the uniform persistence of the system of equations, which means that there exists a positive constant C such that the initial condition $((S (OE (0), I (0) Q (0))$ lies in the interior of the region Ω.

Any solution $\left(S(t), E(t), I(t), Q(t)\right)$

Satisfying $\min\{\liminf_{t\to\infty} S(t), \liminf_{t\to\infty} E(t), \liminf_{t\to\infty} I(t), \liminf_{t\to\infty} Q(t)\} > C$

The preceding condition along with boundedness of the region Ω verifies the endemic equilibrium point interior to Ω.

Theorem 4

The unique endemic equilibrium Ω^* is globally asymptotically stable in the interior of Ω if $R_0 > 1$.

Proof: For the general solution of the system as Equation 7.22.

$$
J = \begin{pmatrix}
-d_1 - \beta I & 0 & -\beta S & 0 \\
\beta I & -(d_1 + \mu) & \beta S & 0 \\
0 & \mu & -(d_1 + d_2 + \alpha) & 0 \\
0 & 0 & \alpha & -(d_1 + d_2 + \delta)
\end{pmatrix}
\tag{7.22}
$$

The second compounded matrix is given by Equation 7.23:

$$J^{[2]} = \begin{pmatrix}
-2d_1 - \beta I - \mu & \beta S & 0 & 0 & 0 & 0 \\
\mu & -2d_1 - d_2 - \alpha - \beta I & 0 & 0 & 0 & 0 \\
0 & \alpha & -2d_1 - d_2 - \delta - \beta I & 0 & 0 & -\beta S \\
0 & \beta I & 0 & -(2d_1 + d_2 + \mu + \alpha) & 0 & 0 \\
0 & 0 & \beta I & \alpha & -(2d_1 + \mu + d_2 + \delta) & \beta S \\
0 & 0 & 0 & 0 & \mu & -(2d_1 + 2d_2 + \alpha + \delta)
\end{pmatrix}$$ (7.23)

To obtain the matrix B in the Bendixson matrices condition, we define the diagonal matrix M as Equation 7.24:

$$M = \text{diag}(1, \frac{E}{I}, \frac{E}{I}, \frac{E}{I}, \frac{E}{I}, \frac{E}{I})$$ (7.24)

We can define a vector field of the system then Equations 7.25 and 7.26:

$$M_f M^{-1} = \text{diag}(0, \frac{E'}{E} - \frac{I'}{I}, \frac{E'}{E} - \frac{I'}{I}, \frac{E'}{E} - \frac{I'}{I}, \frac{E'}{E} - \frac{I'}{I}, \frac{E'}{E} - \frac{I'}{I})$$ (7.25)

$B = M_f M^{-1} + M J^{[2]} M^{-1}$

$$= \begin{pmatrix}
-2d_1 - \beta I - \mu & \dfrac{\beta SI}{E} & 0 \\[2mm]
\dfrac{E\mu}{I} & -2d_1 - d_2 - \alpha - \beta I + \dfrac{E'}{E} - \dfrac{I'}{I} & 0 \\[2mm]
0 & \alpha & \dfrac{E'}{E} - \dfrac{I'}{I} - 2d_1 - d_2 - \delta - \beta I \\[2mm]
0 & \beta I & 0 \\[2mm]
0 & 0 & \beta I \\[2mm]
0 & 0 & 0
\end{pmatrix}$$ (7.26)

$$\begin{pmatrix}
\dfrac{\beta SI}{E} & 0 & 0 \\[2mm]
0 & 0 & 0 \\[2mm]
0 & 0 & -\beta S \\[2mm]
\dfrac{E'}{E} - \dfrac{I'}{I} - (2d_1 + d_2 + \mu + \alpha) & 0 & 0 \\[2mm]
\alpha & \dfrac{E'}{E} - \dfrac{I'}{I} - (2d_1 + \mu + d_2 + \delta) & \beta S \\[2mm]
0 & \mu & \dfrac{E'}{E} - \dfrac{I'}{I} - (2d_1 + 2d_2 + \alpha + \delta)
\end{pmatrix}$$

The preceding matrix can be represented as a block matrix in Equation 7.27:

$$B = \begin{pmatrix} B_{11} & B_{12} \\ B_{21} & B_{22} \end{pmatrix} \tag{7.27}$$

where Equation 7.28

$$B_{11} = (-2d_1 - \beta I - \mu)$$

$$B_{12} = \begin{pmatrix} \dfrac{\beta SI}{E} & 0 & \dfrac{\beta SI}{E} & 0 & 0 \end{pmatrix}$$

$$B_{21} = \begin{pmatrix} \dfrac{E\mu}{I} \\ 0 \\ 0 \\ 0 \\ 0 \end{pmatrix}$$

$$B_{22} = \begin{pmatrix} -2d_1 - d_2 - \alpha - \beta I + \dfrac{E'}{E} - \dfrac{I'}{I} & 0 & 0 \\ \alpha & \dfrac{E'}{E} - \dfrac{I'}{I} - 2d_1 - d_2 - \delta - \beta I & 0 \\ \beta I & 0 & \dfrac{E'}{E} - \dfrac{I'}{I} - (2d_1 + d_2 + \mu + \alpha) \\ 0 & \beta I & \alpha \\ 0 & 0 & 0 \end{pmatrix}$$

$$\begin{pmatrix} 0 & 0 \\ 0 & -\beta S \\ 0 & 0 \\ \dfrac{E'}{E} - \dfrac{I'}{I} - (2d_1 + \mu + d_2 + \delta) & \beta S \\ \mu & \dfrac{E'}{E} - \dfrac{I'}{I} - (2d_1 + 2d_2 + \alpha + \delta) \end{pmatrix} \tag{7.28}$$

The Lozinskii measure of the matrix can be calculated as $\mu(B) \le \sup\{g_1, g_2\}$, where

$$g_1 = \mu(B_{11}) + |B_{12}| = -2d_1 - \beta I - \mu + \dfrac{\beta SI}{E}$$

$$g_2 = |B_{21}| + \mu(B_{22}) = -2d_1 + \mu + \dfrac{E'}{E} - \dfrac{I'}{I} \tag{7.29}$$

This Lozinskii measure (2023) is mieu with respect to the norm of the block matrices B12 and B21, which are the operator norms.

Hence, g_1 and g_2 are reduced to Equation 7.30:

$$g_1 = -d_1 - \beta I + \frac{E'}{E}$$
$$g_2 = -d_1 + \alpha + d_2 + \frac{E'}{E}$$

(7.30)

Thus

$$\mu(B) \leq \sup\{g_1, g_2\} \leq \frac{E'}{E} - d_1 + \sup\{-\beta I, d_2 + \alpha\}$$

Hence, $\mu(B) \leq \frac{E'}{E} - d_1$; thus

$$\int_0^t \mu(B)dt \leq (\log E(t) - d_1 t)$$

Hence, finally as Equation 7.31

$$\bar{q}_2 = \frac{\int_0^t \mu(B)dt}{t} < \frac{\log E(t)}{t} - d_1 < \frac{d_1}{2} - d_1 < 0$$

(7.31)

Therefore, the criteria $\bar{q}_2 < 0$ is satisfied. Hence, the endemic equilibrium of the system is globally stable.

7.6 RESULTS AND DISCUSSION

The analytical results cannot be proven without numerical simulations. These figures explain the behavior of the model in different populations during COVID-19.

The deviation of the populations with respect to time for the parameters is shown in these figures. AI techniques are being developed to detect COVID-19 and to determine the stability of the disease. The major objective is visualization to control infections caused by COVID-19.

Based on different situations during the COVID-19 period, we used different parameters with the help of AI techniques. The COVID-19-free equilibrium point was stable for $R_0 < 1$ (Figure 7.3).

The best uses of the combined effect of isolation and exposure are displayed in the figures. As a result, the quarantine/hospitalization rates after using AI in the exposed population decreased, as shown in Figure 7.4.

This figure indicates that the exposed population quickly survived and went to quarantine if they experienced some illness symptoms, as shown in Figure 7.5.

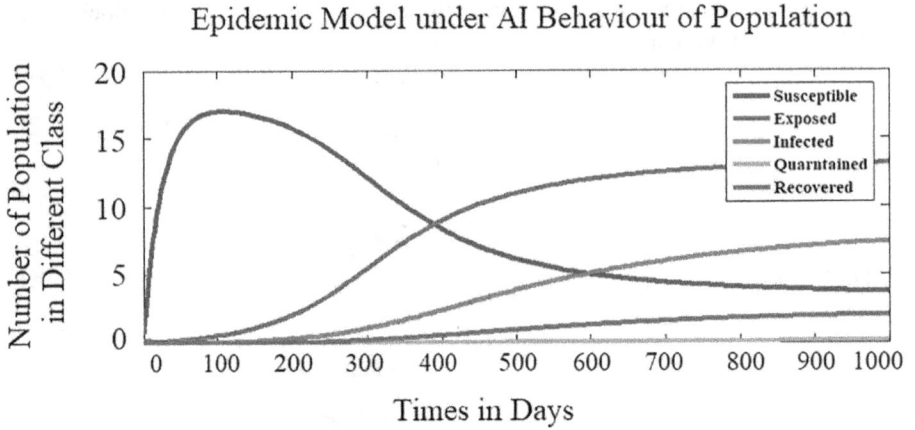

FIGURE 7.3 Dynamic Behavior of Population under AI Effect When $R_0 < 1$.

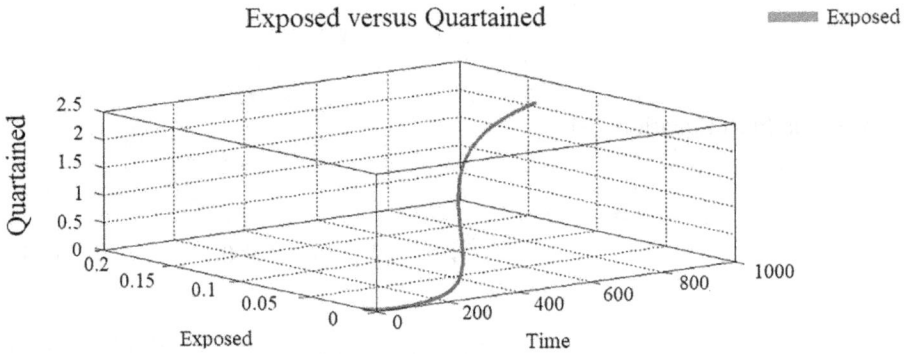

FIGURE 7.4 Exposed Population versus Quarantined Population Model.

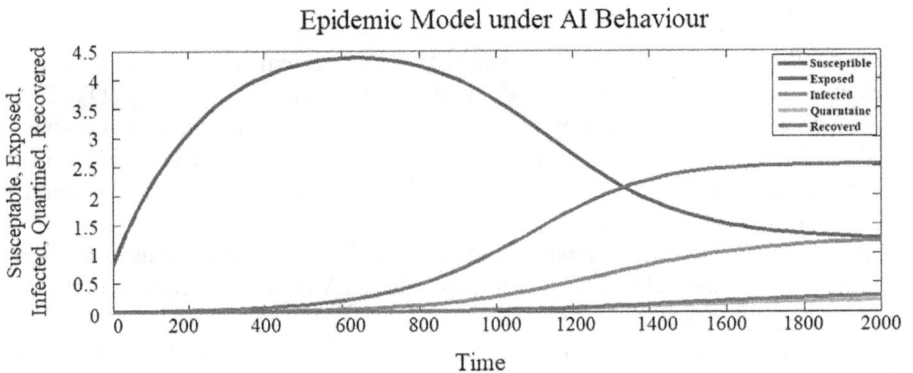

FIGURE 7.5 Dynamic Behavior of the Population under AI Effect for Endemic Equilibrium Point.

At the same time, the development of AI in the field of healthcare systems makes treatment easier. Thus, endemic equilibrium is stable in this region, as shown in Figure 7.6.

As a result, Figure 7.9 indicates that the exposed population increases rapidly upon contact with infected populations and then recovers after proper treatment.

Finally, Figure 7.10 shows that a large number of infected individuals undergo hospitalization by testing them using AI methods.

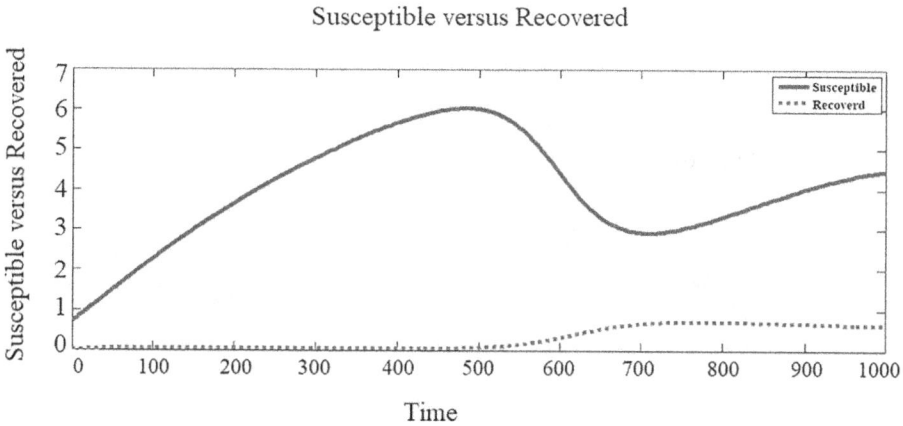

FIGURE 7.6   Behavior of the Susceptible Population versus Recovered Population.

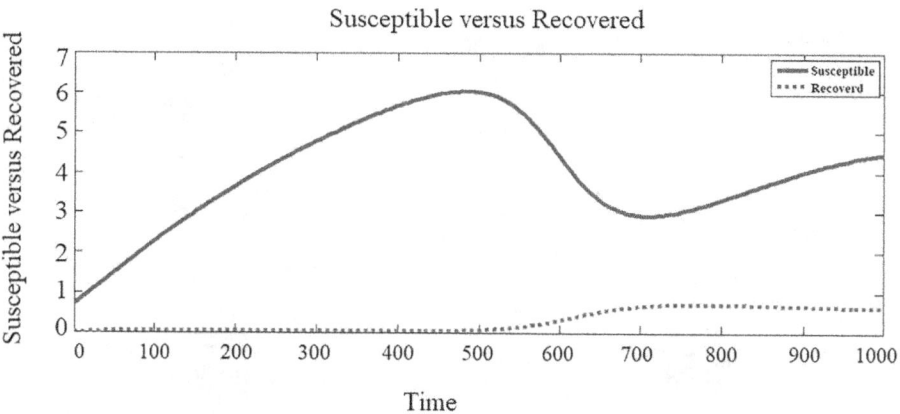

FIGURE 7.7   Infected Population versus Quarantined Population.

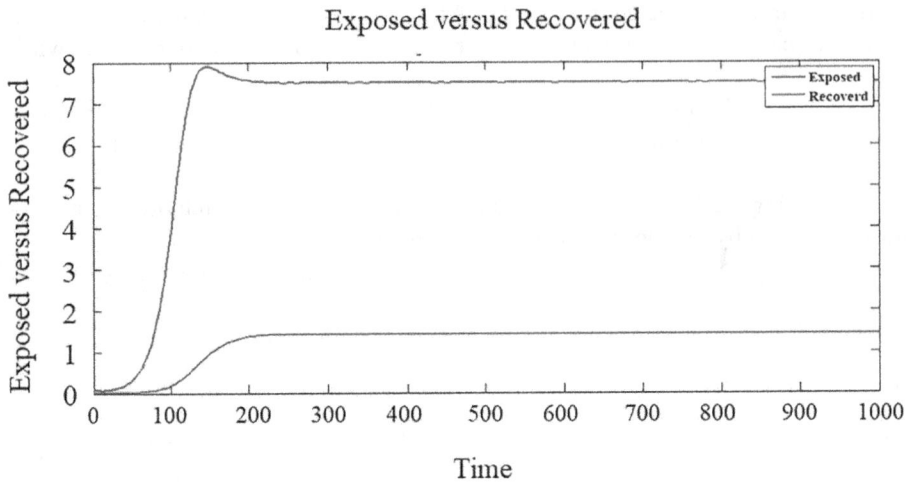

FIGURE 7.8   Exposed Population versus Recovered Population.

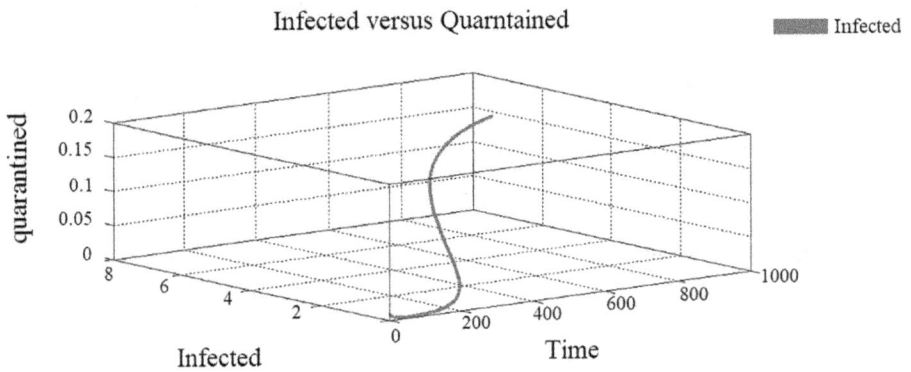

FIGURE 7.9   Infected Population versus Quarantined Population.

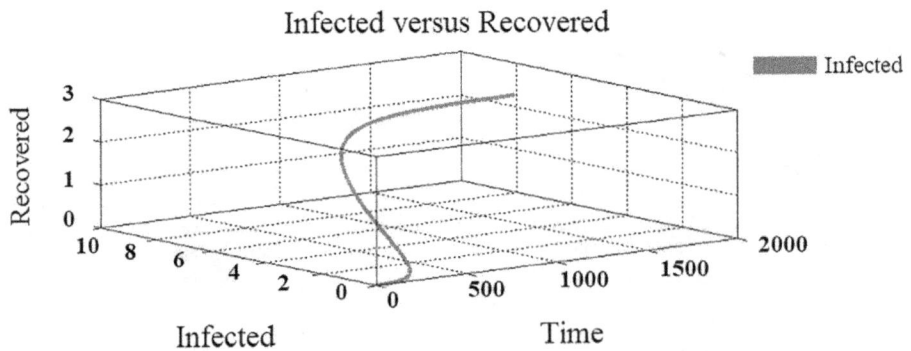

FIGURE 7.10   Impact of Infected Population versus Recovered Population with Respect to Time.

7.7 CONCLUSION

A dynamic epidemic *SEIQR* model for coronavirus was formulated using AI, both numerically and mathematically. An AI technique was used to detect COVID-19.

The existence of equilibrium points in the model was investigated. The observation of the model was stable and depended on the basic reproduction number. The local stability of a disease is asymptotically stable when its number is less than unity.

When the number is greater than unity, the situation becomes pandemic.

- This pandemic can be handled by improving medical facilities and developing vaccines in a short time.
- This AI technique also plays an important role in epidemiology.
- This can help in the diagnosis, detection, prediction, and social control for combating COVID-19 due to latency between the susceptible and infected populations. The *SEIQR* model is more suitable for modeling.

The AI technique was combined with mathematical modeling to speed up the elimination of the virus without human intervention.

To control the spread of the virus, the AI technique rapidly isolated or quarantined people. The developed forecasting model can help in the policymaking and in tackling COVID-19.

Control methods may include booster dose vaccination of susceptible individuals. This model suggested that the healthcare system build up a team with data to control emergency cases.

Our mathematical model shows that quarantine during the pandemic changed the situation. This optimal strategy helped in the recovery of patients from different geographical locations and emphasized the focus of the AI technique on the healthcare system rather than on individuals.

This AI model helps the medical sector move forward to a better solution for the diagnosis and treatment of diseases. Artificial intelligence improves efficiency and productivity by reducing manual errors (Khang et al., 2023c).

Quarantine and social distancing norms play a crucial role that must be addressed with careful consideration.

Artificial intelligence knowledge of equipment may be effectively used to deal with some of these challenges and assist in solving fundamental problems in global disasters. The simulated results are supported by real parametric values.

REFERENCES

Agrebi S., Larbi A., (2020). "Use of artificial intelligence in infectious diseases," In *Artificial Intelligence in Precision Health*, pp. 415–438. Academic Press. www.sciencedirect.com/science/article/pii/B9780128171332000185

Alanazi S. A., Kamruzzaman M. M., Alruwaili M., Alshammari N., Alqahtani S. A., Karime A., "Measuring and preventing COVID-19 using the SIR model and machine learning in smart health care," *Journal of Healthcare Engineering*, 2020, Article ID 8857346, 12. https://doi.org/10.1155/2020/8857346

Behera J., Rauta S. K., Rao Y. S., Patnaik S., (2021). "Mathematical modeling on dou-
ble quarantine process in the spread and stability of COVID-19," In *Sustainability Measures for COVID-19 Pandemic*, pp. 37–57. Springer, E-Book. https://doi.
org/10.1007/978-981-16-3227-3_3

Chawla S., Mittal M., Chawla M., Goyal L. M., (May, 2020). "Corona Virus-SARS-CoV-2:
An insight to another way of natural disaster," *EAI Endorsed Transactions on Pervasive
Health and Technology*. https://eudl.eu/doi/10.4108/eai.28-5-2020.164823

Dihudi B., Rao Y. S., Rout S. K., Panda T. C., (2021). "Transmission modelling on COVID- 19
pandemic and its challenges," *Sustainability Measures for COVID-19 Pandemic*,
pp. 75–90. Springer, E-Book. https://doi.org/10.1007/978-981-16-3227-3_5

Gaur L., Bhatia U., Jhanjhi N. Z., Muhammad G., Masud M., (2021). "Medical image-based
detection of COVID-19 using deep convolution neural networks," *Multimedia Systems*,
1. https://link.springer.com/article/10.1007/s00530-021-00794-6

Khang A., (2021). "Material4Studies," *Material of Computer Science, Artificial Intelligence,
Data Science, IoT, Blockchain, Cloud, Metaverse, Cybersecurity for Studies*. www.
researchgate.net/publication/370156102_Material4Studies

Khang A., Gupta S. K., Hajimahmud V. A., Babasaheb J., Morris G., (2023a). *AI-Centric Modelling and Analytics: Concepts, Designs, Technologies, and Applications* (1st ed.). CRC
Press. https://doi.org/10.1201/9781003400110

Khang A., Hahanov V., Litvinova E., Chumachenko S. T., Hajimahmud V. A., Ragimova
Nazila A., Abuzarova Vusala A., Anh P. T. N., (2023b). "The analytics of hospitality
of hospitals in healthcare ecosystem," In *Data-Centric AI Solutions and Emerging
Technologies in the Healthcare Ecosystem*, p. 4. CRC Press. https://doi.org/10.1201/
9781003356189-4

Khang A., Rana G., Tailor R. K., Hajimahmud V. A., (Eds.). (2023c). *Data-Centric AI Solutions and Emerging Technologies in the Healthcare Ecosystem*. CRC Press. https://doi.
org/10.1201/9781003356189

Khang A., Rani S., Sivaraman A. K., (2022a). *AI-Centric Smart City Ecosystems: Technologies, Design and Implementation* (1st ed.). CRC Press. https://doi.org/10.1201/
9781003252542

Khang A., Ragimova N. A., Hajimahmud V. A., Alyar V. A., (2022b), "Advanced technologies
and data management in the smart healthcare system," In *AI-Centric Smart City Ecosystems: Technologies, Design and Implementation*, Vol. 16, No. 10 (1st ed.). CRC Press.
https://doi.org/10.1201/9781003252542-16

Khanh H. H., Khang A., (2021). "The role of artificial intelligence in blockchain applications,"
Reinventing Manufacturing and Business Processes through Artificial Intelligence,
2(20–40) (CRC Press). https://doi.org/10.1201/9781003145011-2

Lalmuanawma S., Hussain J., Chhakchhuak L., (2020). Applications of machine learning and
artificial intelligence for Covid-19 (SARS-CoV-2) pandemic: A review," *Chaos, Solitons & Fractals*, 110059. www.sciencedirect.com/science/article/pii/S0960077920
304562

Li L., Qin L., Xu Z., Yin Y., Wang X., Kong B., Bai J., Lu Y., Fang Z., Song Q., Cao K.,
Daliang Liu D., Wang G., Xu Q., Fang X., Zhang S., Xia J., Xia J., (2020). "Using
artificial intelligence to detect COVID-19 and community-acquired pneumonia based
on pulmonary CT: Evaluation of the diagnostic accuracy," *Radiology*, 296(2), 65–72
https://pubs.rsna.org/doi/abs/10.1148/radiol.2020200905

Lozinskii, (2023). *Existance of the Lozinskii Measure*. https://math.stackexchange.com/
questions/77017/existance-of-the-lozinskii-measure

McCall B., (2020). "COVID-19 and artificial intelligence: Protecting healthcare workers and
curbing the spread," *Lancet*, 2, e166–e167. www.thelancet.com/journals/landig/article/
PIIS2589–7500(20)300546/fulltext

Mei X., Lee H., Diao K., Huang M., Lin B., Liu C., Xie Z., Ma Y., Robson P. M., Chung M., Bernheim A., Mani V., Calcagno C., Li K., Li S., Shan H., Lv J., Zhao T., Xia J., Long Q., Steinberger S., Jacobi A., Deyer T., . . ., Yang Y., (2020). "Artificial intelligence-enabled rapid diagnosis of patients with COVID-19," *Nature Medicine Letters*. www.nature.com/articles/s41591-020-0931-3

Mishra B. K., Keshri A. K., Rao Y. S., Mishra B. K., Mahato B., Ayesha S., Rukhaiyyar B. P., Saini D. K., Singh A. K., (2020). "COVID-19 created chaos across the globe: Three novel quarantine epidemic models," *Chaos, Solitons & Fractals*, 138, 109928. https://doi.org/1016/j.chaos.2020.109928

Mishra B. K., Keshri A. K., Saini D. K., Ayesha S., Rao Y. S., (2021). "Mathematical model, forecast and analysis on the spread of COVID-19," *Chaos, Solitons & Fractals*, 147, 110995. https://doi.org/j.chaos.2021.110995

Praveen Kumar M., Kumar N., Anuradha M., Khang A., (2023). "Heart disease prediction using logistic regression and random forest classifier," In *Data-Centric AI Solutions and Emerging Technologies in the Healthcare Ecosystem*, p. 6. (1st ed.). CRC Press. https://doi.org/10.1201/9781003356189-6

Rana G., Khang A., Sharma R., Goel A. K., Dubey A. K., (Eds.). (2021). *Reinventing Manufacturing and Business Processes through Artificial Intelligence*. CRC Press. https://doi.org/10.1201/9781003145011

Rani S., Chauhan M., Kataria A, Khang A., (Eds.). (2021). "IoT equipped intelligent distributed framework for smart healthcare systems," In *Networking and Internet Architecture*, Vol. 2, p. 30. CRC Press. https://doi.org/10.48550/arXiv.2110.04997

Rauta A. K., Rao Y. S., Behera J., (2021). "Spread of COVID-19 in Odisha (India) due to influx of migrants and stability analysis using mathematical modeling," *Artificial Intelligence for COVOD-19*, pp. 295–309. Springer Nature, E-Book. https://doi.org/10.1007/978-3-030-69744-0

Rauta A. K., Rao Y. S., Behera J., Dihudi B., Panda T. C., (2021). Algorithms for Intelligent Systems "SIQRS epidemic modelling and stability analysis of COVID-19," In *Predictive and Preventive Measures for Covid-19 Pandemic*, pp. 35–50. Springer. https://doi.org/10.1007/978-981-33-4236-1_3

Sathwik A., Vanshika J., Saravanan S., (2021). "Transforming native epidemic models by using the machine learning approach," *Annals of the Romanian Society for Cell Biology*, 25(4), 1583–6258 (2891–2899). www.annalsofrscb.ro/index.php/journal/article/view/2831

Singh S., (2020). *Cousins of Artificial Intelligence*. https://towardsdatascience.com/cousins-ofartificialintelligence- dda4edc27b55

Thieme Horst R., Lyapunov, (2011). "Global stability of the endemic equilibrium in infinite dimension, Lyapunov function and positive operators," *The Journal of Differential Equations*, 250, 3772–3801. www.sciencedirect.com/science/article/pii/S0022039611000246

Tuli S., Tuli S., Tuli R., Gill S. S., (2020). "Predicting the growth and trend of COVID-19 pandemic using machine learning and cloud computing," *Internet of Things*, 11(100222). https://doi. org/10.1016/j.iot.2020.100222

Vishnu Sai Kumar D., Ritik Ch., Anuradha M., Praveen Kumar M., Khang A., (Aug 24, 2023). "Heart disease and liver disease prediction using machine learning," In *Data-Centric AI Solutions and Emerging Technologies in the Healthcare Ecosystem*, p. 4 (1st ed.). CRC Press. https://doi.org/10.1201/9781003356189-13

WHO, (2021). www.who.int/emergencies/diseases/novel-coronavirus-2019.33–4236–1

Zhao Z., Ma Y., Mushtaq A., et al., (2021). "Applications of robotics, artificial intelligence, and digital technologies during COVID-19: A review," *Disaster Medicine and Public Health Preparedness*. www.cambridge.org/core/journals/disaster-medicine-and-public-health-preparedness/article/applications-of-robotics-ai-and-digital-technologies-during-covid19-a-review/610FFDBD77481D2F8DED4EF3B029AA52

8 Breast Cancer Prediction Using Voting Classifier Model

Sujeet Kumar Jha and Alex Khang

8.1 INTRODUCTION

Breast cancer, mostly targeting the female population, is a major concern for the World Health Organization, as it is persistent worldwide. It is the most common type of cancer in the world (Sung et al., 2021; World Health Organization, 2021).

Depending on the severity of breast cancer symptoms, patients may be exposed to a variety of tests to diagnose the disease, including ultrasonography, biopsy, and mammography. Biopsy is the most evident of these approaches, which entails the removal of sample tissues or cells for analysis.

A fine-needle aspiration (FNA) method is used to take a sample of cells from the breast, which is subsequently sent to a pathology laboratory for microscopic inspection (Khourdifi & Bahaj, 2018).

Numerical properties, such as radius, roughness, perimeter, and area, can be calculated using microscopic images of cells and tissues. The FNA results are then compared to imaging data to determine the possibility of a patient having a malignant tumor (Bhambri et al., 2022).

The fatality rate of breast cancer is comparatively higher than the other types of cancer. Therefore, there seems to be a necessity for a model that can analyze whether a person developing symptoms is vulnerable to cancer or not. If a preliminary scan cannot be conducted in person, this kind of automation is beneficial for both patients and doctors.

In this chapter, the data were extracted from the archive of UC Irvine Machine Learning Repository (Wolberg et al., 2021). Here, several machine learning-based classification algorithms are used to predict whether a patient is susceptible to a malignant tumor or not.

8.1.1 PROBLEM

The main objective of this chapter is to survey the current trends and algorithms in disease prediction systems and propose a model that performs efficient and reliable prediction of patient health at a minimum cost (Khang et al., 2023a).

DOI: 10.1201/9781003400110-8

TABLE 8.1
Stance Category in a Dataset.

Stance	Description	Percentage (%)
0	Benign	356
1	Malignant	212

The specific objectives are listed as follows:

- To compare various classification techniques and build an ensemble model to improve accuracy
- To develop a prediction system for patients with breast cancer with higher accuracy than other systems

8.1.2 REQUIREMENT FOR OUR SYSTEM

The Centers for Disease Control and Prevention, the United States national public health agency, published a dataset known as Wisconsin Breast Cancer Database (WBCD) for public use under CCO license (Praveen et al., 2023).

The dataset contains 569 instances and 32 attributes, where there are two classes (malignant and benign), as shown in Table 8.1. The proportion of malignant people is 32.7% and nondiabetic is 62.7% (Wolberg et al., 2021).

Feature information includes radius, texture, perimeter, area, smoothness (local variation in radius lengths), compactness (perimeter^2/area $-$ 1.0), concavity (severity of concave portions of the contour), concave points (number of concave portions of the contour), symmetry, and fractal dimension ('coastline approximation' $-$ 1).

From Table 8.1, it is clear that the data are slightly imbalanced, which implies that the building model is sufficient, so further treatment is required.

8.2 LITERATURE REVIEW

Depending on the severity of breast cancer symptoms, patients may be exposed to a variety of tests to diagnose the disease, including ultrasonography, biopsy, and mammography (Khang et al., 2022b). Biopsy, the most evident of these approaches, entails the removal of sample tissues or cells for analysis. Fine-needle aspiration (FNA) is a method used to take cell samples from the breast, which are subsequently sent to a pathology laboratory for microscopic inspection (Khourdifi & Bahaj, 2018).

Numerical properties, such as radius, roughness, perimeter, and area, can be calculated using microscopic images of cells and tissues. The FNA results are then compared to imaging data to determine the possibility of a patient having a malignant tumor (Richard O. Duda, 2000).

Kumari et al., (2021) compared AdaBoost, logistic regression, support vector machine, random forest, and other state-of-the-art techniques and base classifiers.

Khuriwa and Mishra (2018) used the voting algorithm approach to evaluate an ensemble of two machine learning algorithms, *artificial neural network* (ANN) and logistic regression, to identify and diagnose breast cancer. They achieved 98% accuracy when using an ANN approach compared with logistic algorithm.

Naji et al. (2021) used individual-based machine learning algorithms such as support vector machine, logistic regression, decision tree, *k*-nearest neighbor, naive Bayes, and well-known majority voting ensemble models.

There are two groups of people in the offered dataset of 569 cases. One group composed of 212 people was diagnosed with possible cancerous tissue termed malignant. The other group, composed of 357 people, was recognized as having benign tumors (Wolberg et al., 2021).

The most important aspect of meaningful evaluation of machine learning algorithms is to ensure that they are all tested on the same data (Khang et al., 2022a). This was accomplished by requiring each algorithm to be assessed using a standardized test harness. Four different categorization techniques were examined using a single dataset:

- Logistic regression
- *K*-nearest neighbors
- Support vector machine
- Decision tree

These models were combined and used in the ensemble model to determine whether the overall accuracy of the classification model could be improved.

These models were then integrated and used in an ensemble model to determine whether the overall accuracy of the classification model could be improved. The following ensemble models were used in this project:

- Hard/majority voting
- Soft voting
- Weighted average

The classifier method was designed to run on 70% training data and 30% test data.

8.3 METHODOLOGY

8.3.1 IMPLEMENTING ENVIRONMENT

The project was implemented in Python version 3.3 using the Jupyter notebook as a standard platform. Libraries, such as Panda, Sklearn, and Pyplot, were used. JupyterLab was used to run the models and analyze the data.

8.3.2 DATASET ANALYSIS AND PRE-PROCESSING

The data used were from the UC Irvine Machine Learning Repository for the breast cancer dataset, known as WBCD. This is a publicly available dataset that consists of 569 instances and 32 attributes (Wolberg et al., 2021).

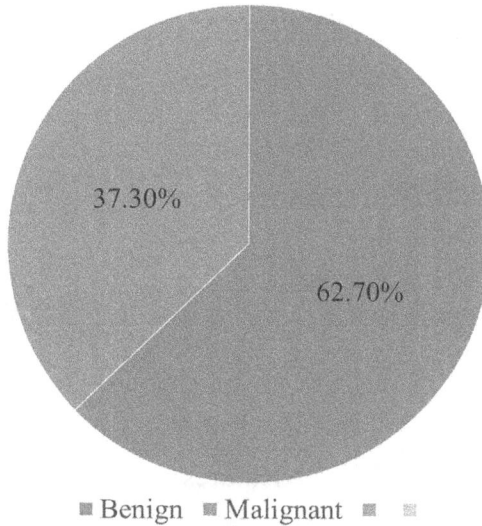

FIGURE 8.1   Distribution Diagnosis Variable.

Data pre-processing is a tried-and-true method for tackling such issues. Data pre-processing is an important and necessary step. The project was accompanied by the following process (Rani et al., 2021).

The pre-processing operations remove irrelevant columns (id), which are referred to as unnamed: 32 because they contain missing values.

To aid in further implementation, prefix the field diagnostic from 'M' and 'B' to a numerical value of 1 and 0, respectively. The numbers of cases of benign and malignant tumors are represented by the graph in Figure 8.1.

Feature selection is a technique that helps select important features that contain maximum information so that it can play an important role in the prediction variable with significant output.

In this feature selection project, a set of methods is applied to minimize the number of features.

8.3.2.1   Univariate Selection

In univariate selection, statistical tests were used to choose the variables/features that have a high correlation with the target variable.

The SelectKBest method selects the feature using chi-square to find the functional association between the feature and the target variable (Vijayvargia, 2018).

8.3.2.2   Feature Importance

We use an extra tree classifier to extract the ranking of the features from the dataset based on score. It calculates the entropy of the features and ranks their importance according to their scores (Vijayvargia, 2018), as shown in Figure 8.2.

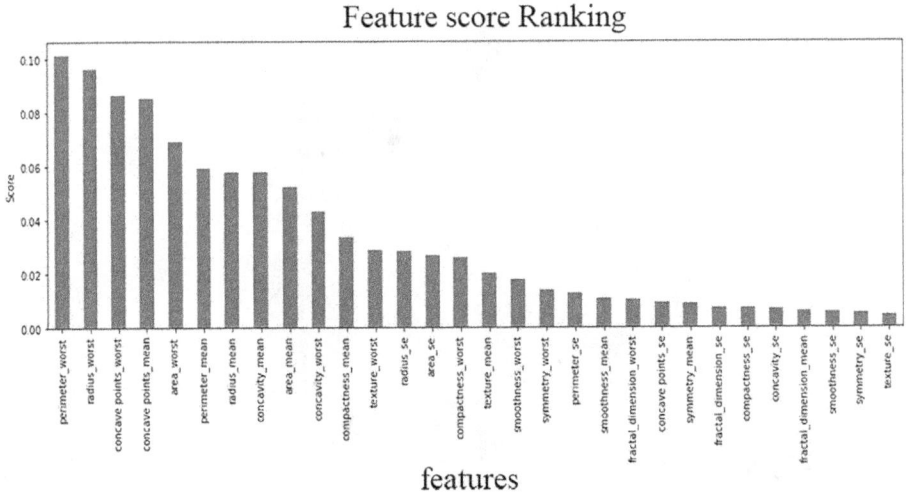

FIGURE 8.2 Top Scoring (Chi-Square) Feature.

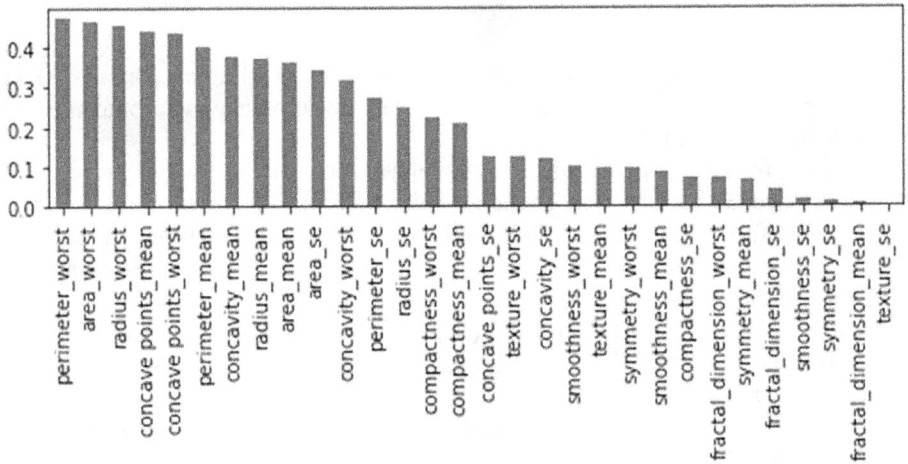

FIGURE 8.3 Score of Mutual Information Parameters.

8.3.2.3 Mutual Information

Mutual information is calculated for fixed categories as in a classification task. The entropy of the variables is used to calculate the rank of the features according to their scores. A value nearer to 0 indicates that the attributes are independent, as shown in Figure 8.3.

From the three feature selection techniques, the features selected to be neglected during training of the models are 'texture se', 'fractal dimension mean', 'symmetry se', 'smoothness se', 'fractal dimension se', and 'symmetry mean'.

8.3.2.4 Correlation Matrix with Heat Map

A correlation matrix is implied, and Pearson correlation is the most commonly used. First, a Pearson correlation heat map was used to visualize the relationship between the independent factors and the output variable diagnosis.

Sklearn was used to generate these charts (Scikit-Learn Developers, 2022). As shown in the heat map in Figure 8.4, area worst was significantly correlated (more than 0.9) with independent variables (radius mean, perimeter mean, area mean, radius worst, and perimeter worst). The worst area was chosen, and all other highly correlated features in the matrix were dropped, as shown in Figure 8.4.

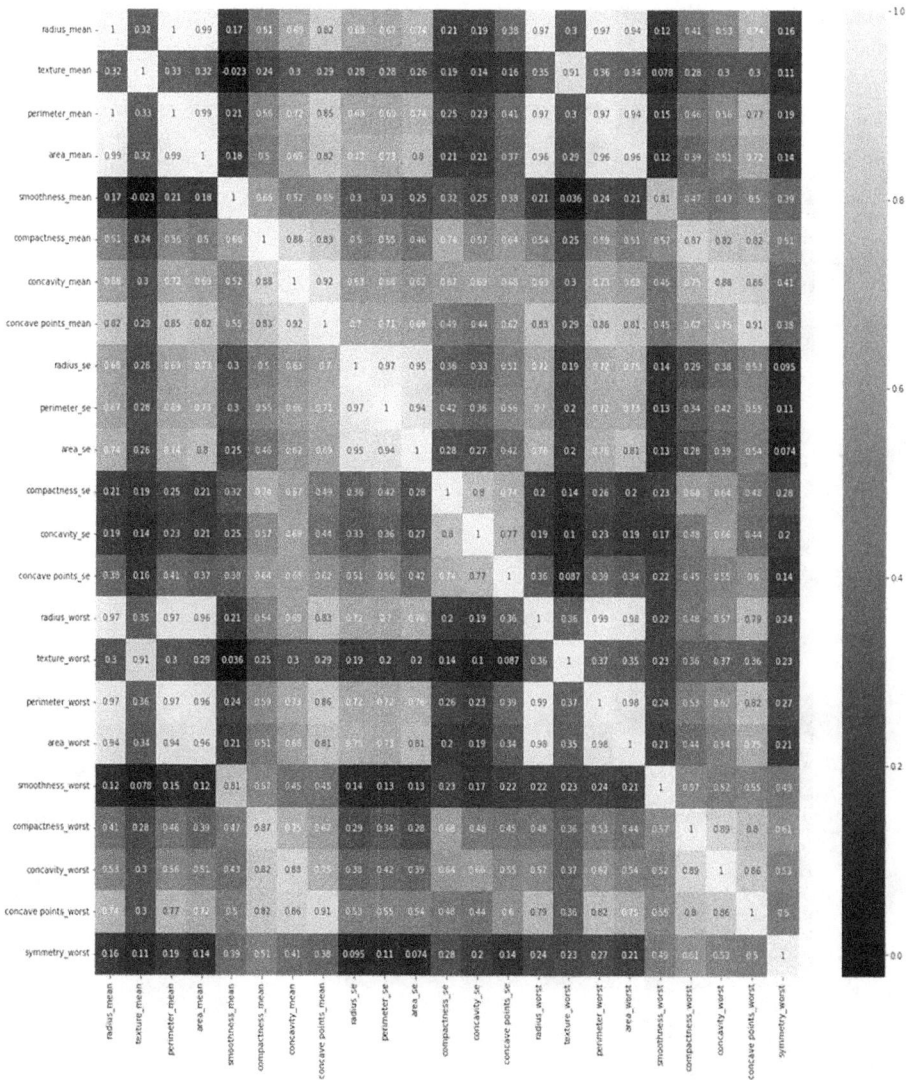

FIGURE 8.4 Heat Map of the Correlated Feature Before Dropping Features.

The reason behind choosing area worst is that it scores the highest in all criteria, so it is wise to choose this. Similarly, the area mean was chosen among perimeter worst, radius worst, perimeter mean, and radius, and area se was selected among concavity mean, concave mean, etc.

Finally, the features that were dropped or neglected from the datasets were 'radius mean', 'perimeter mean', 'radius worst', 'perimeter worst', 'radius mean', 'perimeter se', 'radius se', 'concave points mean', 'texture worst', and 'concavity mean'.

Figure 8.5 represents the heat map after dropping the highly correlated (more than 0.90) features. Only area mean and area worst are present with 0.96 scores, as both carry high importance according to the feature selection mentioned in subsections 1, 2, and 3 in the matrix, as shown in Figure 8.5.

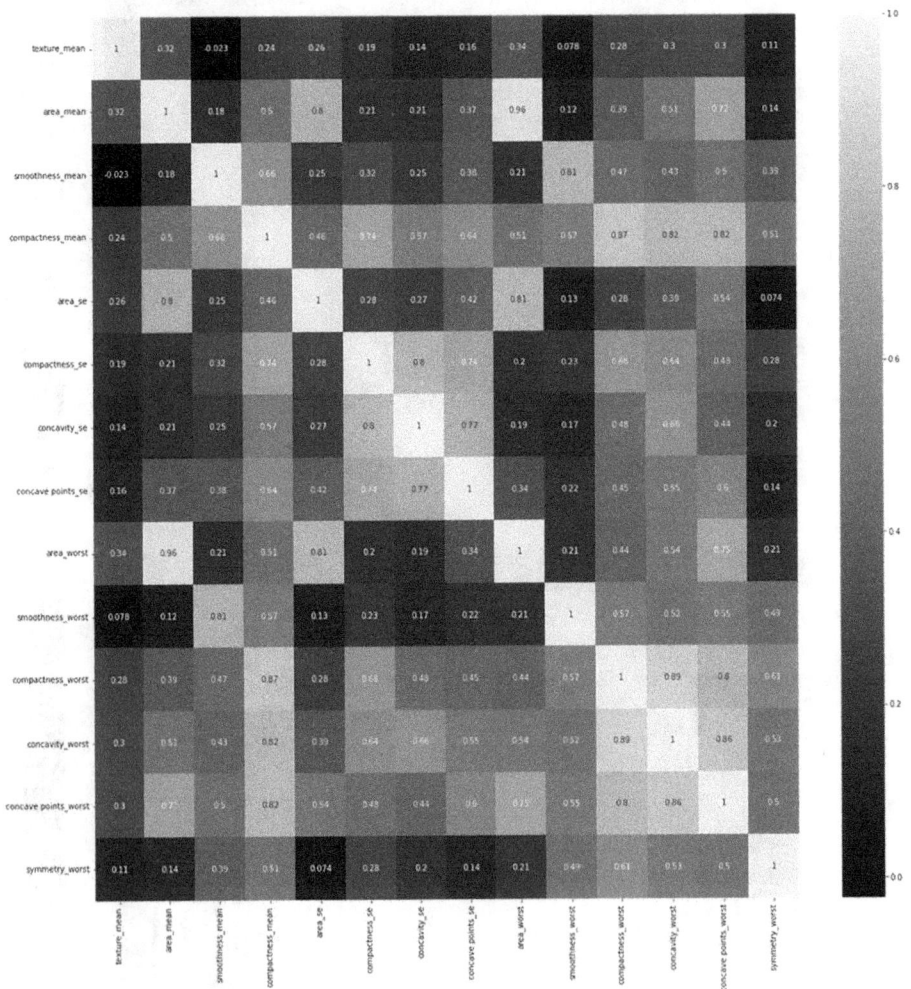

FIGURE 8.5   Heat Map of the Correlated Feature After Dropping Features.

After feature data processing and selection, the shape of the dataset reduced from (569, 32) to (569, 16). The final selected features are 'texture mean', 'area mean', 'smoothness mean', 'compactness mean', 'area se', 'compactness se', 'concavity se', 'concave points se', 'area worst', 'smoothness worst', 'compactness worst', 'concavity worst', 'concave points worst', 'symmetry worst', and 'fractal dimension worst'.

8.3.2.5   Training and Testing Datasets

Using training and testing datasets is an excellent idea. The testing dataset was used exactly toward the end of the project to ensure that the final model was accurate. We used 70% of the dataset for training and 30% of the dataset for testing and validation of the model.

The training and testing datasets were also subjected to standardization, as the obtained data included properties of multiple dimensions and scales (Rani et al., 2023). The evaluation metrics, which were used to measure the performance of the models, are listed as follows:

- Confusion matrix
- Classification accuracy
- Precision and recall score
- K-fold cross validation

8.3.2.6   Confusion Matrix

A confusion matrix is a breakdown of the prediction outputs of a classification task. The total number of correct and incorrect predictions was added and then broken down by class using count values.

From Table 8.2, out of the 171 data points, logistic regression correctly predicted 167 data points, whereas other algorithms, such as the decision tree, had correctly

TABLE 8.2
Confusion Matrix of the Resultant Classification Models.

Classifier algorithm	Confusion matrix		Number of test data
Logistic regression	TN = 103	FP = 2	171
	FN = 2	TP = 64	
Decision tree	TN = 100	FP = 5	171
	FN = 6	TP =60	
Support vector machine	TN = 103	FP = 2	171
	FN = 3	TP = 63	
K-nearest neighbor	TN = 101	FP = 4	171
	FN = 7	TP = 59	
Majority voting	TN = 105	FP = 0	171
	FN = 3	TP = 63	
Soft voting	TN = 105	FP = 0	171
	FN = 3	TP = 63	
Weighted average	TN = 105	FP = 0	171
	FN = 3	TP = 63	

predicted 160 data points. *K*-nearest neighbors also correctly predicted 160 data points, and support vector machines had correctly predicted 166 data points.

Finally, the hard/majority voting ensemble model predicted 168 correct predictions, the soft voting ensemble model identified 167 correct cases, and the weighted average of the soft voting anticipated 169 correct cases out of 171 test cases, which was the highest, compared to others.

8.3.2.7 Classification Report

The classification report displays the precision, recall, F1, and support scores of the model. We can evaluate the effectiveness of the predictive model once it has been built. For this purpose, we evaluated the performance of base-level classifiers and voting-based ensembles based on accuracy, precision, recall, and F1 score values, as shown in Table 8.3. The summarized results are shown in Figure 8.6.

TABLE 8.3
Confusion Matrix of the Resultant Classification Models.

Classifier	Accuracy	Precision	Recall	F1 score
Logistic regression	0.976	0.97	0.97	0.97
Random forest	0.97	0.969	0.955	0.962
Decision tree	0.935	0.923	0.909	0.916
K-nearest neighbor	0.935	0.937	0.894	0.915
Hard/majority voting	0.982	1.00	0.955	0.977
Soft voting	0.976	1.00	0.939	0.969
Weighted average	0.988	1.00	0.97	0.985

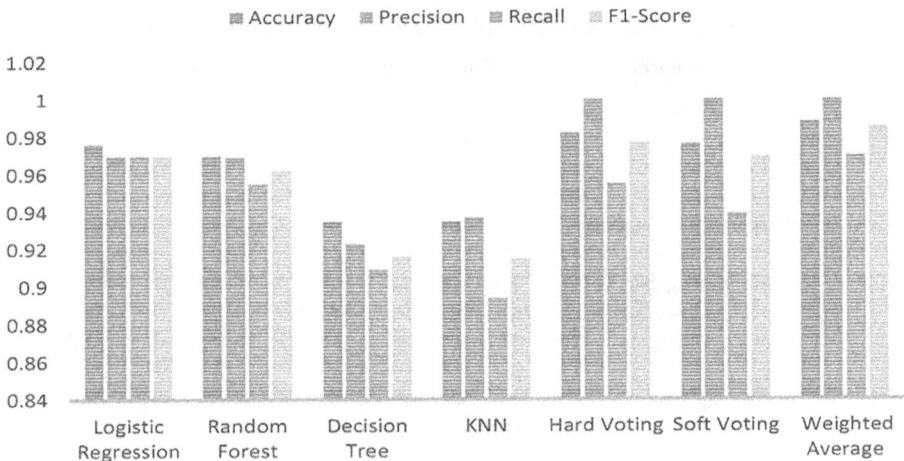

FIGURE 8.6 Models of Accuracy, Precision, Recall, and F1 Score.

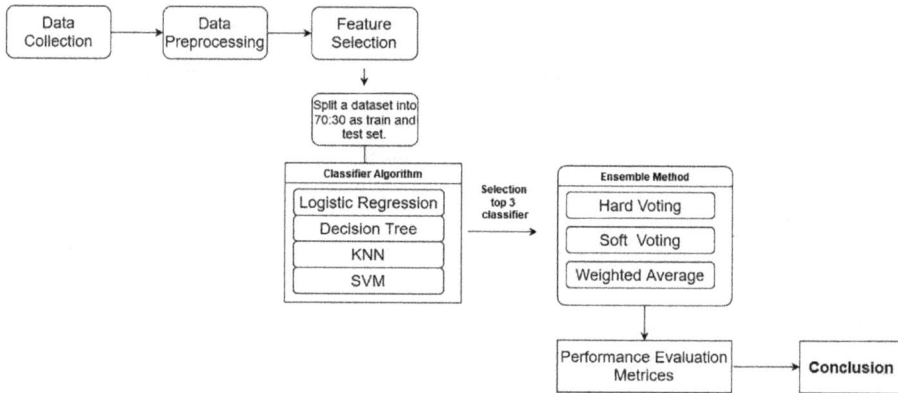

FIGURE 8.7 System Flow Diagram.

8.3.2.8 Classification Chart

The *K*-fold cross-validation approach is a statistical tool for estimating the performance of a prediction model in practice (Wolberg et al., 2021).

The mean accuracy scores of cross validation of hard/majority voting, soft voting, and weighted average soft voting using *K*-fold cross validation ($K = 5$) were 0.97364 ($\pm$0.01110), 0.97188 ($\pm$0.01403), and 0.97365 ($\pm$0.01466).

8.4 PROPOSED WORK

The first step was to identify the problem statement. Then, through a literature review, a research objective was established. The research objective was to develop a prediction system with a higher accuracy.

The dataset used was from the Cleveland Clinic Foundation Dataset available at the UC Irvine Machine Learning Repository. Then, data pre-processing was carried out, and changing the raw data to pre-processed data would have reliable accuracy (Khanh & Khang, 2021).

Feature selection was performed to remove redundant noise from the dataset, leaving only the important attributes, which helped in increasing the accuracy, reducing overfitting, and shortening the training time.

Subsequently, data were split such that 70% of the data were used as a training dataset, and 30% of the data were used as a testing dataset, as shown in Figure 8.7.

8.5 ANALYSIS AND COMPARISON

The performances of the four different classification algorithms were evaluated on the WBCD dataset, as shown in Table 8.4. Among these, logistic regression performed better than the other algorithms. The top three high-performance classification models were selected for the ensemble voting classifier: hard voting, soft voting, and weighted average models.

TABLE 8.4

Comparison with the Latest Research Paper for Breast Cancer Prediction Using WBCD.

Authors	Classifiers	Accuracy (%)
Our work	Weighted average soft voting mechanism	98.83
Kumari et al. (2021)	Soft voting classifier	97.27
Khuriwa and Mishra (2018)	Ensemble voting method	98.50
Naji et al. (2021)	Majority voting ensemble technique	98.10

Among these, the weighted average classifier excels in all metrics. These ensemble voting classifiers are subject to k-cross validation ($k = 5$), with a slight difference in the average accuracy value of cross validation and the model accuracy of all three ensemble classification models.

This illustrates that the suggested system is slightly over-fitted; however, when ample training data are collected, the model becomes more generalized. It also confirms that the proposed approach works with unseen real-world data.

Table 8.1 demonstrates how the proposed weighted average-based voting mechanism compares to previous work on breast cancer prediction using the WBCD. All papers have been published on a well-known platform; thus, they may be trusted.

8.6 CONCLUSION

This project is associated with an intelligent healthcare system. Here, various machine learning classifiers were used in the experiments to determine the best model for use in the ensemble voting mechanism. Thus, the system can predict whether the tumor present in a patient's breast is malignant or benign (Divvela et al., 2023).

The proposed system was built in the following two phases: pre-processing of the dataset and prediction of diseases using an ensemble voting classifier. In this project, 16 attributes were considered, which form the primary basis for the tests and gave more or less accurate results (Khang et al., 2024).

The primary objective of this project was to obtain the optimum accuracy and algorithm for predicting the types of tumors.

An ensemble learning model was developed using a voting classifier and probability. The hard voting/majority voting classifier-based ensemble model yielded an accuracy, precision, recall, and F1 score of 0.9820, 1.00, 0.955, and 0.977, respectively.

The soft voting classifier-based ensemble model yielded an accuracy, precision, recall, and F1 score of 0.976, 1.00, 0.939, and 0.969, respectively, and the average weighted classifier-based ensemble model yielded an accuracy, precision, recall, and F1 score of 0.988, 1.00, 0.970, and 0.985, respectively.

Observations show that the weighted average ensemble model achieves better performance when compared with other classification methods. It was concluded that the best algorithm based on the given dataset is the average weighted ensemble voting classifier (Khang et al., 2023b).

According to Table 8.3, our proposed model scored the highest accuracy compared to the results of the mentioned scientific papers.

Furthermore, it can be inferred that the ensemble voting method for machine learning algorithms can be built into a sophisticated system that decreases human and manual errors when used in medical domain-based projects (Rana et al., 2021).

REFERENCES

Bhambri P., Rani S., Gupta G., Khang A., (2022). *Cloud and Fog Computing Platforms for Internet of Things*. CRC Press. https://doi.org/ 10.1201/9781003213888

Khang A., Rana G., Tailor R. K., Hajimahmud V. A., (Eds.). (2023a). *Data-Centric AI Solutions and Emerging Technologies in the Healthcare Ecosystem*. CRC Press. https://doi.org/10.1201/9781003356189

Khang A., Gupta S. K., Rani S., Karras D. A., (Eds.). (2023b). *Smart Cities: IoT Technologies, Big Data Solutions, Cloud Platforms, and Cybersecurity Techniques*. CRC Press. https://doi.org/10.1201/9781003376064

Khang A., Olena H., Abdullayev V. A., Arvind Kumar S., (2024). *Computer Vision and AI-integrated IoT Technologies in Medical Ecosystem* (1st ed.). CRC Press. https://doi.org/10.1201/9781003429609

Khang A., Ragimova N. A., Hajimahmud V. A., Alyar V. A., (2022a). "Advanced technologies and data management in the smart healthcare system," In *AI-Centric Smart City Ecosystems: Technologies, Design and Implementation*, Vol. 16, No. 10 (1st ed.). CRC Press. https://doi.org/10.1201/9781003252542-16

Khang A., Ragimova N. A., Hajimahmud V. A., Alyar V. A., (2022b), "Advanced technologies and data management in the smart healthcare system," In *AI-Centric Smart City Ecosystems: Technologies, Design and Implementation*, Vol. 16, No. 10 (1st ed.). CRC Press. https://doi.org/10.1201/9781003252542-16

Khanh H. H., Khang A., (2021). "The role of artificial intelligence in blockchain applications," *Reinventing Manufacturing and Business Processes through Artificial Intelligence*, 2(20–40) (CRC Press). https://doi.org/10.1201/9781003145011-2

Khourdifi Y., Bahaj M., (2018). "Applying best machine learning algorithms for breast cancer prediction and classification," In *2018 International Conference on Electronics, Control, Optimization and Computer Science* (ICECOCS), pp. 1–5. IEEE. https://doi.org/10.1109/ICECOCS.2018.8610632

Khuriwa N., Mishra N., (2018). "Breast cancer diagnosis using adaptive voting ensemble machine learning algorithm," *2018 IEEMA Engineer Infinite Conference (eTechNxT)*, pp. 1–5. https://doi.org/10.1109/ETECHNXT.2018.8385355

Kumari S., Kumar D., Mittal M., (2021). "An ensemble approach for classification and prediction of diabetes mellitus using soft voting classifier," *International Journal of Cognitive Computing in Engineering*, 40–46. https://doi.org/10.1016/j.ijcce.2021.01.001.

Naji M. A., Filali S. E., Bouhlal M., Benlahmar E. H., Ait R. A., Debauche O., (2021). "Breast cancer prediction and diagnosis through a new approach based on majority voting ensemble classifier," *Procedia Computer Science*, 481–486. https://doi.org/10.1016/j.procs.2021.07.061

Praveen Kumar M., Kumar N., Anuradha M., Khang A., (Aug 24, 2023). "Heart disease prediction using logistic regression and random forest classifier," In *Data-Centric AI Solutions and Emerging Technologies in the Healthcare Ecosystem*, p. 6. (1st ed.). CRC Press. https://doi.org/10.1201/9781003356189-6

Rana G., Khang A., Sharma R., Goel A. K., Dubey A. K., (Eds.). (2021). *Reinventing Manufacturing and Business Processes through Artificial Intelligence*. CRC Press. https://doi.org/10.1201/9781003145011

Rani S., Bhambri P., Kataria A., Khang A., Sivaraman A. K., (2023). *Big Data, Cloud Computing and IoT: Tools and Applications* (1st ed.). Chapman and Hall/CRC. https://doi.org/10.1201/9781003298335

Rani S., Chauhan M., Kataria A., Khang A., (Eds.). (2021). "IoT equipped intelligent distributed framework for smart healthcare systems," In *Networking and Internet Architecture*, Vol. 2, p. 30. CRC Press. https://doi.org/10.48550/arXiv.2110.04997

Richard O., Duda P. E., (2000). *Pattern Classification.* Wiley. https://doi.org/10.1201/9781003145011

Singh R. P., Javaid M., Haleem A., Vaishya R., Ali S., (2020). "Internet of Medical Things (IoMT) for orthopaedic in COVID-19 pandemic: Roles, challenges, and applications," *Journal of Clinical Orthopaedics and Trauma*, 11(4), 713–717. www.sciencedirect.com/science/article/pii/S097656622030179X

Sung H., Ferlay J., Siegel R. L., Laversanne M., Soerjomataram I., Jemal A., Bray F., (2021). "Global cancer statistics 2020: GLOBOCAN estimates of incidence and mortality worldwide for 36 cancers in 185 countries," *ACS Journal*, 209–249.

Vijayvargia A., (2018). "Feature engineering," In (Machine Learning With Python) A. Vijayvargia (Ed.), *An Approach to Applied Machine Learning*, pp. 41–51. BPB Publications. https://doi.org/10.1201/9781003145011

Vishnu Sai Kumar D., Ritik C., Anuradha M., Praveen Kumar M., Khang A., (Aug 24, 2023). "Heart disease and liver disease prediction using machine learning," In *Data-Centric AI Solutions and Emerging Technologies in the Healthcare Ecosystem*, p. 4. (1st ed.). CRC Press. https://doi.org/10.1201/9781003356189-13

Wolberg W. H., Street W. N., Mangasarian O. L., (Jan 10, 2021). *Breast Cancer Wisconsin (Diagnostic) Data Set.* http://archive.ics.uci.edu/ml; https://archive.ics.uci.edu/ml/datasets/Breast+Cancer+Wisconsin+

WHO, (Dec 21, 2021). *WHO Factsheet.* www.who.int/news-room/fact-sheets/detail/cancer.

9 Privacy Protection for Internet of Medical Things Data Using Effective Outsourced Support Vector Machine Approach

Padmavathi Pragada, Murali Dhar M. S., Rajesh Kumar Rai, Bhargavi Devi P., and Nidhya M. S.

9.1 INTRODUCTION

Increased fraud risk, privacy invasions, uninvited marketing messages, and highly targeted and intrusive marketing messages that disrupt daily routines are just a few of the negative consequences of having widespread access to customer personal information.

Nevertheless, the advantages of information campaigns for consumers are routinely lauded. The smart use of consumer data makes it feasible to provide customized price reductions. Freebies, more relevant marketing messaging, and media materials are all examples of how to improve product offerings and recommendations.

Theoretically, marketers may provide customers with greater benefits because they can operate productively with more information. As a result of these trends, academics, social critics, and regulators have given consumer privacy more attention. However, the costs and advantages to marketers and consumers are substantial and require further study.

We study a large body of marketing literature on consumer privacy and data use to summarize what is known and what is still unknown in this sector (as well as information systems, law, ethics, and other fields) (Papaioannou et al., 2022).

9.1.1 Big Data

Big data is used to describe the enormous volume of data generated from multiple sources in a variety of formats at rapid rates. Big data is a term that refers to

DOI: 10.1201/9781003400110-9

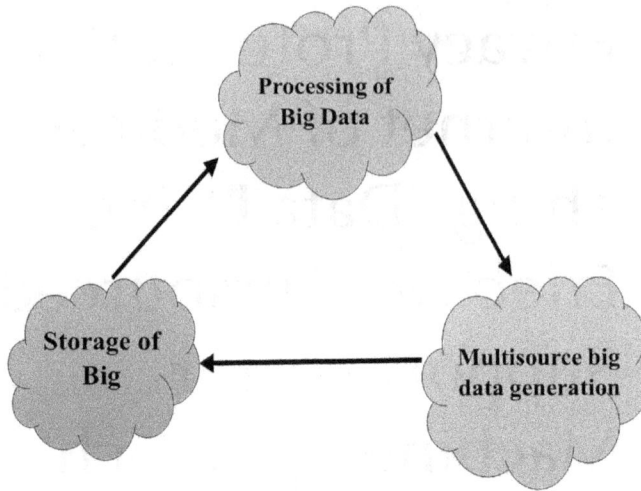

FIGURE 9.1 Big Data Life Cycle Stages.

Source: Khang (2021)

a combination of technologies and architectural methodologies that can enable high-velocity gathering, discovery, and analysis, and are characterized as being "intended to inexpensively separate value from very large volumes of a wide range of data".

Together with volume, velocity, and variety, these three Vs characteristics of big data (Awan et al., 2021).

Later studies revealed that the three Vs definition is insufficient to help elucidate the big data challenges that we currently face. One recurring feature of big data is the diversity of the data, which might include text, music, images, and videos among other things.

Various properties of the data are represented by variety. Recently, various strategies have been developed to protect the privacy of massive data collections. These techniques can be categorized using the phase life cycle of big data, as shown in Figure 9.1, which encompasses data generation, storage, and processing.

To preserve privacy, access restrictions and data falsification procedures are implemented throughout the data-generating phase (Arya et al., 2020).

9.1.2 Privacy-Preserving Data

Encryption techniques are frequently used in data storage privacy-preservation procedures. The data-processing phase includes both data knowledge extraction and privacy-preserving data publishing.

Although clustering and classification are used to partition association rule mining techniques, they classify the input data into categories and find important correlations and patterns in the data. Frameworks can manage different big data metrics in terms of variety, velocity, and volume are necessary. Large amounts of data are entered from various sources at extremely high speeds. There are various stages in the big data life cycle. When combined with external datasets, personal data can be used to infer new user information. Such information can be private and is not intended for public consumption.

Personal information may occasionally be obtained and used to increase commercial value. For instance, the shopping habits of a person can reveal a lot about them. Sensitive data are processed and stored in an unsecure setting, and data leakage can occur throughout this procedure.

9.1.3 INTERNET OF MEDICAL THINGS

An interconnected system of devices, known as the "Internet of Medical Things" (IoMT), offers medical services. As shown in Figure 9.2, IoMT is essentially a networked infrastructure of medical equipment, software programs, and services.

Healthcare organizations can increase the efficiency of their clinical procedures and workflow management, as well as remote patient health monitoring, by connecting equipment and sensors.

The Internet of Things (IoT) connects the digital and physical worlds to improve patient health by accelerating diagnosis and treatment, and modifying patient behavior and health status in real time.

Interconnected medical devices have a significant impact on patients and clinicians (Balica, 2022). The essence of the IoMT healthcare system is illustrated in Figure 9.2.

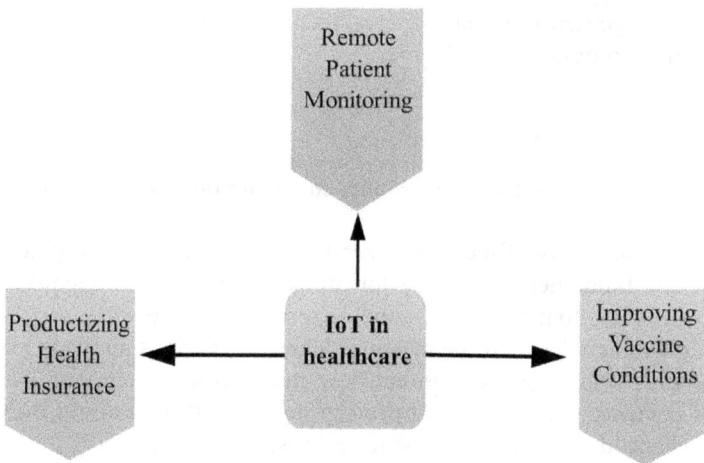

FIGURE 9.2 Essence of IoMT in Healthcare System.

9.1.4 MACHINE LEARNING IN INTERNET OF MEDICAL THINGS

Machine learning (ML) is an additional method for protecting IoMT systems.

In addition, when compared to traditional authentication techniques, a detailed assessment of existing IoMT system security procedures and attacks is required because of advancements in IoMT system security defense techniques, as well as new attack types. Hence, it analyses state-of-the-art security and attacks on IoMT system strategies (Blazek et al., 2022).

Our main contributions are summarized as follows:

- In this chapter, we will look at the security requirements for IoMT systems, as well as a number of methods for gathering, sending, and storing data in a secure manner.
- We discuss the security measures that are now available and how well they stand up to different types of attacks. According to our argument, no method can completely protect against the vast majority of known attacks on these systems.
- We examine the attack surface of IoMT and show how these security mechanisms can thwart various assaults. This chapter lists the current attacks that have been launched against IoMT systems.
- We offer a security architecture for IoMT systems that incorporates some properties of this technique. The security of IoMT systems is addressed by this framework during data collection, transmission, and storage. In our framework, we consider the constraints on these devices.

The remainder of this chapter is organized into five parts.

- Section 1.2 explains the present method and its drawback of data security techniques in IoMT.
- Section 1.3 explains the proposed techniques in data privacy in IoMT with better ML support vector machine (SVM) techniques.
- Section 1.4 presents the experimental results and their graphical analysis.
- Section 1.5 provides a conclusion.

9.2   LITERATURE REVIEW

This section discusses the existing systems and techniques used for data privacy in the IoMT healthcare system.

Various authors have discussed several strategies for securing IoMT. Data encryption is a fundamental strategy for protecting IoMT. The receiver is subsequently given access to the public channel to receive cipher text.

The recipient then decrypts the message. There are various methods for encrypting the data. A lightweight end-to-end key management technique was created because of resource constraints and privacy concerns (Singh et al., 2020), where keys are exchanged while utilizing the fewest resources possible.

The proposed protocol offers both high-security features and limited resources from a security perspective. Given the characteristics of IoT, Aman et al. (2021)

presented a remedy for privacy and security issues in current clever monitoring systems for healthcare.

A prototype system using an effective private homomorphism algorithm and an improved data encryption standard encryption technique was developed.

Jain et al. (2021) proposed a cloud computing technique that uses IoT sensors linked to the mechanism of time stamps, digital signatures, and asymmetric technology for monitoring other personal data. This plan uses fewer medical resources and provides more effective medical services.

Pustokhina et al. (2020) presented a Diffie–Hellman key agreement mechanism and exchange-based secure authentication for a cloud-based wireless body area networks system. When the system participants register, this technique can construct secure routes or means for them. Security and performance analyses showed that the aforementioned solutions are capable of resolving issues with the healthcare system.

9.2.1 Data Mining

The process of extracting knowledge (or useful information) from the data is addressed in knowledge discovery in databases (KDD). In this process, a step called data mining (DM) uses specific algorithms to draw patterns from data.

Nonetheless, a number of publications and business actors refer to the entire KDD process as DM, including the concept of the cross-industry standard procedure for data mining. According to Sugiyarti et al. (2018), the KDD step known as DM deals with the applications of specialized algorithms to extract patterns from data and is denoted by the term "data mining".

9.2.2 Machine Learning

There is a substantial difference between the ML and DM overlap. As they usually use the same methods and have a great deal in common, these two ideas are often confused. The ML pioneer states that a field of research that offers computers the ability to learn without being explicitly taught is ML.

According to Rajkomar et al. (2019), finding fresh characteristics in the data is the basic objective of DM. It places more emphasis on learning new and fascinating information than it does on requiring the domain to have a clear aim. Machine learning can be thought of as the older sibling of DM.

Data mining was initially used in the late 1980s. Some researchers opt to refer to their work as DM (Gao et al., 2021). The techniques used in the papers retrieved by the first and second queries did not differ noticeably.

The guidelines from Alpaydin (2020) imply that a thorough literature search must be valid and trustworthy. In this instance, validity is determined by the databases and publications used, the time period covered, the keywords used, and the usage of a forward and backward search.

The reproducibility of the literature search process is referred to as the reliability (Lalmuanawma et al., 2020). To meet this requirement, the entire search process was meticulously documented. They consisted of security awareness training, security awareness education, security awareness motivation, awareness program, awareness campaign, and security for staff.

To find publications that met at least one of the search criteria in the title, abstract, or keywords, the databases were searched. A full-text search was performed if the search field (such as title, abstract, or keywords) could not be specified in the search query. In total, 4,168 potentially pertinent publications were identified.

The inclusion and exclusion criteria were established to select pertinent publications in the subject research field. We decided to pay attention to not only the highly recommended publications (Roscher et al., 2020) but also to incorporate events or publications that do not receive high marks in global rankings for conferences or journals. This is necessary because a number of conferences or journals (such as computers, security, and information management and computer security) that focus on information security (IS) have many publications that address topics important to this literature study.

White papers and other non-academic publications were not included. Additionally, only articles published after the year 2000 and only publications written in English were considered.

Articles that did not primarily discuss the individual savings account and employee behavior were also excluded. This was done by personally examining each article's title, abstract, and entire text where necessary, and 95 articles were found pertinent to this method.

A forward and backward search was conducted (Lei et al., 2020), and 18 more pertinent articles were identified. In total, 144 publications were found to be pertinent to this study. The number of publications for each journal or conference deemed pertinent is shown in Table 9.1.

Two researchers performed a two-step analysis to reduce errors and subjective biases. First, the applied theory and research methods for each publication were chosen separately by each researcher.

The results were compared with those of other researchers after being categorised in terms of theory and methodology. A consensus on conformance was obtained after discussing divergences. The list of theories was developed inductively while reviewing articles.

We found 54 theories that were used in the under-consideration research field, adhering to the broad meaning of the term "theory" used in contemporary IS literature (Liu et al., 2022). Two or fewer publications used the majority of the identified theories. Seven main theories were identified after considering the frequency of use.

9.3   SYSTEM DESIGN

This session discusses the proposed system methodology and the design of effective outsourced SVM techniques in cybersecurity with data privacy.

- **Patient data to IoMT**: For a range of medical diseases, IoMT systems provide necessary or superior support. There is a need for implantable medical devices (IMDs) for particular medical disorders, such as pacemakers for cardiac conditions. However, assisting devices, such as smartwatches, are mostly wearable for a better healthcare experience.

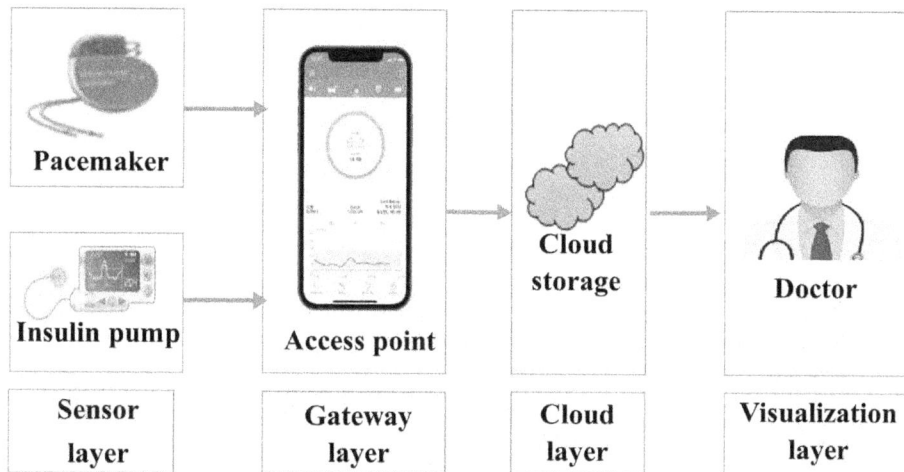

FIGURE 9.3 IoMT System Architecture.

- **Implantable medical devices**: Any implanted device that alters, supports, or improves a biological structure is referred to as an IMD. For instance, a pacemaker is an IMD that can encourage the heart to beat normally if it is beating too quickly or too slowly, helping manage irregular cardiac rhythms. Wireless IMDs have recently been proposed as a remedy to the problems associated with wired IMDs, such as infection and cable breakage.
- **Internet of Wearable Devices**: People wear these devices to track their biometrics, such heart rate, which could enhance their general health. ECG monitors, blood pressure monitors, smartwatches, and fall detection bands are some examples. One of the most popular Internet of Wearable Devices for monitoring biometrics, such as heart rate and movement, are smartwatches. Slow and rapid heartbeats can be detected by monitoring, even when the subject is not moving. Other features of new watches include fall detection and ECG readings for the diagnosis of medical disorders, such as atrial fibrillation (irregular heartbeat). Currently, they frequently serve as noncritical patient monitors (Vrushank et al., 2023).

However, these gadgets are unlikely to play the role of IMDs in life or death circumstances because of sensor accuracy and battery life restrictions. Several well-known IMDs and their locations are shown in Figure 9.3.

Sensor Layer

- A sensing layer represents the type of data that originates from a specific data source, such as web services, traditional wireless sensor networks, and PlayStation Networks.

- Sensors and other data-gathering devices form the foundation of any IoT systems. They serve as the link between the physical and digital worlds, converting analogue signals into digital signals (Khang et al., 2022b).

Data Collection

- Having a backup strategy is helpful when something goes wrong (sensor-hardware manipulation).
- Sensor data are the output of a device that detects and responds to input from the physical environment. The output can be used to inform the user, as input to another system, or as process guidance. Sensors can detect almost any physical element.
- Attacks such as these have the potential to harm patient lives and maintain the functionality of the system.

Gateway Layer

- The processing and storage capacities of IoMT sensors prevent data from being processed before they are sent to the gateway layer, which is the second layer. The devices in this layer could be dedicated access points (APs) or patients' smartphones.
- Access points are frequently more potent than sensors. They are capable of performing pre-processing duties, such as temporary data storage, validation, and basic artificial intelligence-based analysis. They can also transmit sensor data to the cloud via the Internet (Rani et al., 2023).

Data in Transit

- This phase entails communication between the elements in all four layers, such as APs in the gateway layers and IoMT sensors in the sensor layer.
- Attacks in this area can alter or stop the transmission of sensor data. Hence, defending against such assaults would prevent data corruption during the transfer between the four layers.

Cloud Layer

- Internet of Things devices can communicate with each other in real time. Consequently, connected and smart devices can communicate with a variety of trustworthy APIs.
- Cloud computing enables networked technologies such that data processing may be used. The system node IDs and keys are generated by the key generation server. This layer enables remote control and management of sensor access.

Data in Storage

- The sensor and gateway levels collect and transmit patient data, which are then stored in a cloud layer. Attacks on this layer can take the form

of a denial-of-service (DoS), distributed DoS, or account credential theft (Bhambri et al., 2022).
- Data access to both the visualization layer and this layer must be tightly regulated. This is important because the majority of the data in this layer sleep the bulk of the time, in contrast to earlier levels.

Visualization/Action Layer

- To monitor patient health, this layer shows the data to doctors. Doctors' recommendations are also included in this layer, taking into account the patient's medical history (Khang et al., 2022c).
- Examples of actions include writing prescriptions and modifying dosages of various medications.

Support Vector Machine

- With the help of the SVM algorithm, in the future, new data points can be easily categorized by identifying the best line or decision boundary for classifying an *n*-dimensional space.
- The most effective decision boundary is the hyperplane. The SVM decision boundary is shown in Figure 9.4.

Linear SVM

The SVM algorithm can be better understood using an example. Consider a dataset with two features (green and blue) and two tags (green and blue) (x1 and x2). We need a classifier that can distinguish between coordinate pairs in green and blue (x1, x2). Take a look at Figure 9.5.

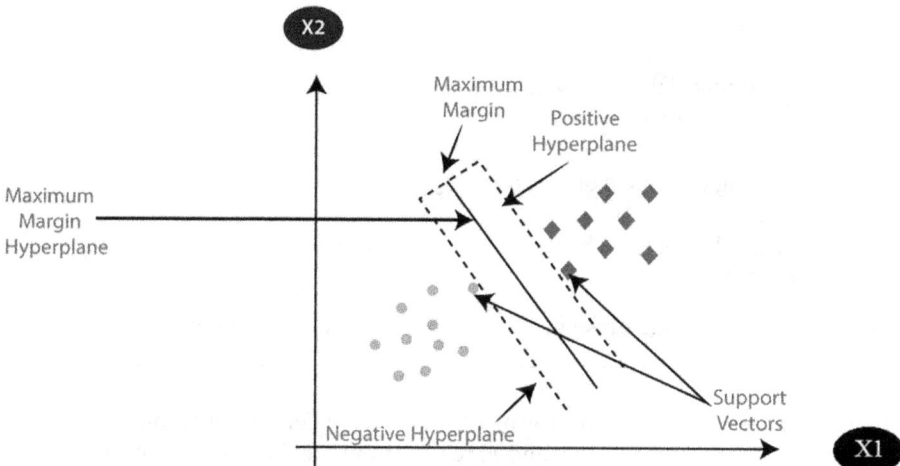

FIGURE 9.4   SVM Decision Boundary.

Green and Blue Coordinate Pairs

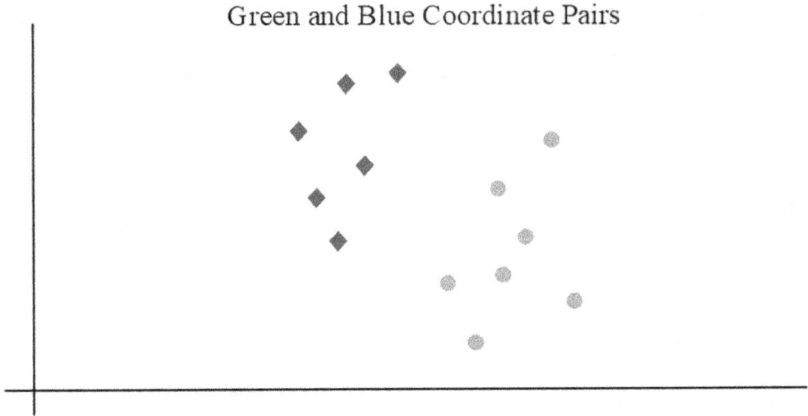

FIGURE 9.5   Green and Blue Coordinate Pairs.

We can simply distinguish between these two classes using a straight line, because the space is only two-dimensional. However, there may be more than one line to divide these classes. Consider Figure 9.6.

A linear or nonlinear classifier, often known as an SVM, is a mathematical function that can distinguish between two different object classes. These items fall under the category of classes and should not be confused with Java classes. The following pseudocode in code block 9.1 shows how to train an SVM.

Train an SVM with the labelled points. $o_i \leftarrow w \cdot x_i + b$.
Assign $y_i \leftarrow 1$ to the largest o_i; -1 to the others.
$C^\sim \leftarrow 10{-}5\ C^*$

    While $C^\sim < C^*$ **do**
    repeat
    Minimizing (1) with $\{y_i\}$ fixed and C^* is replaced by $C^\sim$.
    if $\exists \leftarrow$(i, j) satisfying (6)
    then
    Swapping the labels of y_i and y_j
    end if
    until No labels have been swapped $C^\sim \leftarrow$ min (9.5C, C^*)
    end while

Code block 9.1: Code Presents Algorithm for Training an SVM.

where 'E' denotes the probability of expectation under the probabilities p_u. The lesser of the two losses, $V(1, o_i)$ and V, results from the fact that at optimality with regard to p_u, p_i must concentrate all of its mass on $y_i = $ sign $(wTx_i + b)$ $(-1, o_i)$.

Two-dimensional Space

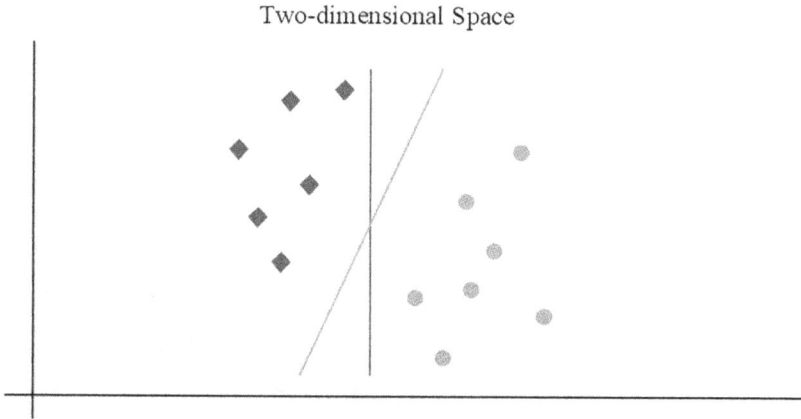

FIGURE 9.6   Illustration of Two-Dimensional Space.

As a result, this relaxing phase does not have any unfavourable effects; it is just a continuous variable-based reformulation of the original target. With *DA—H*, an additional entropy term is added to the target (p_u).

9.4   RESULTS AND DISCUSSION

The performance of the current system was examined in this section and included accuracy ratings of 85.96%, sensitivity ratings of 88.23%, specificity ratings of 86.33%, and precision ratings of 87.1%.

The proposed system is more effective than the current system because of its novel method of disease prediction. Furthermore, the precision of this prediction procedure was improved (Divvela et al., 2023).

- **Confusion matrix**: A confusion matrix was used to evaluate the effectiveness of the post-categorization procedures, as shown in Figure 9.7. It provides examples of correctly categorized TN values that belong to a different class as well as flawlessly classified TP values, FP values that belong to one class but not another, FN values that belong to one class but not another, and FP values that belong to one class but not the other (TN, TP, FP, and FN as mentioned in Figure 9.7).

Accuracy, precision, sensitivity, and real label specificity scores are the most frequently used performance measures for categorization based on these criteria. The confusion matrix is presented in Figure 9.7.

- **Accuracy**: This displays the total accuracy of the model or the percentage of all samples correctly identified by the classifier. Accuracy is determined using Equation 9.1:

$$\text{Accuracy} = (TP + TN)/(TP + TN + FP + FN) \qquad (9.1)$$

Real Label

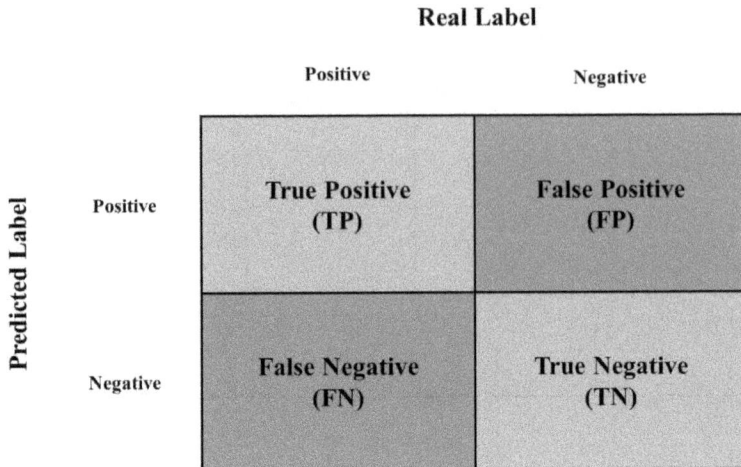

FIGURE 9.7   Confusion Matrix Prediction.

Source: Khang (2021)

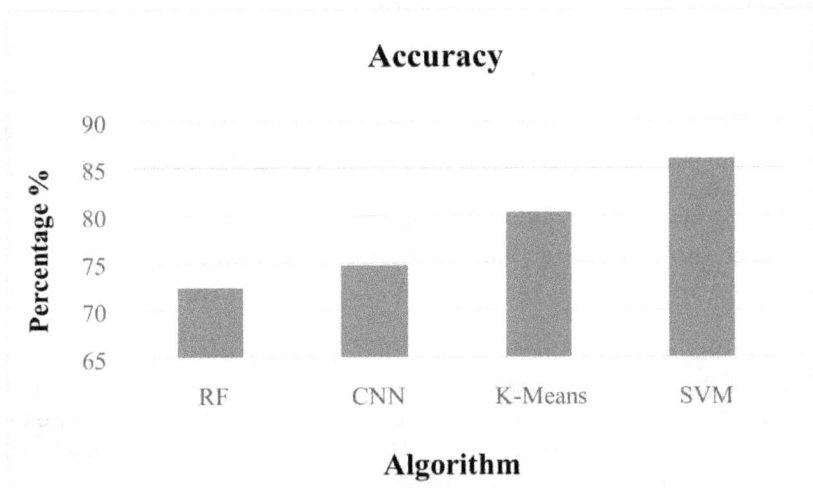

FIGURE 9.8   Accuracy Graph.

TABLE 9.1
Accuracy Result Values for the Proposed and Existing Systems.

Algorithm	Result
Random Forest (RF)	72.36
Convolutional Neural Network (CNN)	74.65
K-means	80.32
SVM	85.96

The accuracy gap is shown in Figure 9.8, and the accuracy result values for the proposed and existing systems are listed in Table 9.1.

- **Precision**: By dividing the total positive predictions (TP + FP) by the total positive predictions that were accurate, one can calculate the precision (TP) using Equation 9.2:

$$\text{Precision} = \text{TP}/(\text{TP} + \text{FP}) \qquad (9.2)$$

The precision graph is shown in Figure 9.9, and the precision values are listed in Table 9.2.

Specificity: The calculation of specificity (N) involves dividing the total number of negatives by the number of accurate negative predictions (TN), as shown in Equation 9.3:

$$\text{Specificity} = \text{TN}/(\text{TN} + \text{FP}) \qquad (9.3)$$

The specificity graph is shown in Figure 9.10, and the specificity values of various algorithms are listed in Table 9.3.

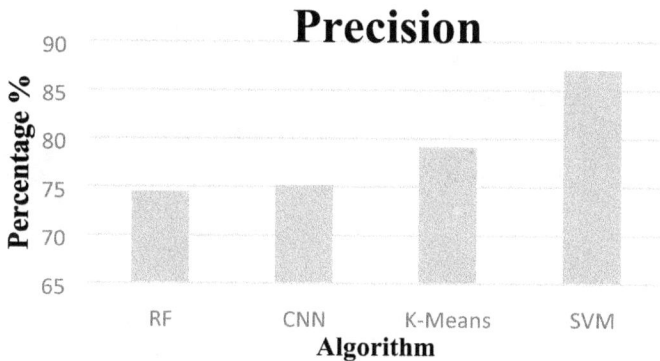

FIGURE 9.9   Precision Graph.

TABLE 9.2
Precision Result Values for the Proposed and Existing Systems.

Algorithm	Result
RF	74.45
CNN	75.12
K-means	79.22
SVM	87.10

Sensitivity: The TP rate is the proportion of positive values among all truly positive events, as shown in Equation 9.4:

$$\text{Sensitivity} = TP/(TP + FN) \qquad (9.4)$$

The sensitivity graph is shown in Figure 9.11, and the sensitivity values of various algorithms are listed in Table 9.4.

Specificity

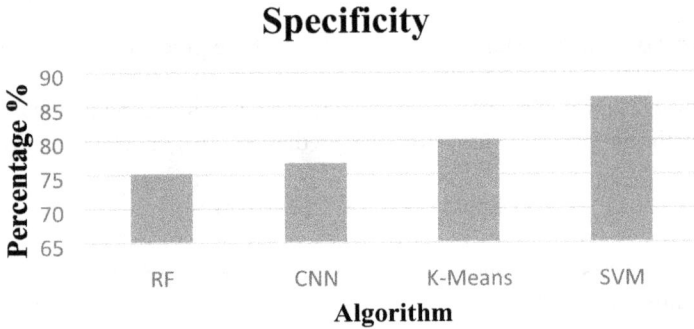

FIGURE 9.10   Specificity Graph.

TABLE 9.3
Specificity Result Values for the Proposed and Existing Systems.

Algorithm	Result
RF	75.15
CNN	76.66
K-means	80.12
SVM	86.33

Sensitivity

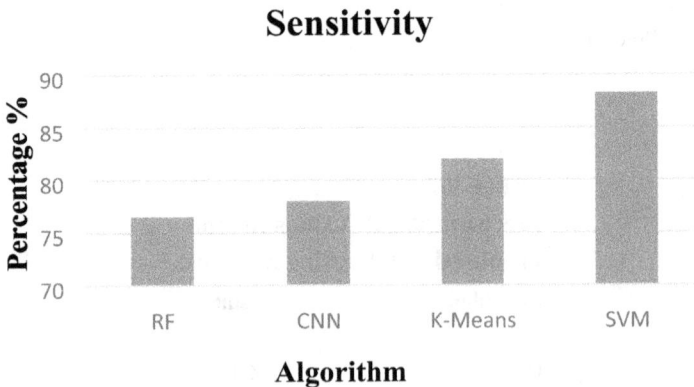

FIGURE 9.11   Sensitivity Graph.

TABLE 9.4
Sensitivity Result Values for the
Proposed and Existing Systems.

Algorithm	Result
RF	76.55
CNN	78.01
K-means	81.97
SVM	88.23

9.5 CONCLUSION

We proposed and tested an approach for intrusion detection using SVMs on a set of selected data. It provides high data accuracy and demonstrates a compatible level of performance.

Everything in the IoT sends data packets, and as the number of linked devices increases exponentially, this communication requires reliable connectivity, storage, and security (Khang et al., 2022a).

An organization must manage, watch over, and secure enormous amounts of data and connections from scattered devices when using IoT (Rani et al., 2021).

A healthy, democratic society requires privacy, which is more than just a concern for individualistic personal data. As technology advances, so will personal and healthcare (Khang et al., 2023a).

We investigated the newly proposed privacy-preserving SVM classification system and discovered that its primary protocol is a secure technique for determining whether the sign of encrypted numbers has a soundness issue or a security leak. Then, we propose a new technique that can be applied to assist privacy-preserving SVM classification and is secure and accurate for finding the signs of encrypted integers (Vrushank et al., 2023).

The experiments show that the MR-SVM model can effectively solve the problem of rapid increases in the SVM cost associated with increase in data volume. However, two nodes were used in the experiment, and the data transfer speed was limited. Multiple nodes can be established for cybersecurity in future studies.

REFERENCES

Alpaydin E., (2020). *Introduction to Machine Learning*. MIT Press. https://mitpress.mit.edu/9780262043793/introduction-to-machine-learning/

Aman A. H. M., Hassan W. H., Sameen S., Attarbashi Z. S., Alizadeh M., Latiff L. A., (2021). "IoMT amid COVID-19 pandemic: Application, architecture, technology, and security," *Journal of Network and Computer Applications*, 174, 102886. www.sciencedirect.com/science/article/pii/S1084804520303490

Arya D., Maeda H., Ghosh S. K., Toshniwal D., Omata H., Kashiyama T., Sekimoto Y., (Dec 2020). "Global road damage detection: State-of-the-art solutions," In *2020 IEEE International Conference on Big Data (Big Data)*. IEEE, pp. 5533–5539. https://ieeexplore.ieee.org/abstract/document/9377790/

Awan U., Shamim S., Khan Z., Zia N. U., Shariq S. M., Khan M. N., (2021). "Big data analytics capability and decision-making: The role of data-driven insight on circular economy performance," *Technological Forecasting and Social Change*, 168, 120766. www.sciencedirect.com/science/article/pii/S0040162521001980

Balica R. S., (2022). "Networked wearable devices, machine learning-based real-time data sensing and processing, and Internet of Medical Things in COVID-19 diagnosis, prognosis, and treatment," *American Journal of Medical Research*, 9(1), 33–48. www.ceeol.com/search/article-detail?id=1038875

Bhambri P., Rani S., Gupta G., Khang A., (2022). *Cloud and Fog Computing Platforms for Internet of Things*. CRC Press. https://doi.org/ 10.1201/9781003213888

Blazek R., Hrosova L., Collier J., (2022). "Internet of Medical Things-based clinical decision support systems, smart healthcare wearable devices, and machine learning algorithms in COVID-19 prevention, screening, detection, diagnosis, and treatment," *American Journal of Medical Research*, 9(1), 65–80. www.ceeol.com/search/article-detail?id=1038880

Gao Y., Li J., Zhou Y., Xiao F., Liu H., (2021). "Optimization methods for large-scale machine learning," *SIAM Review*, 60(2), 223–311. https://ieeexplore.ieee.org/abstract/document/9674150/

Jain S., Nehra M., Kumar R., Dilbaghi N., Hu T., Kumar S., Li C. Z., (2021). "Internet of Medical Things (IoMT)-integrated biosensors for point-of-care testing of infectious diseases," *Biosensors and Bioelectronics*, 179, 113074. www.sciencedirect.com/science/article/pii/S0956566321001111

Khang A., (2021). "Material4Studies," *Material of Computer Science, Artificial Intelligence, Data Science, IoT, Blockchain, Cloud, Metaverse, Cybersecurity for Studies*. www.researchgate.net/publication/370156102_Material4Studies

Khang A., Hahanov V., Abbas G. L., Hajimahmud V. A., (2022a). "Cyber-physical-social system and incident management," In *AI-Centric Smart City Ecosystems: Technologies, Design and Implementation*, Vol. 2, No. 15 (1st ed.). CRC Press. https://doi.org/10.1201/9781003252542-2

Khang A., Ragimova N. A., Hajimahmud V. A., Alyar V. A., (2022b). "Advanced technologies and data management in the smart healthcare system," In *AI-Centric Smart City Ecosystems: Technologies, Design and Implementation*, Vol. 16, No. 10 (1st ed.). CRC Press. https://doi.org/10.1201/9781003252542-16

Khang A., Rani S., Sivaraman A. K., (2022c). *AI-Centric Smart City Ecosystems: Technologies, Design and Implementation* (1st Ed.). CRC Press. https://doi.org/refb10.1201/9781003252542

Khang A., Rana G., Tailor R. K., Hajimahmud V. A., (Eds.). (2023a). *Data-Centric AI Solutions and Emerging Technologies in the Healthcare Ecosystem*. CRC Press. https://doi.org/10.1201/9781003356189

Khang A., Olena H., Abdullayev V. A., Arvind Kumar S., (2023b). *Computer Vision and AI-integrated IoT Technologies in Medical Ecosystem* (1st ed.). CRC Press. https://doi.org/10.1201/9781003429609

Lalmuanawma S., Hussain J., Chhakchhuak L., (2020). "Applications of machine learning and artificial intelligence for Covid-19 (SARS-CoV-2) pandemic: A review," *Chaos, Solitons & Fractals*, 139, 110059. www.sciencedirect.com/science/article/pii/S0960077920304562

Lei Y., Yang B., Jiang X., Jia F., Li N., Nandi A. K., (2020). "Applications of machine learning to machine fault diagnosis: A review and roadmap," *Mechanical Systems and Signal Processing*, 138, 106587. www.sciencedirect.com/science/article/pii/S0888327019308088

Liu X., Zhao J., Li J., Cao B., Lv Z., (2022). "Federated neural architecture search for medical data security," *IEEE Transactions on Industrial Informatics*, 18(8), 56285636. https://ieeexplore.ieee.org/abstract/document/9684972/

Papaioannou M., Karageorgou M., Mantas G., Sucasas V., Essop I., Rodriguez J., Lymberopoulos D., (2022). "A survey on security threats and countermeasures in internet of medical things (IoMT)," *Transactions on Emerging Telecommunications Technologies*, 33(6), e4049. https://onlinelibrary.wiley.com/doi/abs/10.1002/ett.4049

Pustokhina I. V., Pustokhin D. A., Gupta D., Khanna A., Shankar K., Nguyen G. N., (2020). "An effective training scheme for deep neural network in edge computing enabled internet of medical things (IoMT) systems," *IEEE Access*, 8, 107112–107123. https://ieeexplore.ieee.org/abstract/document/9109259/

Rajkomar A., Dean J., Kohane I., (2019). "Machine learning in medicine," *New England Journal of Medicine*, 380(14), 1347–1358. www.nejm.org/doi/full/10.1056/NEJMra1814259

Rani S., Bhambri P., Kataria A., Khang A., Sivaraman A. K., (2023). *Big Data, Cloud Computing and IoT: Tools and Applications* (1st ed.). Chapman and Hall/CRC. https://doi.org/10.1201/9781003298335

Rani S., Chauhan M., Kataria A, Khang A., (Eds.). (2021). "IoT equipped intelligent distributed framework for smart healthcare systems," In *Networking and Internet Architecture*, Vol. 2, p. 30. CRC Press. https://doi.org/10.48550/arXiv.2110.04997

Roscher R., Bohn B., Duarte M. F., Garcke J., (2020). "Explainable machine learning for scientific insights and discoveries," *IEEE Access*, 8, 42200–42216. https://ieeexplore.ieee.org/abstract/document/9007737/

Singh R. P., Javaid M., Haleem A., Vaishya R., Ali S., (2020). "Internet of Medical Things (IoMT) for orthopaedic in COVID-19 pandemic: Roles, challenges, and applications," *Journal of Clinical Orthopaedics and Trauma*, 11(4), 713–717. www.sciencedirect.com/science/article/pii/S097656622030179X

Sugiyarti E., Jasmi K. A., Basiron B., Huda M., Shankar K., Maseleno A., (2018). "Decision support system of scholarship grantee selection using data mining," *International Journal of Pure and Applied Mathematics*, 119(15), 2239–2249. www.academia.edu/download/57052927/DECISION_SUPPORT_SYSTEM_OF_SCHOLARSHIP_GRANTEE.pdf

Vishnu Sai Kumar D., Ritik C., Anuradha M., Praveen Kumar M., Khang A., (Aug 24, 2023). "Heart disease and liver disease prediction using machine learning," In *Data-Centric AI Solutions and Emerging Technologies in the Healthcare Ecosystem*, p. 4. (1st ed.). CRC Press. https://doi.org/10.1201/9781003356189-13

Vrushank S., Vidhi T., Khang A., (2023). "Electronic health records security and privacy enhancement using blockchain technology," In *Data-Centric AI Solutions and Emerging Technologies in the Healthcare Ecosystem*, p. 1 (1st ed.). CRC Press. https://doi.org/10.1201/9781003356189-1

10 Robotics in Real-Time Applications Using Bayesian Hyper-Tuned Artificial Neural Network

Arvind Kumar Shukla, Poongodi S., Alex Khang, and Shashi Kant Gupta

10.1 INTRODUCTION

A robot must adapt to a limited set of measurements while functioning in real-world robotics contexts where it may encounter unexpected visual features. A broad shift toward system robotics capable of responding to complicated dynamic settings and tackling a variety of uncertainties has recently taken place (Haleem et al., 2022).

The employment of increasingly smart robots in a broader variety of tasks has been made possible by recent developments in robotics and automation systems that have increased their competence (Hajimahmud et al., 2022).

The ability to grasp multiple distinct morphological features is crucial for robotics to achieve more versatile functions. The spatial capabilities of public robotics can be implemented to identify mechanical end-sensor layouts and to directly and securely hold and carry heavy objects without sliding (Mayoral-Vilches, 2022).

The robotics sector has emerged over the past 50 years. However, the creation of robots frequently lacks cybersecurity risks, which is typically a sign that a system remains in its infancy. Automated cybersecurity is often confused with automated protection.

The majority of these devices, from commercial to customers to expert robots, are vulnerable to security assaults. Security is a fundamental priority for both designers and the current regulations. Privacy is not considered significant (Tuomas, 2022).

A single safety and protection risk framework was created as a consequence of the convergence of these sectors from a value perspective. Cybersecurity aims to protect networks from threats offered by bad attackers, while robotic safety is often defined as building preventive procedures for incidents or defects, as shown in Figure 10.1.

Another viewpoint is that privacy concerns defending the robots from a different area, while protection concerns defending the surroundings from a specific robot (Bhardwaj et al., 2022).

 DOI: 10.1201/9781003400110-10

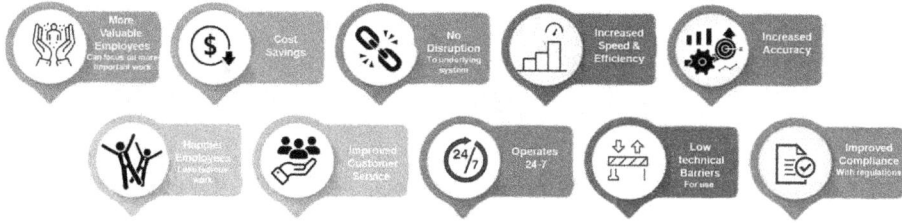

FIGURE 10.1   Benefits of Using Robotic Automation.

Cybersecurity is not just a product. The scale and complexity make this important. The tremendous functionality of modern robotic applications often results in broad security attacks and several possible attack routes, rendering the use of standard approaches challenging.

Consequently, this creates the following research questions: What is the state of robotics cybersecurity? How can robot security be effectively enhanced?

By applying basic functionalities, distributing computational capabilities to enable trainees to distribute the processing of complicated processes over a network, and establishing programmer groups around them to exchange scripts and equipment, robotics and automated testing conceal the difficulty of controlling varied devices.

Initially, the robotics data are gathered and pre-processed using min-max normalization. Feature extraction is performed using kernel linear discriminant analysis (KLDA). Feature detection is then achieved by our novel Bayesian hyper-tuned artificial neural network (BH-ANN).

Finally, the performance of the system is evaluated, and the parameters are compared with those of existing approaches.

The remainder of this chapter is organized as follows. Section 10.2 presents the problem statements. The proposed method is presented in section 10.3. Section 10.4 presents the performance analysis, and section 10.5 offers the conclusion.

10.2   PROBLEM STATEMENT

Interactions involving robots/machines and people are unsafe and vulnerable to different assaults because of a lack of private connections. Cyberattacks that utilize common passwords and login credentials result from improper validation, which is readily breached by a specific hacker.

Creating automaton clusters, enhancing mobility and discovery, and creating artificial intelligence (AI) that can acquire understanding and deploy ordinary sense to render ethical and societal judgments are several tasks that concentrate on basic issues in robotic systems.

The working mechanism, movement, and human-centered automation are the three main obstacles, with communication being the most difficult component needed for such automated systems.

Detection, location, cognitive, and control system are the four primary competency components in navigation. Automated robotic functionality and effectiveness

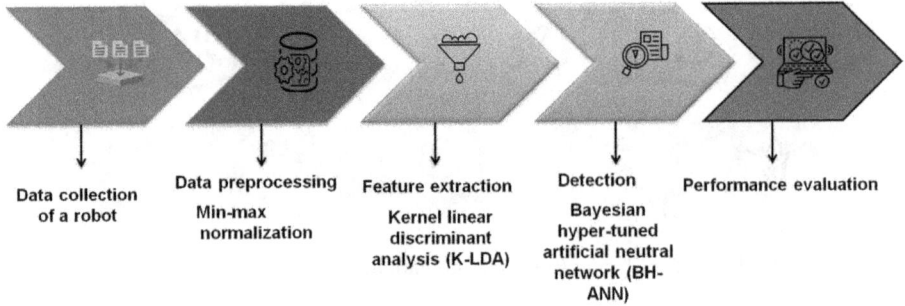

FIGURE 10.2   Working Principle of the Proposed Work.

are affected by network and management weaknesses, which can interrupt manufacturing and commercial operations and lead to economic shortfalls.

More specifically, these may cause network blocking, communication eavesdropping, report generation, and bodily harm (Jin et al., 2022).

10.3   PROPOSED WORK

Robotic automation has swiftly emerged as one of the most important strategies for helping firms improve their efficiency for main contributions. People can no longer need to perform hazardous tasks, even though robotic systems can operate under these conditions (Tailor et al., 2022).

People are capable of handling routine tasks, dangerous substances, and heavy loads. This has caused companies to avoid numerous hazards while also protecting their time and money. Figure 10.2 represents the working principle of the proposed system.

10.3.1   DATA COLLECTION

The multimodal human–robot interaction dataset is intended for human-guided engaging material perception. This dataset depicts a scenario in which a person instructs the robots regarding modeling techniques by using a limited range of mission actions.

Instead of researching actual human–robot cooperation, this data collection aims to educate robots by using a human's natural teaching capacity (Khang et al., 2023c).

Thirteen women out of the 24 candidates were chosen from the Pittsburgh region. The ages of 17 people ranged from 18 to 24, four from 25 to 30, one from 31 to 35, and two from 41 to 45.

The candidate sample was vetted to ensure that none of the participants had performed the task before using this robotic system in other experiments.

The research was conducted in the Human and Robot Partners Laboratory of Carnegie Mellon University. For their waking hours of one-and-a-half hours, the participants received $15 (Newman et al., 2022).

10.3.2 PRE-PROCESSING USING MIN-MAX NORMALIZATION

Data preparation for automated computation often uses a normalization procedure. The objective of the normalization process is to transform the characteristics of the quantitative fields in the dataset into a standard format in a manner that is insensitive to the inherent differences in the significance thresholds.

A pre-processing, matching, or scalability strategy may be used for normalization. It may prove to be highly useful for modeling or estimating. As a result, we use z-score normalization in this chapter.

In min-max normalization, Equation 10.1 is used to normalize the features that fall within the range [0, 1]:

$$u' = \frac{u - \min_B}{\max_B - \min_B} \tag{10.1}$$

The minimum and maximum possible values of characteristic B are denoted by the symbols "$\min_B$ and $\max_B$" accordingly in this text.

Both the original values of the attribute and its normalized values are indicated by the values u and u', respectively. As can be seen from the equation, both the highest and lowest possible feature values are reduced to 1 and 0, respectively.

10.3.3 FEATURE EXTRACTION USING KERNEL LINEAR DISCRIMINANT ANALYSIS

They used KLDA to extract the characteristics because it is the most important component of an operating system. To get things started, a general introduction to KLDA was ordered.

A training sample exists in the form of a column vector for each class that may be used. For computing, the between and total scatter matrices of the KLDA (Equations 10.2 and 10.3) are used:

$$S_b = \sum_{i=1}^{C} \left(a_i^j - a_i \right)\left(a_i^j - a_i \right)^T \tag{10.2}$$

$$S_t = \sum_{i=1}^{C} \sum_{j=1}^{n} \left(a_i^j - \overline{a} \right)\left(a_i^j - \overline{a} \right)^T \tag{10.3}$$

where a_i^j represents the i th class training sample with the j th training sample, $\overline{a}$ represents the mean for all the training samples, and a_i denotes a mean of an i th class, as shown in Equation 10.4:

$$S_b x = \lambda S_t x \tag{10.4}$$

If all eigenvalues of Equation 10.4 are $\lambda_1 \geq \lambda_2 \geq \therefore N \ldots 1\,2$ and similar eigenvectors are . . . The LDA eigen-equation is as follows. Because of the way it works, KLDA uses the top two or three largest eigenvectors to create d-dimensional vectors out of references. The dimensionality of the reference vectors is denoted by N.

Feature extraction and selection are both applicable to dimension reduction. The given sample and an eigenvector of a KLDA covariance matrix are $x = [x_1 \ldots x_N]^T$ and $a = [a_1 \ldots a_N]^T$. Because of x, the result of the KLDA-based extracting features of a is Equation 10.5:

$$z = a^T x = \sum_{i=1}^{n} a_i x_i \qquad (10.5)$$

It is exactly that the size of $x_i (i = 1, 2, \ldots,) N$ statistically mirrors the contribution of the i th component of the sample to the result of the feature extraction. It is exactly that the i th component of a sample contributes to reducing the smaller absolute value for x_i.

Removing $a_k x_k \sum_{i=1}^{N} a_i x_i$ from = N I I I an x 1 will only slightly change the output of the feature extraction if an absolute value of x_k is small enough. This means that they can disregard the significance of the k th element of the sample when x_k is of a modest absolute value.

When assessing the relevance of a single component, many eigenvectors should be considered because there are almost multiple eigenvectors. Using this method, they were able to identify the best features.

Phase 1: KLDA among the total scatter matrices was generated using the initial training samples. Finally, the eigenvectors and eigenvalues of Equation 10.4 are solved.

Phase 2: They chose the eigenvectors with the greatest eigenvalues in the first m, designating them with the letters $V_{1,\ldots0,}$ and V_m in turn.

Phase 3: The impact of the j th component on feature extraction is calculated as follows (Equation 10.6):

$$c_j = \sum_{p=1}^{m} |V_{pj}| \qquad (10.6)$$

where V_{pj} denotes the j th element of the vector V_p, where $j = 1, 2, \ldots, N$ and $p = 1, 2, \ldots, m$, and $|a| V_{pj}$ represents an absolute value of V_{pj}.

Phase 4: Where $j = 1, 2 \ldots N$, they sort c_j in descending order and use $d_{\square}$ to record an order.

For instance, if the s th and t th components of the original samples are the first and second most significant features, respectively, between all the c_j, where $j = 1, 2 \ldots N$, then they should let $d_1 = s$ and $d_2 = t$ instead.

If n-dimensional features are required, the d_1 th, d_2 th $\ldots$ and d_n th components will be the outcome of the feature extraction. The feature extraction data are placed in Internet of Things (IoT).

10.3.4 FEATURE DETECTION USING BAYESIAN HYPER-TUNED ARTIFICIAL NEURAL NETWORK

A theoretical model called an artificial neural network simulates several aspects of the biological brain waves. The proposed method performs well in robotic technology in a similar manner (Rana et al., 2021).

More technically, it often refers to a computer network that, in response to a particular type of input, produces some useful insights. Although they are cognitive systems, they are not designed using any connection between input and outcome.

Instead, BH-ANNs are simply installed in a permanent feature and subsequently taught to perform their functions, emulating neurological characteristics.

There are several varieties of neural networks, each with a unique goal, teaching methods, and transmission networks (Khang et al., 2023a).

10.3.4.1 Feature of Bayesian Hyper-tuned Artificial Neural Network

Many regular and quasi-actuators use BH-ANNs.

- Bayesian hyper-tuned artificial neural networks can manage a single intake with various outcomes.
- The statistical challenge of translating spatial values into strategy and plans is difficult for BH-ANNs to solve.
- The majority of extensively employed control strategies, including the regulation of nonlinearity events, involve BH-ANNs.
- Several different sensors have been effectively incorporated into BH-ANNs.
- The denial of service neural system model is most commonly employed in robotics.

10.3.4.2 Bayesian Hyper-tuned Artificial Neural Network Operations

Because it decreases the dimensions of the training data to simplify the relationship between the various layers involved in network defense, the method employs a thick network of fully linked layers as its outcome. Algorithm 1.1 shows how the BH-ANN operates.

Algorithm 1.1: BH-ANN Operations

Input: Neurons, Activation, Optimizer, Learning Rate, Batch Size, Epochs, Layers, Normalization, Drop Out, Dropout Rate
Output: Optimized Values for all the estimators
Begin:
Step1: Create a sequential layer.
Step 2: Define a set of dictionaries that consist of all possible optimizers, such as Adam, SGD, RMSProp, and others, for both convolution and dense layers.
Step 3: Define activation functions.

Step 4: Define neurons in the range of 10 to 1000, and batch size in the range of 200 to 1000.

Step 5: Define three dense layers and drop out layers with a threshold greater than 0.5.

Step 6: Define a dense output layer with a sigmoid activation function.

Step 7: Compute the cross-validation score by applying the stratified folding technique.

End

10.4   PERFORMANCE EVALUATION

We proposed the BH-ANN. The experiment used security, sensing level, robustness, and implementation cost. Existing methods, such as IoT, AI, and cloud computing, are compared to the proposed work (Khang et al., 2023b).

10.4.1   SECURITY

Because of the ongoing development of robotic systems and the rise in security and safety threats and issues, the interdisciplinary field of robotics security is becoming more relevant and significant (Khang et al., 2022).

According to the local police, robots can patrol the streets. They received training to recognize and apprehend all street criminalities, as shown in Figure 10.3.

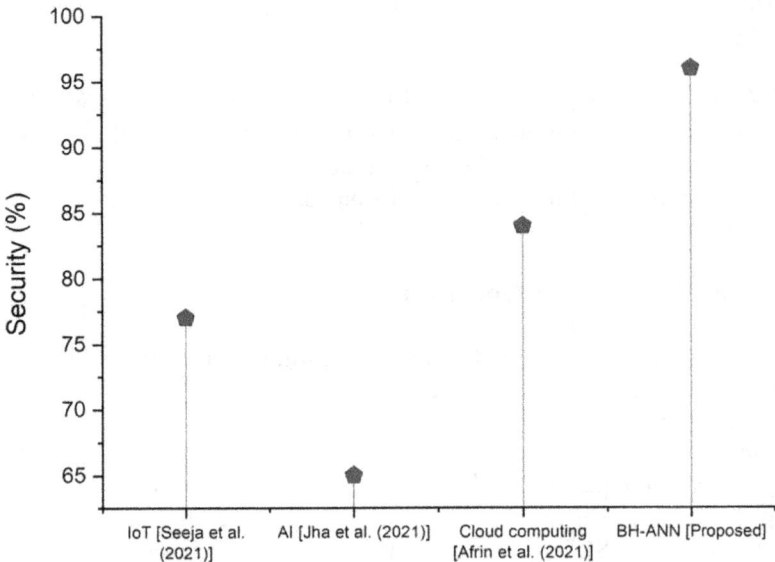

FIGURE 10.3   Comparison of Security.

10.4.2   SENSING LEVEL

Robotic sensing mostly employs algorithms that require environmental input or sensory data to enable robots to see, touch, hear, and move (Hahanov et al., 2022).

The three main sensor types (visual, force/torque, and tactile) identify the three most common robotic control strategies: visual serving control, force control, and tactile control. They are often the foundation of robotic controllers that interpret and evaluate sensory data, as shown in Figure 10.4.

In contrast to existing methods, such as IoT (65%), AI (75%), cloud computing (88%), and BH-ANN (93%), the proposed work is evaluated.

10.4.3   ROBUSTNESS

The control policy desired actions (such as those brought on by a step) and undesirable actions (such as those brought on by a fall) may be used to classify a robot's robustness, as shown in Figure 10.5.

10.4.4   IMPLEMENTATION COST

The costs associated with developing and implementing an implementation strategy that focuses on one or more particular evidence-based treatments are known as implementation costs. The approach will have a direct impact on the next intervention, perhaps affecting its effectiveness, use, or quality, as shown in Figure 10.6.

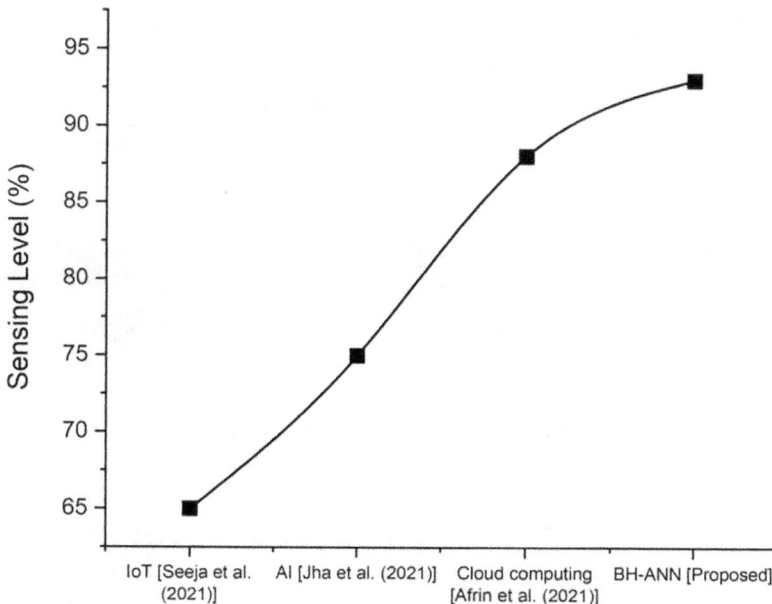

FIGURE 10.4   Comparison of Sensing Level.

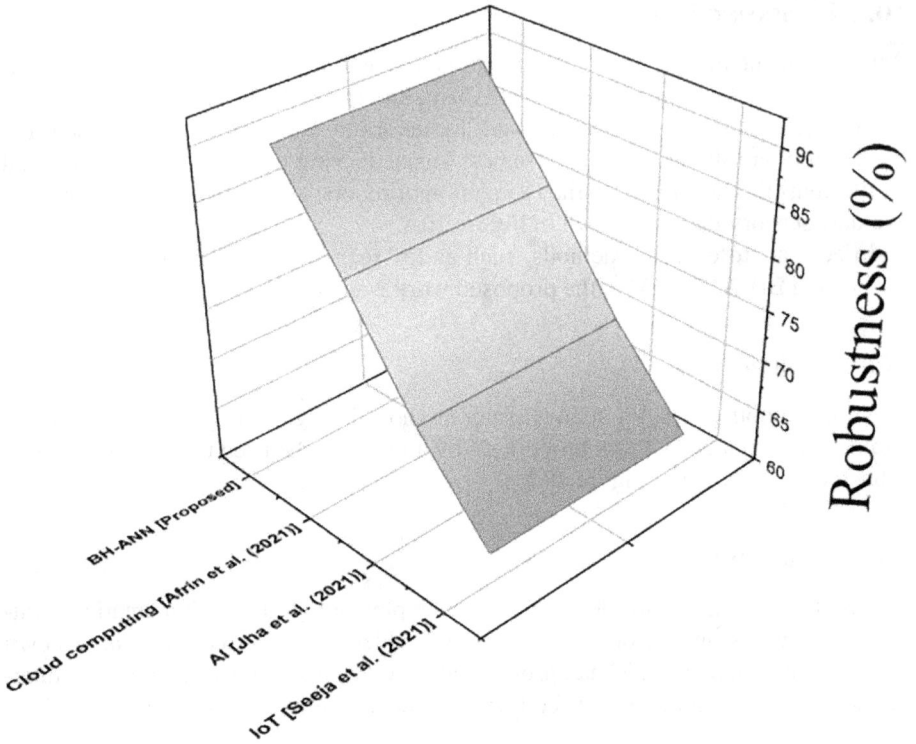

FIGURE 10.5   Comparison of Robustness.

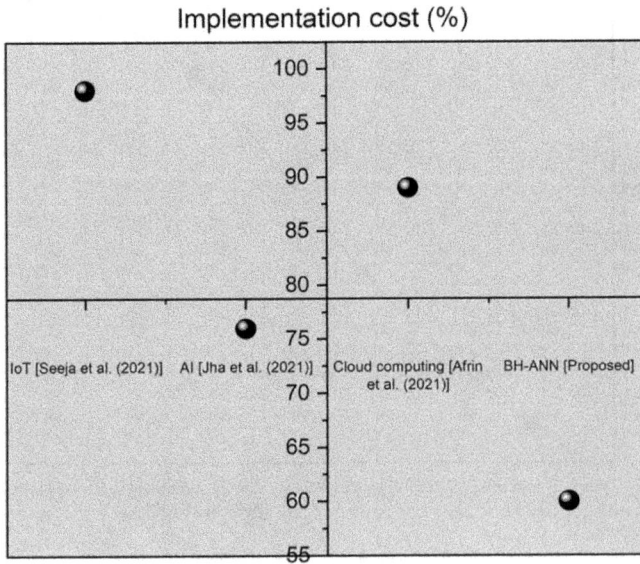

FIGURE 10.6   Comparison of Implementation Costs.

10.5 CONCLUSION

In this chapter, BH-ANN is suggested for use in flexible industrial sectors with robotics in real-time applications. With the robotic aspect as a guideline for future smart factories, the new cybersecurity scheme comprises neural networks that communicate with other aspects in a completely integrated system.

An industrial context dataset was collected to illustrate a manufacturing assembly process employing human–robot cooperation to determine the viability of the proposed system.

The feature extraction phase of this study uncovered the true difficulties that pose a severe threat to robotics. To demonstrate the usefulness of the suggested strategy, its aspects are presented.

A security-compromised physical asset may result in a severe safety hazard in automation, owing to the increased connection, as shown by the simulation.

Future industrial processes, including human operators and robots, will be created within the framework of revolution (Khang, 2024).

REFERENCES

Bhardwaj A., Alshehri M. D., Kaushik K., Alyamani H. J., Kumar M., (2022). "Secure framework against cyber-attacks on cyber-physical robotic systems." *The Journal of Electronic Imaging*, 31(6), 061802. https://doi.org/10.1117/1.JEI.31.6.061802

Hahanov V., Khang A., Litvinova E., Chumachenko S., Hajimahmud V. A., Alyar V. A., (2022). "The key assistant of smart city—sensors and tools," In *AI-Centric Smart City Ecosystems: Technologies, Design and Implementation*, Vol. 17, No. 10 (1st ed.). CRC Press. https://doi.org/10.1201/9781003252542-17

Hajimahmud V. A., Khang A., Hahanov V., Litvinova E., Chumachenko S., Alyar V. A., (2022). "Autonomous robots for smart city: Closer to augmented humanity," In *AI-Centric Smart City Ecosystems: Technologies, Design and Implementation*, Vol. 7, No. 12 (1st ed.). CRC Press. https://doi.org/10.1201/9781003252542-7

Haleem A., Javaid M., Singh R. P., Rab S., Suman R., (2022). "Perspectives of cybersecurity for ameliorative Industry 4.0 era: A review-based framework," *Industrial Robot: The International Journal of Robotics Research and Application*, 49(3), 582–597. https://doi.org/10.1108/IR-10-2021–0243

Jin S., Lian W., Wang C., Tomizuka M., Schaal S., (2022). "Robotic cable routing with spatial representation." *IEEE Robotics and Automation Letters*, 7(2), 5687–5694. https://doi.org/10.1109/LRA.2022.3158377

Khang A., (Eds.). (2024). *(AIoCF) AI-Oriented Competency Framework for Talent Management in the Digital Economy: Models, Technologies, Applications, and Implementation*. CRC Press. https://doi.org/10.1201/9781003440901

Khang A., Gupta S. K., Hajimahmud V. A., Babasaheb J., Morris G., (2023a). *AI-Centric Modelling and Analytics: Concepts, Designs, Technologies, and Applications* (1st ed.). CRC Press. https://doi.org/10.1201/9781003400110

Khang A., Vrushank S., Rani S., (2023b). *AI-Based Technologies and Applications in the Era of the Metaverse* (1st ed.). IGI Global Press. https://doi.org/10.4018/978166 8488515

Khang A., Rani S., Gujrati R., Uygun H., Gupta S. K., (Eds.). (2023c). *Designing Workforce Management Systems for Industry 4.0: Data-Centric and AI-Enabled Approaches*. CRC Press. https://doi.org/10.1201/9781003357070

Khang A., Hahanov V., Abbas G. L., Hajimahmud V. A., (2022). "Cyber-physical-social system and incident management," In *AI-Centric Smart City Ecosystems: Technologies, Design and Implementation*, Vol. 2, No. 15 (1st ed.). CRC Press. https://doi.org/10.1201/9781003252542-2

Mayoral-Vilches V., (2022). "Robot cybersecurity, a review," *International Journal of Cyber Forensics and Advanced Threat Investigations*. https://orcid.org/0000-0001-8308-3363

Newman B. A., Aronson R. M., Srinivasa S. S., Kitani K., Admoni H., (2022). "Harmonic: A multimodal dataset of assistive human-robot collaboration," *The International Journal of Robotics Research*, 41(1), 3–11. https://doi.org/10.1177/02783649211050677.

Rana G., Khang A., Sharma R., Goel A. K., Dubey A. K., (Eds.). (2021). *Reinventing Manufacturing and Business Processes through Artificial Intelligence*. CRC Press. https://doi.org/10.1201/9781003145011

Tailor R. K., Ranu P., Khang A., (Eds.). (2022). "Robot process automation in blockchain," In Khang A., Chowdhury S., Sharma S., Eds., *The Data-Driven Blockchain Ecosystem: Fundamentals, Applications, and Emerging Technologies*, Vol. 8, No. 13, pp. 149–164 (1st ed.). CRC Press. https://doi.org/10.1201/9781003269281-8

Tuomas T., (2022). "Cybersecurity Testing Automation." 202205026731. https://urn.fi/URN:NBN:fi:amk

11 Quantitative Study on Variation of Glaucoma Eye Images Using Various EfficientNetV2 Models

*Aniverthy Amrutesh, Asha Rani K. P.,
Amruthamsh A., Gowrishankar S.,
and Gowtham Bhat C. G.*

11.1 INTRODUCTION

A group of eye conditions, known as glaucoma, can cause blindness. All forms of glaucoma cause damage to the optic nerve, which connects the eye to the brain, mostly because of the elevated blood pressure in the arteries that supply blood in the eyes.

There are five distinct forms of glaucoma:

- Open-angle glaucoma
- Acute-angle closure glaucoma
- Normal-tension glaucoma
- Glaucoma in children
- Pigmentary glaucoma

Open-angle glaucoma is the most prevalent type of glaucoma (Weinreb et al., 2014). The primary symptoms include severe headache, eye pain, nausea, and redness in the eyes.

The later stages may lead to a dull or cloudy eye, halo, or coloured rings around light, as well as the final stages of vision loss (Rani et al., 2023).

How is glaucoma diagnosed?

Specific tests are performed to determine:

- Inner eye pressure
- Shape and colour of the optic nerve
- Total field of vision
- The angle in the eye where the iris meets the cornea
- Corneal thickness

DOI: 10.1201/9781003400110-11

If any of these symptoms are seen, refer to an ophthalmologist (a doctor whose specialization includes eye care unlike optometrists and opticians). Ophthalmologists have specific training and experience in the diagnosis and treatment of vision conditions (Weinreb et al., 2014).

- **Tronometry**: This device measures intraocular pressure (IOP) or pressure inside the eyes. Before the test, numbing eye drops are applied to expose the eye, increasing the comfort of the test. The corneas are mildly indented by the device, and the resistance to this indentation is measured using the measuring equipment.
- **Ophthalmoscopy**: This examination is used to determine the size and shape of the optic nerve, retina, optic disc, and blood vessels.
- **Perimetry**: This is an example of visual field testing. The visual field describes the distance the eyes can see while focusing on a single location. This test determines whether a patient's vision has any blind spots.
- **Gonioscopy**: This is performed to determine the precise angle at which the iris and cornea meet. A numbing drop is applied to the eye, followed by a customised contact lens containing a mirror to show the angle between the iris and the cornea. If the angle is blocked, there is a dramatic increase in IOP. This examination also reveals aberrant blood vessels and damage caused by earlier trauma.
- **Pachymetry**: This is performed to measure the corneal thickness. Because of its ability to alter IOP, a probe measures its thickness. This is necessary because corneal thickness can interfere with an accurate reading of IOP, resulting in misinterpretation of internal pressure and possibly misdiagnosis.

The following are glaucoma medications and treatments (Weinreb et al., 2014):

- Medication: Eye drops, such as Xalatan, Lumigan, and beta blockers
- Surgery: laser surgery
- Medical procedure: trabeculoplasty

To detect the presence of glaucoma, we considered various publicly available binary classifiable datasets obtained from tonometry examinations. The behaviour of the eye dataset is observed on various EfficientNetV2 models with varying activation functions to improve the baseline results, and the behaviour of the same is analysed.

Older age, family background, African heritage, high IOP, and specific medical disorders, including blood pressure and diabetes, are other risk factors for glaucoma (Weinreb et al., 2014). People with these risk factors should undergo frequent eye examinations to check for early indications of glaucoma.

What are the benefits of transfer learning (TL) in the health sector (especially in glaucoma detection)? Transfer learning has the potential to be highly beneficial in the health sector (Khang et al., 2023c), specifically in the detection of glaucoma. The following are some of the benefits of TL in glaucoma detection:

- **High accuracy**: TL can improve the performance of models for glaucoma detection because pre-trained models can be fine-tuned on a smaller dataset of medical images, which can detect specific diseases, such as glaucoma. This can lead to a higher accuracy in detecting glaucoma compared to training a model from scratch.

- **Limited data availability**: In healthcare, limited amounts of data are often available because of privacy and ethical concerns. Transfer learning can help overcome this limitation by allowing models to be fine-tuned on smaller datasets, making it possible to train models in scenarios in which data are scarce (Vrushank et al., 2023).
- **High computational resources**: The detection of glaucoma requires a large amount of computational resources and memory. Transfer learning can help reduce the computational resources required by using pre-trained models that are already trained on large datasets.
- **Continual learning**: Glaucoma detection is a dynamic field, and new treatments and drugs are constantly emerging. Transfer learning can update models with new data, allowing the model to continually learn and adapt to new information.
- **Explainable AI**: Transfer learning can also improve the interpretability of models, as the pre-trained models have already learned features from large amounts of data, and fine-tuning on task-specific data can help understand the decision-making process of the model.

The upcoming sections in the paper contain dataset descriptions that provide information about number of glaucoma datasets available and considered, and the number of images employed for the training and testing of the TL models available in the Keras module.

In the next section, the methodology contains the procedure that is employed and the process of implementation results and discussions. Finally, the conclusion provides a clear idea about the model to be used for the glaucoma-related dataset.

11.2 LITERATURE SURVEY

Ambreen et al. (2019) conducted a meta-analysis of 42 academic articles, mostly from the IEEE, ACM, and Springer digital libraries, on automated glaucoma systems, suggested in the previous 10 years. Thirty-seven of them were devoted exclusively to automated glaucoma detection methods.

The anatomical and functional aspects of both eyes that were used in the automatic glaucoma diagnosis were also included in the study. For more precise glaucoma detection, a hybrid feature set encompassing the combined structural and textural aspects of retinal pictures is suggested.

Jaja et al. (2020) collected data from 50 pictures of human eye optical coherence tomography (OCT) and fundus images, including control and glaucomatous images, which can be useful for developing an automated system for glaucoma diagnosis.

To diagnose glaucoma, Shinde et al. (2021) suggested an offline computer-aided diagnosis approach based on images of the retinal fundus.

Le-Net utilised the brightest spot approach to extract the return on investment and achieved 98.67% accuracy for input photo validation. They also employed the U-Net design, which, for such segmentation of the optic disc and optic cup, yielded dice coefficients of 0.93 and 0.87, respectively.

The optic disc region was first divided using the DeepLabv3+ architecture proposed by Sreng et al. (2020), which is a fully automated glaucoma screening approach.

However, the encoder module was replaced with several deep convolutional neural networks (CNNs). For their three ideas, they used pre-trained deep CNNs: (1) TL, (2) learning feature descriptors with support vector machines (SVMs), and (3) constructing an ensemble of methods from (1) and (2).

These methods were evaluated on five datasets comprising 2787 retinal images. DeepLabv3+ and MobileNet were combined to yield the best results for optic disc segmentation, with an accuracy of 99.7%, a dice coefficient of 91.73%, and an intersection over union of 84.89%.

Kumar et al. (2016) proposed a collection of image processing techniques for identifying glaucoma. They found that using a glaucoma screening approach that employed super pixel classification on a set of 650 pictures resulted in overlapping errors of 9.5% and 24% in the optic disc and optic cup, respectively.

They found that glaucoma detection without segmentation had an 86% success rate on 200 real-world photographs using three different classifiers: naive Bayes, k-nearest neighbour, and SVM.

Zhuo et al. (2010) proposed an approach that uses deep neural network (DNN)-based modules, such as a retinal sickness diagnosis and clinical description generator, and a DNN visual explanation module.

They trained a DNN-based model on the DeepEyeNet dataset to effectively learn resilient features of retinal images. Their method can generate therapeutically useful descriptions and explanations of retinal images.

To categorise glaucoma, Hafsa Ines et al. (2014) suggested extracting certain traits from retinal fundus images. The cup-to-disc ratio (CDR), a crucial physiological parameter for the diagnosis of glaucoma, and the neuroretinal rim ratio inside the inferior, superior, nasal, and temporal quadrants (ISNT quadrants), a measure used to confirm the ISNT rule, are examples of these features.

Three different databases – Digital Mean Elevation Data, FAU (An online collection that may include articles, statistics, images, music and/or other types of information), and Methods to Evaluate Segmentation and Indexing Techniques in the field of Retinal Ophthalmology —were used to assess the suggested approaches. When evaluated on 80 retinal images, the method exhibited a 97.5% accuracy rate and required an average computation time of 0.8141 seconds.

Fathima and Subhija (2019) discovered that when collaborating with the Center of Prevention and Attention of Glaucoma in Bucaramanga, Colombia, the success rate of glaucoma identification using fundus photographs was 88.5%.

Ophthalmologists estimated the CDR for each image, and measurements were obtained across a group of 26 fundus photographs of both unhealthy and healthy eyes. The error percentage was 19.2%, while the absolute error was 8.6%.

11.3 TECHNICAL APPROACH

11.3.1 DATASET DESCRIPTION

The following are the publicly available datasets considered for this work:

- **ACRIMA** (Fan et al., 2020) is a public dataset of fundus images that contains annotations of glaucoma made by two experts. The dataset contains 705 images, with 396 images labelled as glaucomatous and 309 images labelled as normal. It is crucial to remember that no additional clinical information was taken into account; the comments were made simply on the basis of how the fundus pictures appeared.
- The **Drishti-GS** (Sivaswamy et al., 2014) dataset is a publicly accessible dataset for glaucoma evaluation that contains 101 monocular fundus pictures, 70 of which are glaucoma images and 31 are normal images. It is split into two sets: testing and training. Four professional segmentations of an optic disc and cup are included in the training package.
- The **High-Resolution Fundus (HRF) Image Database** is a collection of images of the retina, or the inner lining of the eye, which are used for research and medical purposes (Li et al., 2019). The images in the database are high resolution, meaning that they have a high level of detail and clarity. The HRF Image Database is often used by researchers and medical professionals to study the structure and function of the retina and to develop and evaluate new technologies and methods for diagnosing and treating eye conditions. The HRF dataset, consisting of 45 images that were split into three groups and labelled as diabetic retinopathy, glaucomatous, and healthy, is a dataset for fragmenting retinal vessels. In the present study, only glaucomatous and healthy subjects were considered (Khang et al., 2022).
- **ORIGA** is an online repository of retinal fundus photographs developed for glaucoma examinations and studies. Photographs of the retina, optic disc, and blood vessels at the back of the eye are called retinal fundus images. Glaucoma is a group of eye conditions that can damage the optic nerve and cause vision loss (Zhuo et al., 2010). The ORIGA-light dataset is anticipated to be a helpful resource for researchers looking into the underlying biological reasons for the condition, as well as academics and clinicians focusing on glaucoma diagnosis and treatment. ORIGA30 includes 168 photographs of glaucoma patients and 482 photographs of healthy people, in addition to disc and cup segments.
- **Multi Eye Disease Dataset** is a collection of eye-related medical photographs that were divided into four groups: normal, cataract, glaucoma, and retinal disease. The dataset was developed in an Indian hospital and has 500 photographs in total, with 100 images for each of the three illness groups and 300 photographs for the healthy category. Because the data have been anonymised, it is unlikely that the dataset contains any private information about the people in the photographs. The dataset appears to have been built using the data fusion procedure, and it was developed for research on

selecting features based on optimization. Only healthy and glaucomatous individuals were included in this study.

11.3.2 EfficientNetV2 Models

Evolution of TL:

- **Early TL**: In the early days of artificial intelligence (AI) research, TL was proposed as the concept of "learning to learn," where a system could learn from one task and transfer the acquired knowledge to improve performance on a different task. However, the field of AI was still in its infancy and the computational resources are limited, so the idea of TL remained theoretical (Rana et al., 2021).
- **Transfer learning with feedforward neural networks**: As computational resources improved and large datasets became available, researchers began to explore practical applications of TL. Early works in this area focused on transferring knowledge between different but related tasks, such as between different languages or between different image classification tasks. These TL approaches are mainly based on feedforward neural networks, which are simple and efficient architectures that are well-suited for the limited computational resources of the time.
- **Convolutional neural networks and TL**: With the introduction of CNNs, TL has received considerable attention. Convolutional neural networks can learn feature representations from images and have been used in various image-related tasks, such as object detection, semantic segmentation, and image classification (Khang et al., 2023a).
- **Transfer learning using pre-trained DNNs**: As deep learning has gained popularity, TL has also gained popularity. Deep neural networks that have already been trained on big datasets can perform better after being fine-tuned on small, task-specific datasets. This approach, known as fine-tuning, has become a popular method for TL in various applications, such as natural language processing, computer vision, and speech recognition.
- **Transfer learning with EfficientNetV2 models**: EfficientNetV2 models represent the latest evolution of TL. EfficientNetV2 is an improved version of the original EfficientNet model, which is a family of CNNs designed to be more efficient in terms of computational resources while maintaining high accuracy. EfficientNetV2 models were trained using a combination of TL and machine learning techniques. They are based on the EfficientNet architecture and are pre-trained on a large dataset, allowing them to be fine-tuned on specific tasks with smaller datasets (Khang et al., 2023b).

The evolution of TL has progressed from early theoretical proposals to the practical application of feedforward neural networks, CNNs, pre-trained DNNs, and finally to the latest EfficientNetV2 models.

With each step, the models become more complex and powerful, capable of learning from larger and more diverse datasets, and can fine-tune specific tasks to achieve improved performance.

11.4   WHY EFFICIENTNETV2?

The architecture of EfficientNetV2 is based on the idea of scaling up CNNs in a "smart" way (Luke et al., 2023). This is achieved using a compound scaling method, which scales up the dimensions of the network in a more structured manner than simple scaling. The compound scaling method adjusts the following network dimensions:

- **Width**: The number of channels in the convolutional layers is increased.
- **Depth**: The number of layers in the network is increased.
- **Resolution**: The input image resolution is increased.

In addition to the compound scaling method, EfficientNetV2 also uses several other techniques to improve efficiency, such as lightweight convolutional layers, global depth-wise convolution, and a linear bottleneck structure.

Overall, the goal of the EfficientNetV2 architecture was to achieve the best performance with the smallest number of parameters and computational resources.

The main differences between EfficientNetV2 and the original EfficientNet architecture are as follows:

- EfficientNetV2 adjusts the model dimensions (depth, width, and resolution) in a more balanced manner using a different technique termed compound scaling.
- EfficientNetV2 introduces a new version of the Swish activation function, called memory-efficient Swish, which is more memory efficient than the original Swish activation function.
- EfficientNetV2 introduces a new version of the bottleneck building block, called the EfficientNetV2 block, which is more efficient than the original bottleneck block used in EfficientNet.

The models considered under EfficientNetV2 series are as follows.

11.4.1   EFFICIENTNETV2B0

EfficientNet-V2B0 is a variant of the EfficientNet-B0 architecture that is designed to be efficient and scalable. The architecture is based on MobileNetV2 architecture and is trained using a combination of manual and automated methods.

The EfficientNet-V2B0 architecture is characterized by its use of depth-wise separable convolutions and a lightweight residual block called mobile inverted residual bottleneck.

It also uses a compound scaling method that scales the dimensions of the network, along with the number of channels and the resolution of the input images. This allows the network to maintain a high level of accuracy while being more efficient in terms of computational resources.

EfficientNetV2B0 architecture consists of a stack of EfficientNetV2 blocks, with varying number of filters and layers in each block.

The input to the model is a 224 × 224 image, and the output is a 1000-dimensional vector showing the class probabilities of the image. The model also includes a fully linked layer at the end as well as a global average pooling layer.

With only a few parameters, EfficientNet-V2B0 performs well on a range of tasks, including semantic segmentation, object identification, and picture classification. It is suitable for use in areas with limited resources, such as portable devices or edge computing settings.

11.4.2 EFFICIENTNETV2B1

EfficientNet-V2B1 is one of the models in this family, and it is a variant of the EfficientNet-V2 model.

The architecture of the EfficientNet-V2B1 model relies on the MobileNetV2 model, which is CNN developed for efficient picture categorisation. It employs depth-wise separable convolutions and residual connections to decrease the number of parameters and calculations needed while maintaining high performance (Subhashini & Khang, 2023).

EfficientNet-architecture V2B1 consists of a succession of convolution layers followed by batch normalisation layers and the rectified linear unit (ReLU) activation functions.

A global average pooling layer lowers the spatial dimension of the feature maps to a single number per channel, and a fully connected layer with only a Softmax activation function delivers the final class probabilities. The cross-entropy loss function is used to train the model.

EfficientNet-V2B1 is a relatively small model with only 3.2 million parameters, but it is still able to achieve good performance on image classification tasks. It is well suited for use on mobile devices and other resource-constrained platforms, where computational efficiency is important.

Some of the key features of EfficientNetV2B1 are as follows:

* It is an efficient architecture that is designed to be smaller and faster than other models while still achieving good performance.
* It uses a compound scaling method to scale the dimensions of the network, which helps improve the efficiency and accuracy of the model.
* It has a balanced number of parameters and computational cost, making it a good choice for a wide range of applications.
* It employs a mix of depth-wise and pointwise convolutions, which lowers the model's parameter count and increases its effectiveness.
* It includes a number of other design innovations, such as an inverted residual block and a linear bottleneck, which help to further improve the efficiency and accuracy of the model.

11.4.3   EfficientNetV2B2

The EfficientNetV2B2 model is a variant of the EfficientNetV2 architecture with a depth of 2 (hence the "B2" in its name). It is a relatively small model, with around 0.5 million parameters. It is well suited for image classification tasks with limited computational resources, such as mobile devices.

The EfficientNet V2B2 architecture is based on the EfficientNet architecture, which uses a compound scaling method to scale up the network size and capacity while also improving its efficiency.

The specific details of the V2B2 architecture may vary; however, in general, it is likely to have a larger number of layers and more filters per layer than the smaller models in the EfficientNet family. It may also use techniques, such as depth-wise separable convolutions and squeeze-and-excitation (SE) blocks to improve the model's efficiency.

11.4.4   EfficientNetV2B3

EfficientNet V2B3 is a specific member of the EfficientNet family, characterised by its use of the "B3" size configuration.

The EfficientNet architecture is based on the idea of scaling up neural network models by carefully balancing the width, depth, and resolution of the network. The model hyper-parameters employed, which include the number of layers, total number of filters for each layer, and the resolution of an input picture, are referred to as the "B3" size configuration.

The EfficientNet-V2 B3 variation is a larger model with a higher computational complexity to obtain better performance. It is composed of 10 blocks that alternate convolutional and maximum pooling layers, must have a dimensionality of the data of 224 × 224 pixels, and is completely linked after the global average pooling layer.

As the network develops, the convolutional layers contain more filters, and the model employs skip connections to enhance information flow across the network.

The B3 variation is frequently used for tasks, such as image classification, object identification, and semantic segmentation. It comprises more than 25 million parameters.

A CNN, particularly created for image classification applications, is called EfficientNet V2B3. Other EfficientNet models often comprise many convolution and pooling layers, followed by one or much more fully connected (dense) layers; therefore, it is probable that this model has a similar overall architecture.

To enhance model performance, skip connections, batch normalisation, and other methods can also be used.

11.4.5   EfficientNetV2S

EfficientNetV2S is a member of the EfficientNet family. It is a smaller version of EfficientNet, with fewer parameters and lower computational complexity. It is designed for scenarios in which the model size and computational resources are limited, such as on mobile devices or edge devices.

One key aspect of the EfficientNet V2S architecture is its use of compound scaling, which involves scaling the network dimensions (width, depth, and resolution) in a compound manner to achieve better performance. This allows the network to maintain a good performance even when the input size is reduced.

The implementation of a depth-wise separable convolution layer and the use of a global average pooling layer at the end of the network are two more crucial components of the EfficientNet V2S design. These design decisions aid in reducing the number of parameters inside the network, increasing the efficiency and simplifying the training process.

The architecture of EfficientNet-V2s is based on the MobileNetV2 architecture, which uses inverted residual blocks with linear bottlenecks and width multipliers to control model complexity.

The EfficientNet-V2s model also includes additional enhancements, such as an SE block, which helps improve the model's representational power by adaptively recalibrating the channel-wise feature responses, and a Swish activation function, which helps improve the model's performance and training speed.

In summary, the architecture of EfficientNet-V2s consists of a series of inverted residual blocks with linear bottlenecks and SE blocks followed by a classification head. The model is designed to be flexible and can be modified by adjusting the width multipliers and resolution of the input image to achieve a trade-off between performance and efficiency (Khanh & Khang, 2021).

11.4.6 EfficientNetV2M

The EfficientNet architecture, which was created to increase the effectiveness of CNNs for computer vision applications, has a variation called EfficientNet-V2M. The study "EfficientNet: Rethinking Model Scaling for Convolutional Neural Networks" presented EfficientNet-V2, which it extends.

The publication "EfficientNet-EdgeTPU: Hardware-aware Neural Network Design for Edge TPUs" debuted the V2M version of EfficientNet. It is particularly suited for use in edge devices that have limited resources, such as Google's Edge Vector Processing Elements, because it was created to be more efficient in terms of computing and memory needs tensor processing units (TPUs).

The main differences between EfficientNet-V2M and EfficientNet-V2 are as follows:

- EfficientNet-V2M has fewer filters per layer than EfficientNet-V2.
- EfficientNet-V2M uses depth-wise separable convolutions in the middle layers instead of regular convolutions, which reduces the number of required parameters and computations.
- EfficientNet-V2M uses a modified version of the Swish activation function, which is more computationally efficient than ReLU.

Overall, EfficientNet-V2M is a highly efficient CNN architecture that is well suited for deployment on edge devices with limited resources.

11.4.7 EFFICIENTNETV2L

EfficientNet-V2L is a variant of the EfficientNet family of models, which is designed to be more efficient and accurate than other models in their class. The "V2L" in the name stands for "Variant 2, Large."

The following is a brief summary of the architecture of EfficientNet-V2L:

1. Input size: 224×224
2. Number of layers: 32
3. Filter expansion: 3.1
4. Depth-wise convolution expansion: 6
5. Width expansion: 2.6
6. Resolution expansion: 2.2
7. Output stride: 16

EfficientNet-V2L uses a combination of depth-wise separable and traditional convolutions to build its network. It also uses a number of techniques to increase the efficiency of the model, including using Swish activation functions, using a mobile inverted bottleneck (MBConv) block as its building block, and using a grid search to optimise the architecture.

The EfficientNet v2L model was built using the EfficientNet architecture, which is based on a combination of the following key ideas:

1. Scaling up model size: The model is scaled up systematically, starting from a baseline model and increasing the model size and complexity in a way that is designed to improve accuracy.
2. Using effective convolutional layers: To decrease the number of parameters and increase the model performance, the model employs depth-wise separable convolution and a global average pooling layer.
3. Reducing the resolution of input images: The model uses an input resolution that is smaller than the typical input size for image classification models, which reduces the amount of computation required and allows the model to run faster.

11.4.8 EFFICIENTNETV2XL

EfficientNetV2-XL is a version of the EfficientNet model, which is a family of CNNs designed to be more efficient and accurate than previous CNN architectures.

EfficientNetV2-XL is the largest and most powerful version of the EfficientNet family, with the highest number of parameters and the best performance in image classification tasks. It is designed to be used on large-scale image datasets and is trained using the latest techniques in neural architecture search (NAS) and TL, as shown in Table 11.1.

In Table 11.1, the models are listed in increasing order of accuracy.

TABLE 11.1

Characteristics of EfficientNetV2 Models.

Model	Image resolution	Depth	Width	FLOPS (G)	Top-1 Acc	Top-5 Acc
B0	224	18	24	0.5	76.3	92.9
B1	240	19	40	1	77.8	93.4
B2	260	20	52	1.8	79.2	93.9
B3	300	22	88	5.3	80.8	94.4
S	380	24	64	3.3	81.3	94.6
M	456	26	112	7.8	82.6	94.9
L	528	28	184	15.4	83.3	95.2
XL	600	30	328	30.4	84	95.4

- **Image resolution**: The scale of the input picture on which the model is trained is referred to as image resolution. It is measured in pixels and is typically represented by the width and height of the image. A higher image resolution implies that the model can process more detailed information from the image, leading to improved performance.
- **Depth**: The depth of the model is the number of layers. Because a deep model has more layers, it can learn characteristics with more complexity from the input data. Deep models are challenging to train and require more computational power.
- **Width**: The number of channels inside the model is referred to as width. A broader model may learn more characteristics from the input data, because it has more learning channels. However, wider models are both computationally demanding and challenging to train.
- **FLOPS (G)**: FLOPS stands for floating-point operations per second and is measured in billions (G). This refers to the computational resources required to run a model. A higher FLOPS value indicates that the model requires more computational resources to run, which may be a limiting factor when deploying the model in certain scenarios.
- **Top-1 Acc**: Top-1 Acc refers to the accuracy of the model in image classification tasks in terms of the percentage of images that are correctly classified in the top 1 predictions. A higher Top-1 Acc value indicates that the model is better at correctly identifying objects in the images.
- **Top-5 Acc**: Top-5 Acc refers to the accuracy of the model in image classification tasks in terms of the percentage of images that are correctly classified in the top 5 predictions. A higher Top-5 Acc value indicates that the model is better at correctly identifying the objects in the images and can provide more accurate results in cases where the objects in the images are not certain.

How are excellent accuracies achieved even though the efficientv2 models contain the smallest number of parameters?

EfficientNetV2 models can achieve good accuracy despite having fewer parameters than the other models because of their efficient architecture. They are more efficient in terms of computational resources while maintaining good performance.

The secret behind this efficiency is the use of a compound scaling method, where the depth, width, and resolution of the model are scaled together in a balanced manner. This method allows the model to learn more complex features from the input data while reducing the number of parameters required.

EfficientNetV2 models use a technique called automated machine learning (AutoML) to search for the optimal architecture and hyper-parameters, which can lead to better performance.

AutoML is an approach that automates the process of selecting the best model architecture and hyper-parameters for a given task, and it is based on a combination of NAS and hyper-parameter optimisation techniques.

EfficientNetV2 models also use a technique called EfficientNet-Edge, which applies a specific set of operations to the model's architecture, which further improves the efficiency of the model.

11.5 HYPER-PARAMETERS USED

11.5.1 ADAM (OPTIMIZER)

Adam is a stochastic gradient descent version that employs moving averages of the parameters to give a running estimate of the mean and variance of the second raw moments of the gradients. Adam is the name of the algorithm, which uses an adaptive approach to stochastic optimisation.

The adaptive moment estimation is abbreviated as "Adam" (Kingma & Ba, 2015). Compared to the common stochastic gradient descent technique, the Adam optimiser often uses less memory and computes the update step for each parameter more quickly.

Equation 11.1 is used by the Adam optimiser to modify the parameters of the neural network:

$$\text{param} = \text{param} - \text{lr} \frac{m}{\sqrt{v + \text{epsilon}}} \tag{11.1}$$

where

- param is the current value of the parameter that we want to update
- lr (learning rate) is a hyper-parameter that controls the step size of the update (here, it is set to 0.001)
- m is the first moment of the gradient (i.e., the mean)
- v is the second moment of the gradient (i.e., the variance)
- epsilon is a small value added to the denominator to prevent division by zero

The moving average of the first and second moments are computed using Equations 11.2 and Equation 11.3:

$$m = \text{beta } 1 * m + (1 - \text{beta } 1) * \text{gradient} \qquad (11.2)$$

$$v = \text{beta } 2 * v + (1 - \text{beta } 2) * \text{gradient}^{2} \qquad (11.3)$$

where beta 1 and beta 2 are the two hyper-parameters, commonly set as 0.9 and 0.999, respectively.

It should be noted that the Adam optimiser uses an exponentially weighted moving average; therefore, the most recent gradient has a higher weight than the older gradients.

11.5.2 Sigmoid (Activation Function)

The sigmoid activation function is a specific type of activation function that uses the sigmoid function as its mathematical formula (Papers with Code, 2023). The function $f(x) = \dfrac{1}{1 + e^{-x}}$ is called sigmoid function. When the output of a neural network reflects a probability or a binary choice, this function is frequently utilised at the output layer of the network, as shown in Figure 11.1.

The output of the sigmoid function, which ranges from 0 to 1, may be considered the likelihood that the input belongs to a certain class. The output value is also helpful when attempting to train the model using the backpropagation approach because it is simple to calculate the gradient of the sigmoid function.

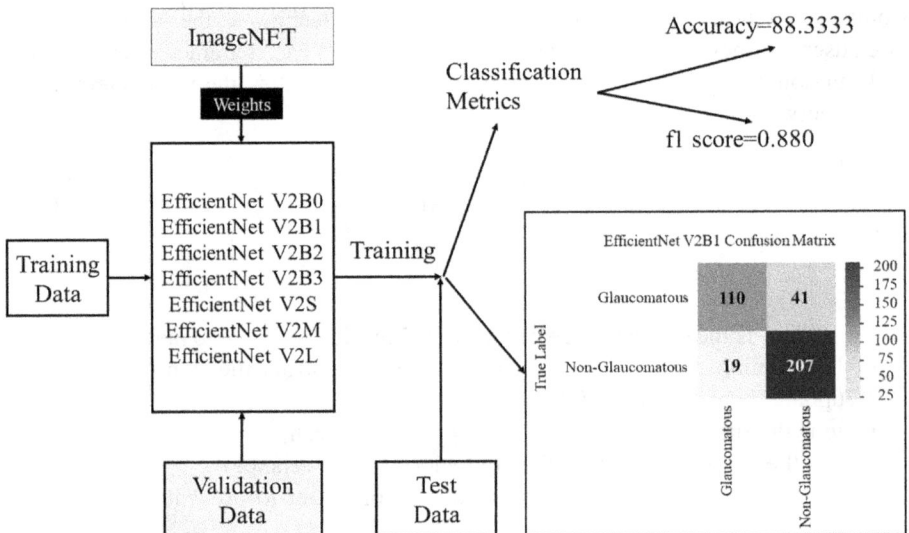

FIGURE 11.1 Data Flow Diagram of the Proposed Approach for Glaucoma Detection.

Source: Khang (2021)

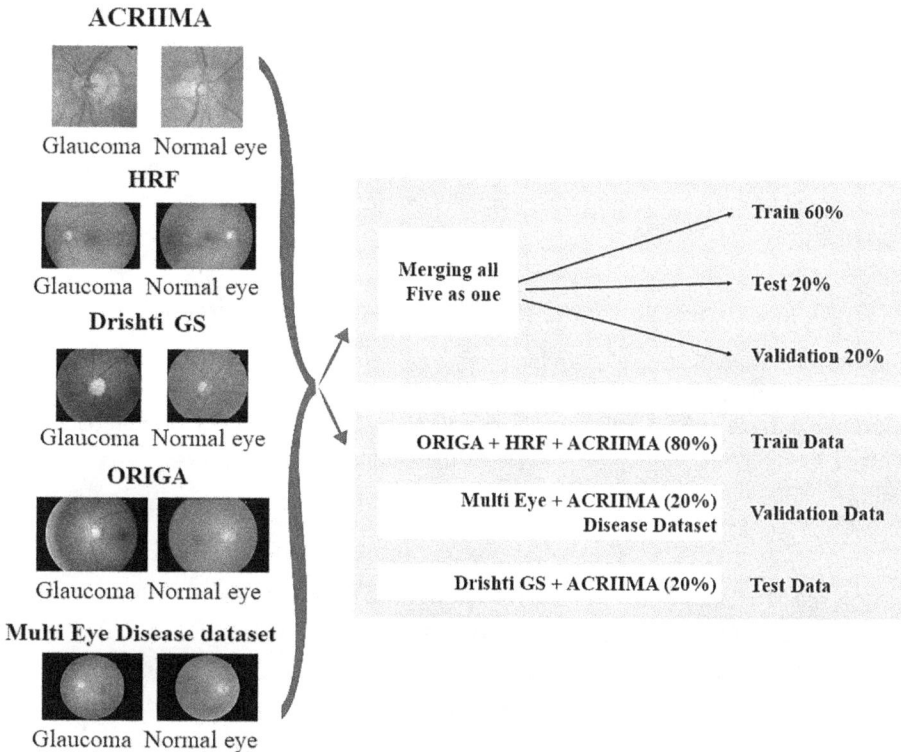

FIGURE 11.2   Use of the Five Datasets Considered.

11.6   PROPOSED METHODOLOGY

The suggested technique involves training models using all five datasets combined and then dividing them into three separate datasets: training, testing, and validation datasets, and by using a combination of training, testing, and validation datasets, with certain datasets serving as the former, as shown in Figures 11.1 and 11.2.

- **Case 1**: For implementation purposes, all five datasets are merged as one and divided into training, testing, and validation datasets.
- **Case 2**: ORIGA, HRF, and 80% of ACRIMA datasets are considered to be training datasets; Multi Eye Disease datasets and 20% of ACRIMA datasets are considered to be validation datasets; and Drishti-GS and 20% of ACRIMA datasets are considered to be testing datasets.

The ACRIMA dataset contains particular variation data in the photographs of fundus, allowing it to be considered in all three datasets.

There are numerous different options for selecting the training, testing, and validation datasets, but this one is chosen because it meets the 80:20:20 ratio (training:testing:validation).

11.7   IMPLEMENTATION AND RESULTS

The implementation of five datasets, namely, ORIGA, HRF, Drishti-GS, and ACRIMA with EfficientNetV2B0, EfficientNetV2B1, EfficientNetV2B2, EfficientNetV2B3, EfficientNetV2S, EfficientNetV2M, EfficientNetV2L, and EfficientNetV2XL with Adam as optimiser and sigmoid as activation function considering merging of all the datasets as one and setting different datasets for different parameters is shown in Tables 11.2 and 11.3.

11.7.1   EVALUATION OF A MODEL USING CLASSIFICATION METRICS

The performance of several EfficientNetV2 models is shown in Tables 11.2 and 11.3 on two distinct datasets: a separate dataset (created by merging all five datasets) and a merged dataset (where training, testing, and validation data are different datasets). The F1 Score and accuracy are the performance indicators employed.

TABLE 11.2
Classification Metrics Glaucoma Dataset where Training, Testing, and Validation Data Are Different Datasets.

	Sl no.	Model	F1 score	Accuracy
Merged	1	EfficientNetV2B1	0.830	84.0849
dataset	2	EfficientNetV2B2	0.815	82.2281
	3	EfficientNetV2B0	0.810	81.9629
	4	EfficientNetV2B3	0.805	81.9175
	5	EfficientNetV2L	0.800	80.9019
	6	EfficientNetV2M	0.780	79.0451
	7	EfficientNetV2S	0.775	78.5146
	8	EfficientNetV2XL	0.668	68.9098

TABLE 11.3
Classification Metrics Glaucoma Dataset where Dataset Is Obtained from Merging All Five Datasets.

	Sl no.	Model	F1 score	Accuracy
Separate	1	EfficientNetV2S	0.880	88.3333
dataset	2	EfficientNetV2B3	0.855	86.6667
	3	EfficientNetV2L	0.850	86.6667
	4	EfficientNetV2B1	0.850	86.5876
	5	EfficientNetV2B0	0.835	84.5833
	6	EfficientNetV2B2	0.835	83.7500
	7	EfficientNetV2M	0.745	74.5833
	8	EfficientNetV2XL	0.710	69.9990

The F1 score is used to assess a model's accuracy and recall. It is calculated as the harmonic mean of the accuracy and recall. A higher F1 score indicates better performance.

The ability of a model to classify events is a measure of accuracy. It is calculated by dividing the number of forecasts by the proportion of accurate forecasts. A higher accuracy is a sign of better performance.

The EfficientNetV2S model has the highest F1 score and accuracy in the case of the independent dataset, suggesting that it is the top-performing model on such a dataset.

However, the EfficientNetV2B1 model must have the highest F1 score and accuracy when the same models are assessed on the combined dataset, showing that it is the top-performing model on this dataset.

In the separate dataset, the EfficientNetV2S model had the highest F1 score (0.880) and accuracy (88.3333) among the compared models.

The EfficientNetV2B3 model had the second highest F1 score (0.855) and accuracy (86.6667), followed by EfficientNetV2L (0.850; 86.6667) and EfficientNetV2B1 (0.850; 86.5876). The rest of the models had lower F1 scores and accuracy.

In the merged dataset, the EfficientNetV2B1 model had the highest F1 score (0.830) and accuracy (84.0849) among the compared models.

The EfficientNetV2B2 model had the second highest F1 score (0.815) and accuracy (82.2281), followed by EfficientNetV2B0 (0.810; 81.9629) and EfficientNetV2B3 (0.805; 81.9175). The rest of the models had lower F1 scores and accuracy.

The reason for the difference in performance between the two datasets and across the models is likely because the merged dataset contains a larger and more diverse set of examples. This can make it more difficult for the models to generalise and perform well because they may encounter examples in which they were not specifically trained.

On the other hand, the separate dataset may have a more homogenous distribution of examples, making it easier for the models to perform well.

In addition, the distribution of the classes in the two datasets is different, which affects the performance of the model.

For example, if one class is over-represented in one dataset and under-represented in another, models trained on the first dataset will perform better on that class than models trained on the second dataset.

The models were trained or fine-tuned differently on the two datasets, leading to performance variations. EfficientNetV2S model, for example, was trained or fine-tuned differently on a separate dataset, which could explain its high performance on that dataset.

It is also worth noting that the EfficientNetV2XL model had the lowest F1 score and accuracy in both datasets, which could be because it is the largest and most computationally expensive among the models.

The model may require more data and computational resources for training and fine-tuning, which might make it less practical to use in real-world scenarios.

In conclusion, the F1 score and accuracy were used to gauge the effectiveness of the model. The type of data in both datasets, as well as their quantity and quality, are the causes of performance differences.

11.7.2 EVALUATION OF MODELS USING CONFUSION MATRIX

Models are evaluated using confusion matrix, loss vs. epoch curve, accuracy vs. epoch curve, classification report, and model training time.

11.7.2.1 Confusion Matrix

A table called a confusion matrix is used to describe the performance of a classification system. It is used to quantify the accuracy of model predictions (Mohan & Manas, 2015).

Typically, the matrix is shown in a 2 × 2 arrangement, with the columns representing the anticipated values and the rows indicating the real values. The diagonal members of the matrix reflect the number of accurate forecasts, whereas the off-diagonal components represent the opposite.

When many cells of the matrix are colour-coded to reflect various values, a confusion matrix is also referred to as a heat map. The heat map will help understand the performance of the model and spot trends in the data.

For instance, in a binary classification job, a heat map may be made with true positives (TPs) represented by a green colour, false negatives (FNs) by a red colour, false positives (FPs) by a yellow colour, and true negatives (TNs) by a blue colour.

This type of visualisation makes it easy to identify where the model is making errors, such as a large number of red cells, indicating a high number of FNs. It can also help identify patterns in the data, such as a large number of yellow cells, indicating that the model is biased towards one class.

11.7.2.2 Loss vs. Epoch Curve

It is a curve used to assess how well a model performs during training (Gershoni, 1979). The loss function value is plotted on the y-axis, and the number of training iterations (or "epochs") is plotted on the x-axis. Loss reduction over time demonstrates model learning and improvement.

11.7.2.3 Accuracy vs. Epoch Curve

The accuracy vs. epoch curve is a plot used, like loss vs. epoch curve, to assess how well a model performs during training (Gershoni, 1979). The y-axis shows the accuracy of the model, while the x-axis shows the number of training steps (or "epochs"). Accuracy improvements over time show that the model is being developed and learned.

11.7.2.4 Classification Report

The results of the classification model on a test dataset for which the real values are known are summarised in a classification report. Every class in the dataset offers several assessment measures, including the accuracy, recall, and F1 score.

Precision measures the ratio of accurate positive forecasts to all the positive predictions. Recall is the ratio of correctly predicted positive outcomes to all positively observed outcomes. The F1 score, which measures the harmony between accuracy and recall, is the harmonic mean of these two metrics.

Additionally, the micro- and macro-averages of both metrics among all classes are included in the categorisation report. Each sample was given an equal weight by the micro-average, but each class was given an equal weight by the macro-average.

11.7.2.5   Model Training Time

The length of time required to train a model on a certain dataset is referred to as the model training time. The quantity of the dataset, computing resources available, and complexity of the model architecture can all have an impact on this (Ajayi & John, 2023).

While deploying models for real-world settings or when working with large datasets, it is crucial to consider the training time because models with longer training time may not be suitable for certain applications.

A model's efficiency may be increased by making improvements after analysing the training duration to detect bottlenecks in the model's design or training procedure.

The evaluation of models using a confusion matrix, loss vs. epoch curves, accuracy vs. epoch curve, feature maps, and model training time can provide valuable insights into the performance of a model.

11.7.2.6   Confusion Matrix

By analysing the confusion matrix, we can determine the number of TPs, TNs, FPs, and FNs. This can help us understand the model's performance in terms of precision, recall, and overall accuracy, as shown in Figure 11.3.

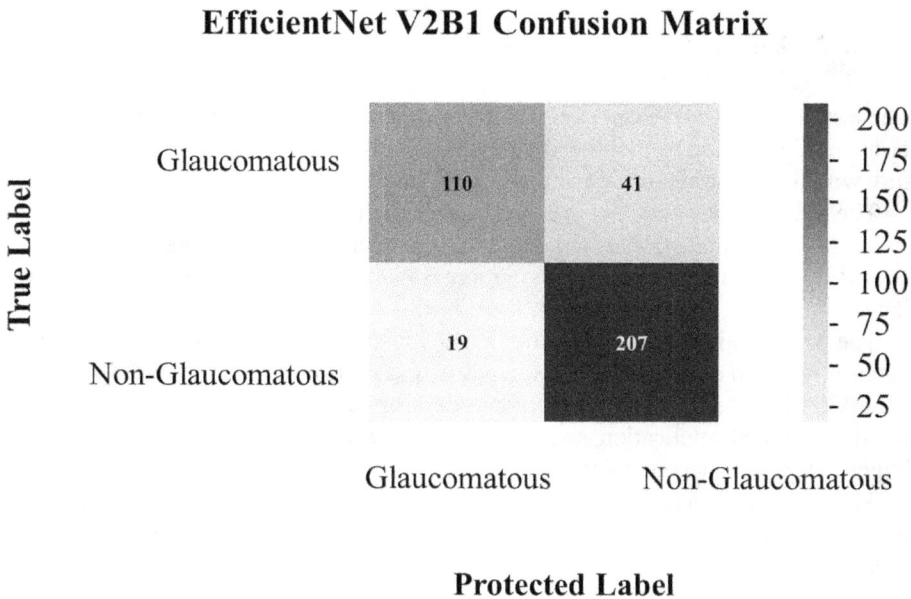

EfficientNet V2B1 Confusion Matrix

FIGURE 11.3   Confusion Matrix of EfficientNetV2B1 Model Trained for Merged Dataset.

Source: Khang (2021)

The confusion matrix in Figure 11.3 represents the performance of the model on a binary classification task for a glaucoma dataset, where class 1 is labelled as glaucomatous and class 2 is labelled as non-glaucomatous. The matrix is organised as follows:

- [[True positives (TPs) | false negatives (FNs)],
- [False positives (FPs) | true negatives (TNs)]]

The number of cases in which the model successfully identified glaucoma was measured as TPs. In this instance, 110 patients were appropriately classified as glaucomatous by the model.

The number of situations in which the model misclassified a situation as not having glaucoma is shown by the FNs. The model misclassified 41 instances as not being glaucomatous.

The number of cases in which the model misdiagnosed glaucoma is known as FPs. The model misidentified 19 patients as having glaucoma.

The TNs show the number of occurrences of the model properly identified as non-glaucomatous. In total, 207 instances were accurately classified as non-glaucomatous by the model.

This matrix provides insight into the performance of the model. However, it does not provide information on class imbalance and the relative importance of each class.

To address this, the matrix can be normalised and the ratios of misclassification can be presented in terms of precision and recall.

11.7.2.7 Loss vs. Epoch Curve, Accuracy vs. Epoch Curve, and Model Training Time

The loss vs. epoch curve can gauge how well the model is trained. The model's over- or under-fitting may be identified by examining the curve, and the model architecture with hyper-parameters can be changed accordingly.

The accuracy vs. epoch curve may assess the model's performance throughout training in a manner similar to the loss vs. epoch curve. We can determine whether the model is developing and achieving a reasonable degree of accuracy by examining the curve.

The model training time is an important consideration when working with large datasets or when deploying models in real-world scenarios, as shown in Figure 11.4. By evaluating the training time, we can determine whether the model is practical for the intended application and make adjustments to the model architecture or hyper-parameters accordingly.

The loss vs. epoch curve shows the change in the loss function over time as the model was trained. Because the model was trained over multiple epochs, the loss function should decrease as the model becomes better at predicting the correct output.

As the training goes on, the loss in the example reduces, starting at a value of 1.3 and ending at 0.1468 at the conclusion of the 30th epoch. The accuracy against the epochs graph shows how the model's accuracy changes as it is trained over time.

EfficientNetV2B1 Accuracy

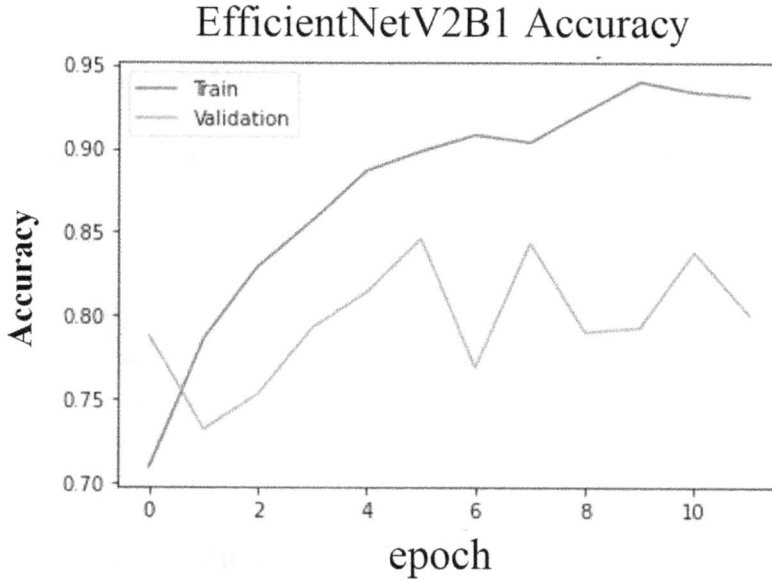

FIGURE 11.4 Accuracy versus Epoch Graphs of EfficientNetV2B1 Model Trained for Merged Dataset.

EfficientNetV2B1 Loss

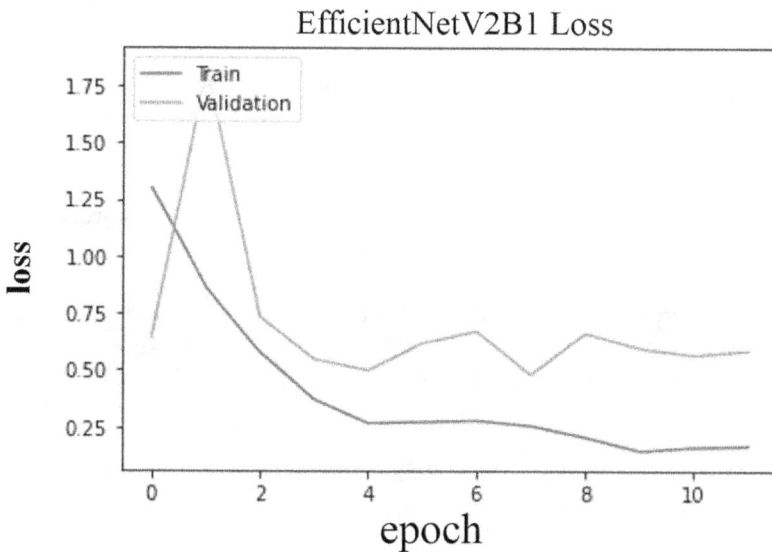

FIGURE 11.5 Loss versus Epoch and Accuracy versus Epoch Graphs.

The accuracy should increase when the model is trained over a number of epochs, because it improves the prediction of the right output. In the example, we can see that the accuracy increases over time, beginning at a value of 70.96% and ending at 92.23% at the end of the 30th epoch.

TABLE 11.4

Classification Report of EfficientNetV2B1 Model Trained for Merged Dataset.

		Precision	Recall	F1 score
Classification report	Glaucomatous	0.85	0.73	0.79
	Non-glaucomatous	0.83	0.92	0.87
	Accuracy			0.84
	Macro-average	0.84	0.82	0.83
	Weighted average	0.84	0.84	0.84

The trade-off between the two where we can see the link between the two is loss vs. accuracy. The accuracy of the model increased with a decrease in loss. In the example, as the loss decreases, the accuracy of the model increases, and vice versa.

11.7.2.8   Classification Report

The effectiveness of the model was assessed using a classification report (Table 11.4). The report provides several measures for each category in the dataset, including accuracy, recall, and F1 score.

As shown in Table 11.4, the dataset has two classes: 0 (glaucoma) and 1 (non-glaucoma). The report shows the following metrics for each class:

- **Precision**: The precision for class 0 is 0.85, which means that 85% of the samples predicted as class 0 are actually class 0. The precision for class 1 is 0.83, which means that 83% of the samples predicted as class 1 are actually class 1.
- **Recall**: For class 0, the recall is 0.73, which means that 73% of the samples that belong in class 0 were properly predicted to be in class 0. The recall for class 1 is 0.92, which means that 92% of the samples that are truly in class 1 were forecasted as class F1 score in a correct manner.

The harmonic mean of accuracy and recall is the F1 score. The F1 scores for classes 0 and 1 are 0.79 and 0.87, respectively.

The model successfully predicted 84% of the samples, with an accuracy of 84%.

The average accuracy, recall, and F1 score for all classes comprised the macro-average and weighted average.

Overall, the model is performing well in predicting class 1 (non-glaucoma) samples with a high recall of 0.92 and a high F1 score of 0.87. However, it can be improved in predicting class 0 (glaucoma) samples with a lower recall of 0.73 and F1 score of 0.79.

11.8   CONCLUSION

This chapter describes a TL approach for glaucoma detection. There are two vertices by which the five different datasets, namely, Drishti-GS, ACRIMA, ORIGA, HRF, and Multi Eye Disease datasets, which includes retinal (fundus) and pupil images, are trained on 'EfficientNetV2' series of TL models.

First, all five datasets are merged together to evaluate performance. The second approach involves combining the five datasets differently, wherein ORIGA and HRF datasets are merged as the training dataset, Multi Eye Disease Dataset as the testing dataset, and Drishti-GS dataset as the validation dataset.

In particular, the ACRIMA dataset is divided into the same testing, training, and validation and merged with the above. This approach is beneficial in when inputs with different variations are considered.

The study used an early stopping technique with a maximum of 10 to 16 epochs and a patience of 6, preventing overfitting. The results indicate that among the EfficientNetV2 models, EfficientNetV2M and EfficientNetV2XL are the baseline models, and EfficientNetV2B1, EfficientNetV2B3, and EfficientNetV2L are good models for glaucoma detection.

This study emphasizes the importance of using TL for glaucoma detection, as it allows for better performance compared to training models from scratch, and it shows the potential of using publicly available datasets for training models in medical imaging.

The use of EfficientNetV2 models for glaucoma detection provides several advantages:

1. High accuracy: The EfficientNetV2 models can achieve a high classification accuracy for glaucoma detection, which is crucial for the early diagnosis and treatment of the disease.
2. Transfer learning: Using TL allows for better performance compared with training models from scratch, and it enables the use of the knowledge of pre-trained models, which can be especially useful when the amount of data is limited.
3. Model selection: The study shows that EfficientNetV2M and EfficientNetV2XL are the baseline models, and EfficientNetV2B1, EfficientNetV2B3, and EfficientNetV2L are good models for glaucoma detection.
4. Early stopping technique: The use of an early stopping technique helps prevent overfitting and improves model performance.

REFERENCES

Ajayi O. G., John A., (2023). "Effect of varying training epochs of a faster region-based convolutional neural network on the accuracy of an automatic weed classification scheme," *Smart Agricultural Technology*, 3, 100128 (Elsevier BV). https://doi.org/10.1016/j.atech.2022.100128

Ambreen T., Saleem A., Park, C. W., (1 Mar 2019). "Numerical analysis of the heat transfer and fluid flow characteristics of a nanofluid-cooled micropin-fin heat sink using the Eulerian-Lagrangian approach." *Powder Technology*, 345, 509–520. https://doi.org/10.1016/j.powtec.2019.01.042

Fan R., Bowd C., Brye N., Christopher M., Weinreb R. N., Kriegman D. J., Zangwill L. M., (2020). *One-Vote Veto: Semi-Supervised Learning for Low-Shot Glaucoma Diagnosis. ArXiv: Computer Vision & Pattern Recognition*. http://export.arxiv.org/pdf/2012.04841

Fathima C. S., Subhija E. N., (2019). "Glaucoma detection using fundus images & OCT images," *SSRN Electronic Journal* (Elsevier BV). https://doi.org/10.2139/ssrn.3445912

Gershoni H., (1979). "Learning Curves as a Measure of Progress in Training," *Education + Training*, 21(2), 38–42 (Emerald). https://doi.org/10.1108/eb002026

Hafsa I., Bernard C., Su J. K., Alain L. B., Thierry R., Sylvie C., Powder Technology, (April 2014). *Description of Internal Microstructure of Agglomerated Cereal Powders Using X-Ray Microtomography to Study of Process–Structure Relationships*. Vol. 256, 512–521. https://doi.org/10.1016/j.powtec.2014.01.073

Khang A., (2021). "Material4Studies," *Material of Computer Science, Artificial Intelligence, Data Science, IoT, Blockchain, Cloud, Metaverse, Cybersecurity for Studies*. www.researchgate.net/publication/370156102_Material4Studies

Khang A., Gupta S. K., Hajimahmud V. A., Babasaheb J., Morris G., (2023a). *AI-Centric Modelling and Analytics: Concepts, Designs, Technologies, and Applications* (1st ed.). CRC Press. https://doi.org/10.1201/9781003400110

Khang A., Vrushank S., Rani S., (2023b). *AI-Based Technologies and Applications in the Era of the Metaverse* (1st ed.). IGI Global Press. https://doi.org/10.4018/9781668488515

Khang A., Rana G., Tailor R. K., Hajimahmud V. A., (Eds.). (2023c). *Data-Centric AI Solutions and Emerging Technologies in the Healthcare Ecosystem*. CRC Press. https://doi.org/10.1201/9781003356189

Khang A., Ragimova N. A., Hajimahmud V. A., Alyar V. A., (2022). "Advanced technologies and data management in the smart healthcare system," In *AI-Centric Smart City Ecosystems: Technologies, Design and Implementation*, Vol. 16, No. 10 (1st ed.). CRC Press. https://doi.org/10.1201/9781003252542-16

Khanh H. H., Khang A., (2021). "The role of artificial intelligence in blockchain applications," *Reinventing Manufacturing and Business Processes through Artificial Intelligence*, 2(20–40) (CRC Press). https://doi.org/10.1201/9781003145011-2

Kingma D. P., Ba J., (2015). "Adam: A method for stochastic optimization," *Cornell University—ArXiv*. https://doi.org/10.48550/arxiv.1412.6980

Kumar B., Naveen, et al., (2016). "Detection of glaucoma using image processing techniques: A critique," In *Seminars in Ophthalmology, Informa UK Limited*, pp. 1–9. Crossref. https://doi.org/10.1080/08820538.2016.1229801.

Li R., Li M., Li J., (2019). "Connection sensitive attention U-NET for accurate retinal vessel segmentation," *ArXiv: Computer Vision & Pattern Recognition*. http://export.arxiv.org/pdf/1903.05558

Luke J., Khang A., Vadivelraju C., Antony R. P., Sriram K., (Eds.). (2023). "Smart city concepts, models, technologies and applications," In *Smart Cities: IoT Technologies, Big Data Solutions, Cloud Platforms, and Cybersecurity Techniques* (1st ed.). CRC Press. https://doi.org/10.1201/9781003376064-1

Mohan P., Manas R. P., (2015). "A novel approach to compute confusion matrix for classification of n-class attributes with feature selection," *Transactions on Machine Learning & Artificial Intelligence*. Scholar Publishing (Crossref). https://doi.org/10.14738/tmlai.32.1108

Papers with Code—Sigmoid Activation Explained (2023). https://paperswithcode.com/method/sigmoid-activation

Raja Hina, et al. "Data on OCT & Fundus Images for the Detection of Glaucoma." *Data in Brief*, vol. 29, Elsevier BV, Apr. 2020, p. 105342. https://doi.org/10.1016/j.dib.2020.105342

Rana G., Khang A., Sharma R., Goel A. K., Dubey A. K., (Eds.). (2021). *Reinventing Manufacturing and Business Processes through Artificial Intelligence*. CRC Press. https://doi.org/10.1201/9781003145011

Rani S., Bhambri P., Kataria A., Khang A., Sivaraman A. K., (2023). *Big Data, Cloud Computing and IoT: Tools and Applications* (1st ed.). Chapman and Hall/CRC. https://doi.org/10.1201/9781003298335

Shinde Rutuja et al. "Glaucoma Detection in Retinal Fundus Images Using U-Net & Supervised Machine Learning Algorithms." *Intelligence-Based Medicine*, vol. 5, Elsevier BV, 2021, p. 100038. https://doi.org/10.1016/j.ibmed.2021.100038.

Sivaswamy J., Krishnadas S. R., Datt Joshi G., Jain M., Syed Tabish A. U., (2014). "Drishti-GS: Retinal image dataset for optic nerve head (ONH) segmentation," *2014 IEEE 11th International Symposium on Biomedical Imaging (ISBI)*. https://doi.org/10.1109/isbi.2014.6867807

Sreng S., et al., (2020). "Deep learning for optic disc segmentation & glaucoma diagnosis on retinal images," *Applied Sciences*, Vol. 10, No. 14, p. 4916. MDPI AG (Crossref). https://doi.org/10.3390/app10144916

Subhashini R., Khang A., (Eds.). (2023). "The role of Internet of Things (IoT) in smart city framework," In *Smart Cities: IoT Technologies, Big Data Solutions, Cloud Platforms, and Cybersecurity Techniques*. CRC Press. https://doi.org/10.1201/9781003376064-3

Vrushank S., Vidhi T., Khang A., (2023). "Electronic Health Records Security and Privacy Enhancement using Blockchain Technology," *Data-Centric AI Solutions and Emerging Technologies in the Healthcare Ecosystem*. P (1). (1st ed.). CRC Press. https://doi.org/10.1201/9781003356189-1

Weinreb R. N., Aung T., Medeiros F. A., (2014). "The pathophysiology & treatment of glaucoma: A review," *JAMA*, 311(18), 1901–1911. https://doi.org/10.1001/jama.2014.3192

Zhuo Z., Feng Shou Y., Jiang L., Wing Kee W., Ngan Meng T., Beng Hai L., Jun C., Tien Yin W., (2010). "ORIG-light: An online retinal fundus image database for glaucoma analysis & research," *2010 Annual International Conference of the IEEE Engineering in Medicine & Biology*. https://doi.org/10.1109/iembs.2010.5626137

12 Disaster Management System for Forest Fire Prediction
Fog and Cloud Data-Driven Analytical Compatible Model

Radhika Kumari, Kiranbir Kaur, and Salil Bharany

12.1 INTRODUCTION

A disaster is a crisis that causes widespread damage that is beyond the ability to recover. In 2015, 346 natural disasters caused 22,764 deaths, 110.3 million victims worldwide, and economic damages of US$ 70.3 billion (Guha-Sapir et al., 2016).

Humans are unable to prevent disasters, especially natural disasters, but assessment and planning can mitigate the various harmful effects and damage caused by such events to a great extent (Zhao et al., 2014). Therefore, international agencies and governments of vulnerable countries are transitioning towards disaster management.

Disaster management is one of the most promising research fields because of its economic and social implications (Hibino & Shaw, 2014; Anderson, 2015). Many studies have been conducted on disaster management for early warning and timely intervention to reduce disruption and loss of life.

A disaster is a natural or human-made recurring calamitous affair that quantitatively and qualitatively affects the socioeconomic life of human society and may have an adverse impact on the environment (Monacelli et al., 2005).

Disasters have the potential to cause large-scale financial and environmental damage that is beyond the reach of society to recover its own resources. Forests encompass nearly one-third of the earth's land surface area and are a vital resource for humanity (Trivedi & Srivastava, 2014).

Forests are economically and socially significant in addition to being ecologically important. Forest fires are a global threat. Wildfires, also known as wildland fires or forest fires, burn 4 to 5 million acres (1.6 to 2 million hectares) of land in the United States on average every year (Yoon et al., 2012).

As a result of global warming, wildfires are expected to become more frequent and intense. Therefore, accurate wildfire forecasting and detection are important for successful mitigation and control (Bolourchi & Uysal, 2013). Disasters can be

DOI: 10.1201/9781003400110-12

of any volume, leading to devastating effects on human life, environment, and the economic conditions of the country.

Disasters can be categorized as either natural or generated through human activities (IFRC, 2013). To this end, dedicated efforts by researchers have yielded mechanisms and models for early detection and prediction of forest fires.

Thein et al. (2020) conducted a survey of forest fires in Myanmar. Real-time monitoring and early warning systems have been developed using a machine learning-based approach.

The prediction of Vinck et al. (2013) was based on parameters, such as moisture levels within the soil and temperature. Juyal and Sachin (2021) discussed a forest fire susceptibility using machine learning approach. The predictor variables used for detection included only the moisture levels. The classification accuracy using this approach was lower.

Hartomo et al. (2017) proposed an exponential smoothing method using Google API for the early prediction of forest fires. Applications of fog computing have rarely been used to store information regarding forest fires and generate appropriate warnings for relevant authorities (Sun et al., 2015; Alam, 2016).

This study proposes a fog-based model for the early detection and prediction of forest fires, ensuring the least loss in terms of financial and human resources (Ayalew et al., 2004).

The United Nations Office for Disaster Risk Reduction (ESCAP, 2015) has cautioned that wildfires would become more recurrent and destructive as global temperatures soar, drought conditions plague, and temperature changes in many regions of the world (Saoudi et al., 2016; Sinha et al., 2017).

Therefore, to combat these wildfires, it is essential to adopt a comprehensive multifaceted approach that enables continuous monitoring of wildfire-susceptible terrains as well as reporting of time-critical and latency-sensitive data and events for early prediction and forecasting of wildfires (Loomba & Anderson, 2018).

Over the years, various techniques have been deployed to monitor wildfire-susceptible regions and fight wildfires (Jha et al., 2015). Wildfire detection systems based on satellite imagery have been extensively used, but they are not suitable for real-time applications because of various issues, such as lengthy scanning cycles, poor resolution, and high cost (Giuntini et al., 2017).

Furthermore, systems based on short-range optical infrared and thermal images have been proposed; however, these systems suffer from sensitivity to outliers, such as direct and strong sunlight, inadequate light, or smoke.

In addition, these systems are extremely prone to false alarms (Bornmann & Leydesdorff, 2014). To address the aforementioned challenges, there has been a shift towards the use of wireless sensor networks (WSNs) for time-critical applications (Jan et al., 2018), such as real-time wildfire prediction and detection, to minimize the destruction caused by them.

Several wildfire detection systems centred on WSNs have been proposed; however, the majority of these systems are designed to detect wildfires upon their outbreak (Dhar et al., 2017). However, once wildfires break out, they become extremely difficult to contain because of their uncontrolled and unpredictable rapid growth and behaviour.

Kabenge et al. (2017) highlighted the pressing need for early prediction and forecasting of wildfires for their effective mitigation and management.

Internet of Things (IoT) provides a mechanism to integrate various diverse components (sensors) to work in a synchronized manner. A system of smart interconnected components can be used to monitor abnormal situations in which existing resources or infrastructure fails (Kansal et al., 2015; Mouradian et al., 2017).

This chapter aims to study the techniques and framework used to analyse forest fire disasters. The primary objective of this study is to provide energy conservation using idle state battery power elimination techniques associated with fire detection in considerably larger premises, such as forests. A fire detection system and an alert system were employed in this case.

This work provides a mechanism to conserve energy and compares it with existing approaches to determine better approaches that can be enhanced on the parameters of energy conservation and fault tolerance in future work. The proposed method was partitioned into multiple layers.

In the first layer, noise-handling mechanisms are applied to handle missing values and outliers. Normalized data are fed into the second layer (Rau et al., 2013). The second layer contains a mechanism for reducing the size of the extracted features.

Exploratory data analysis (EDA) is applied to this layer. The cloud layer is used to store the results produced through the fog layer (Komac, 2006). The rest of the chapter is organized as follows:

- Section 12.1 presents an analysis of the mechanisms used for the prediction of forest fires along with the definition of the proposed mechanism.
- Section 12.2 gives an in-depth analysis of existing mechanisms used for the prediction of forest fires at an early stage. The datasets used are also explored through this section.
- Section 12.3 gives the methodology of the proposed work along with an explanation of each phase.
- Section 12.4 presents the performance analysis and results.
- Section 12.5 gives implementation and demonstration forms.
- Section 12.6 gives the conclusion and future scope.

12.2 RELATED WORK

This section highlights the different techniques used to detect and predict forest fires at an early stage. Dai et al. (2021) proposed an ensemble-based approach for the prediction of forest fires.

The ensemble-based approach used K-nearest neighbour (KNN), random forest, support vector machine, and decision tree for the prediction process. The overall process detected the maximum true-positive values predicted by the classifiers.

The highest prediction was the result. The classification accuracy through this approach was in the range of 90s. A real-time dataset was employed for the detection and prediction processes.

Azmoon et al. (2021) proposed an image-based temperature stability analysis using deep learning mechanism. The layer-based approach works on real-time datasets. The prediction of forest fires depends on the clarity of the extracted images. The results were presented in the form of prediction accuracy.

Amit et al. (2017) proposed a disaster detection method using aerial images. A spatial mechanism was employed to address noise from images. The boundary-value analysis accurately detected the image boundary, and the rest of the image segment was eliminated. The results of the proposed approach were expressed in terms of classification accuracy.

Jana and Singh (2022) discussed the impact of climate and the environment on natural disasters in various countries. The official datasets available on government websites were explored for this purpose.

Sarwar and Muhammad (2022) proposed a mechanism to explore the issue of forest fires within the Chittagong City in Bangladesh. A real-time dataset corresponding to the Hill Tracts of Bangladesh was presented in this analysis.

Marjanović et al. (2011) discussed forest fire susceptibility detection and prediction using support vector machine. Only two hyper-planes were used in this case. The prediction was oriented towards forest fire detection. The classification accuracy through this approach was poor because of the high degree of misclassification.

Lee (2005) discussed the applications of logistic regression in the detection and prediction of forest fire. The prediction model used a real-time dataset, and a high degree of misclassification causes this model to perform adversely in the case of a large dataset collection.

Lee (2005) proposed a fuzzy-based model for the early detection of forest fires using a benchmark dataset derived from Kaggle. The results of the system were expressed in the form of classification accuracy.

The fire detection system proposed by Kaur et al. (2019) was designed using heat and smoke sensors. A humidity sensor was also used for detecting abnormal situations.

However, the problem of energy conservation is missing, and hence, fire detection could be delayed. Reddy et al., (2011) proposed a wireless sensor-based fire detection system.

In this study, the nodes are distributed. Nodes with the maximum energy are referred to as monitoring nodes. The fire detection system detects abnormalities or intruders in terms of heat and flames. The fire detection system then sends a signal to the alarm system to indicate critical situations.

A Zigbee-oriented fire monitoring system was used for the detection of abnormal situations within the monitored area. Fire detection using Zigbee and general packet radio services (GPRS) system. Forest fire detection using Zigbee and GPRS has been addressed in the literature. The forest fire detection proposed in the literature includes algorithms for detecting humidity and temperature changes. The hardware circuitry of the proposed solution was based on Arduino board with an ATmega328 microcontroller, temperature sensor, humidity sensor, Zigbee, and GPRS modules.

Mobin et al. (2018) proposed an intelligent fire detection system. This chapter presents a system in which dissipative fires, such as citrates and welding smoke, are eliminated using a fusion algorithm.

During a fire hazard, small form factor notifies the fire service and others through text messages and telephone calls. Along with the ringing fire alarm, it announces fire-affected locations and severity. To prevent the fire from spreading, it breaks the electric circuits of the affected area and releases the extinguishing gas pointing to the exact fire locations.

This chapter presents how this system is built, its components and connection diagrams, and implementation logic. Room temperature control using the IoT and MQTT was proposed by Kang et al. (2017).

Amazon Web Service is considered for evaluation through message-queue telemetry transportation. A broker is used in this case, which senses the room temperature with the help of IoT. Threshold values are maintained when a violated alarm blows. This alarm allows temperature monitoring within the room.

A fire detection system using IoT applications was proposed by Shinde et al. (2017). This study divided the entire fire detection system into three parts. The first part involves the detection of smoke. The second part involves the detection and monitoring of flames, and the third part includes temperature monitoring.

If any of these three cases are violated, an alarm is blown. An automatic WSN is considered in this case. The 'automatic' word signifies minimal human interaction.

Once this system is in place, fire is detected, and an alarm is blown to safeguard the place and humans where it is installed. Johnson et al. (2019) investigated the disconnection of utility-scale converters in photovoltaic power facilities caused by transmission problems during wildfires.

This research focuses on tripping commands produced by phase-locked loops and determines the dynamics, which are not frequently studied.

According to Mitchell (2013), transmission line failures that can spark wildfires fall into two categories:

- Elastic extension of surrounding objects (e.g., tree branches) or conductors causing electrical contact and arcing, and
- Fatigue failure of objects (e.g., vegetation) or system components (e.g., poles and conductors) under high-strain conditions.

Because both failure types are strongly influenced by increased wind speed, outage data from regular grid operations may be utilized to assess system susceptibility to extreme fire weather conditions (Khang et al., 2023a).

Additional power system research has examined how heat stress affects the mechanical strength of aluminum conductor steel-reinforced transmission cables in wildfires. One approach for reducing the possibility of wildfire ignition near transmission lines in urban and rural regions is vegetation management. Inspection, herbicide treatment, trimming, and vegetation removal are examples of such tasks.

One purpose of vegetation management is to eliminate vegetation within and outside the transmission line rights-of-way to produce a vegetation-free control buffer where plants can grow for a certain amount of time before approaching the danger zone of the transmission lines.

Vegetation management research may help energy businesses improve their vegetation management performance. Sun et al. (2015) examined the wildfire probability along electricity lines using meteorological, topographic, and remote sensing data. The findings show that precipitation, normalized difference vegetation index, and vegetation cover are all important factors in wildfires. Methods for enhancing current wildfire forecasting systems are presented, including the integration of meteorological data from the Automated Surface Observing Systems network into geographic information system (GIS) applications for improved geographical coverage.

Liang et al. (2019) developed a model to estimate the magnitude of wildfires using a mix of meteorological data, fire size, and fire duration. The authors evaluated multiple neural networks to determine the one with the highest accuracy.

High temperatures, strong winds, and extremely low humidity can contribute to the ignition of significant wildfires. In such instances, a Public Safety Power Shutoff (PSPS) or the selective de-energization of power lines may be required for public and infrastructure safety. Although a PSPS event can be beneficial in minimizing power line ignition, dimerization can be disruptive to the public.

For over a week, from October 26 to November 1, 2019, the greatest PSPS incident in history disrupted the electricity supply of 38 California counties and about 968,000 consumers.

The PSPS research includes the work of Rhodes et al. (2020), which provides an optimization model for the power shutdown problem that optimizes the quantity of power given while minimizing power consumption grid components to reduce wildfire risk ignition. Measurements of live fuel moisture and plant water content have been considered as major markers of wildfire danger.

In particular, fuel moisture content (FMC) is an important component in determining wildfire risk and the pace of fire spread. Because of their spatially continuous and comprehensive coverage of wide surface areas, satellite data may be utilized to monitor live fuel moisture (Khang et al., 2023b

Rau et al. (2013) explored methods for predicting live fuel moisture from satellite images and proposed a nonlinear vegetation water content model based on stem factor (the highest amount of water residing in plant stems) that outperformed previous models. Transmission lines are the backbone of electricity infrastructure, and inspection is critical to finding risks or defects that might spark wildfires, among other problems. An unmanned aerial vehicle (UAV) (drone) technology may now be used for overhead transmission line inspections and patrols. Based on GIS concentrated control, enormous image identification, fault diagnosis, centralized monitoring, and distributed database cluster technology, this study proposed an intelligent central control for transmission line inspections.

The turn of events and utilization of the new model have shown the benefit of considering both anthropogenic and normal fire drivers. The proposed direct model is satisfactory for locations where anthropogenic fires start as often as possible, yet it is challenging to apply in locales where these starts happen unpredictably.

More extensively, the formation of widespread fire expectation models with sensible transient and spatial accuracy continues to be intense, on the off chance that certainly feasible, issue because of the characteristically irregular nature of the underlying start of fierce blazes.

Barmpoutis et al. (2020) discussed the re-examined approach as an additional material for modern and apartment machines. However, only legitimate assessment news is expected for the mentioned framework. As a likely future irritation, a multi-choice partnership utilizing IoT landing is examining an item and completing an important investigation (Rani et al., 2021).

At that point, the fog layer utilizes this model to conjecture the organization's method of activity in view of current climate conditions. While a high expected probability of a wood fire makes the WSN more dynamic, a low fire likelihood makes the organization less dynamic (Bhambri et al., 2022).

Accordingly, the energy utilization of the WSN was enhanced by our recommended model, and the time taken to identify woodland fires was expanded. Beraja Christian (2021) discussed that the current system will be put into practice on a broad scale with a variety of sensor nodes to power and augment the dataset to increase the precision and collaboration of data between various nodes.

Throughout the operation, the system learning technique can be used continuously. In our future research, we intend to employ wind-direction sensors to accurately determine and pinpoint the origin of the fire. It also improves the decision-making process and provides real-time composite burn index-based fire risk forecasts, considering the local weather at each individual WSN deployment site (Rani et al., 2023).

Targeted alert notifications are automatically generated in emergency situations to raise awareness and prompt greater mobilization, and systematic logs are maintained to successfully correlate the data and enable stakeholders to effectively monitor the status and health of the monitored lands and the energy consumption levels of the system modules (Khang et al., 2022).

Rajasekaran Sathish et al. (2012) have proposed a method of early detection of forest fires to reduce the harm caused by forest fires and to regulate the initiation of fires and their spread. Three steps make up this approach.

The overall danger rating of the forest was calculated, the likelihood of flames in various locations was evaluated and forecasted, and the place declared to be burning with the help of a UAV. For smoke detection in outdoor fires, a sophisticated learning object recognition model was implemented based on the Detectron2 stage.

The advanced learning model was created by moving the pre-built RetinaNet and Faster R-CNN models for object discovery through move learning. The datasets used to retrain the models were combined based on several systems (Khang, 2021).

The two top models achieved a dynamic treatment regimes of more than 86% and an F1 score of more than 80% at the fourth image for a free test. Both models were created using a dataset that was labelled with an increase and mid- and low-smoke classifications.

Another structure was distinguished by combining WSN-based observation frameworks with ground identification frameworks to carry out early discovery with greater vigour and steadfast quality. This depends on the strategy of sound vision computation and the on-the-fly application of a clever imager with minimal power.

A model vision-powered WSN hub contains these two essential elements. The results of the finished field testing demonstrated that the suggested strategy contains a strong justification for the advancement of fine-grained spatio-worldly detection frameworks in the future, as shown in Table 12.1.

Such future development should consider a number of previously unconsidered factors, such as how the presentation of the vision computation alters for sensor-tuft distances outside the considered reach.

Benzekri (2017) discussed that the engineering was then completely analyzed, with its information stream capabilities—beginning with the field hubs and finishing with the moment the data were provided to the ideal individuals—being analyzed exhaustively, as shown in Table 12.2.

Furthermore, the ongoing work provides details regarding the plan and execution of a demo model that, in light of the proposed engineering, may promptly adjust its usefulness to deal with significant ecological concerns.

To accomplish this, an equipment/programming arrangement that utilizes profoundly adaptable parts and regulators equipped for system administration,

TABLE 12.1
Different Forest Fire-Related Phenomena.

No.	Forest fire-related phenomena	Measured variables	IoT technology used	Types
1	Heat	Rate of heat rise	Heat sensors	Thermistor and liner sensors
2	Smoke	Particulates	Smoke sensors	Optical smoke sensor and ionization smoke sensor
3	Temperature	Rate of rise in temperature	Temperature sensors	Thermocouples, infrared sensors, and Resistance Temperature Detector (RTD)
4	Gas	Gases, such as carbon monoxide	Gas sensors	Gas detector and carbon monoxide sensor
5	Flames	Light of fire	Flame sensors	Optical detectors and infrared detectors

TABLE 12.2
Types of Sensor.

Type of sensor	Wildfire phenomena-related attributes	Meteorological attributes	Location attributes
Temperature sensor	Heat	Temperature	Latitude
Potential applications: detect change in temperature with respect to change in humidity			
Humidity sensor	Smoke	Relative humidity	Longitude
Potential applications: detect change in humidity with respect to change in temperature; smoke sensor: detects the presence of smoke, which typically indicates occurrence of fire; carbon monoxide (CO) sensor: detects CO gas released from fire			
Infrared (IR) sensor	Atmospheric CO	Precipitation	Longitude
Potential applications: IR sensor is used as a flame indicator; barometric pressure sensor: detects change in pressure in environment in case of fire			
Passive microwave imaging sensor	Atmospheric CO_2	Wind speed	Longitude
Potential applications: microwaves can easily penetrate dense smoke			

correspondence, and estimation is recommended. It is also inexpensive and appropriate for fast prototyping handling and detection.

12.3 DISCUSSION AND RESULTS

The methodology of the proposed work starts with dataset acquisition. The dataset corresponding to states of Jammu and Kashmir was collected.

The dataset was collected corresponding to the states of Jammu and Kashmir in India. The structure of the dataset is presented in Table 12.3.

The data acquisition layer will receive this dataset and perform an initial analysis. The details of the used layers are given as follows.

12.3.1 Data Acquisition Layer

This layer is critical for the operation of a fog-based forest fire prediction model. This layer receives the dataset and removes noise from the dataset. The noise in terms of missing and unnamed values is addressed through replacement with '0.'

Outliers indicating extreme values will be tackled using the box plot method. The values inside the box plot are retained, and the remaining values are outliers. These outliers are handled using the median values. The pre-processed dataset is fed into the fog layer.

TABLE 12.3
Dataset Description.

Field	Description
Event_Date	Date at which a forest fire occurred
Category	Type of disaster
Forest Fire_trigger	Cause of a forest fire
Size	Size of destruction
Setting	Location of an event
Latitude	Latitude of a location
Longitude	Longitude of a location
Dew/Frost point at 2mtrs	Amount of water vapour present in the air
Earth skin temperature	Temperature of the earth
Temperature 2mtrs range	Water vapour temperature
Specific humidity	Humidity present in the air
Relative humidity	Relative humidity of environment
Precipitation	Amount of precipitation attributable to temperature
Surface pressure	Pressure on the surface where an event occurred
Wind speed	Wind speed during an event
Surface soil wetness	Wetness could be critical for forest fires
Root zone soil wetness	Zone at which a disaster occurred
Profile soil moisture	Soil moisture compared against the threshold

12.3.2   FOG LAYER

The primary purpose of this layer is to conserve energy of sensors. This is possible only if a dimensionality reduction mechanism is in place.

Principal component analysis (PCA) is used for dimensional reduction. An exploratory data analysis is used to determine the highest correlated values. The highest correlated values are used as predictor variables (Khanh & Khang, 2021).

Thus, the fog layer has two tasks: the first task is associated with dimensionality reduction and then identifying predictor variables with EDA.

12.3.3   CLOUD LAYER

The cloud layer stores the generated predictions. To generate the predictions, we first apply KNN clustering and then the ARIMA model for forecasting.

The forecasted results are accessed with the help of accounts within the cloud.

Early prediction can help governments initiate preventive steps to reduce financial loss and loss of human life.

The algorithm corresponding to KNN clustering is given as KNN_Clustering:

- Receives the dataset with the predictor variables
- Set the value of $K = P$, where K is the distance metric, and P is the static value corresponding to the distance.
- Repeat the following steps until all the values within the dataset are checked for inclusion within cluster.
- If (distance $<K$) include within cluster end of if.
- Move to the next value within the dataset.
- End of loop
- Return clusters

The clustering mechanism provides groups corresponding to parameters possessing a similar nature. Clustering causes faster propagation of results. The obtained clusters, which are fed into the ARIMA model to generate predictions corresponding to the forest fire, are given as ARIMA_Prediction (Clusters):

- Stores clusters
- Repeat the following steps corresponding to test datasets for predictions.
- Perform regression analysis.
- Perform integration by obtaining difference with raw observations to make the time series become stationary.
- Calculate the moving averages by evaluating the error by subtracting observations from actual values.
- Generate a prediction.
- End of loop

The flow of the proposed model is given in Figure 12.1.

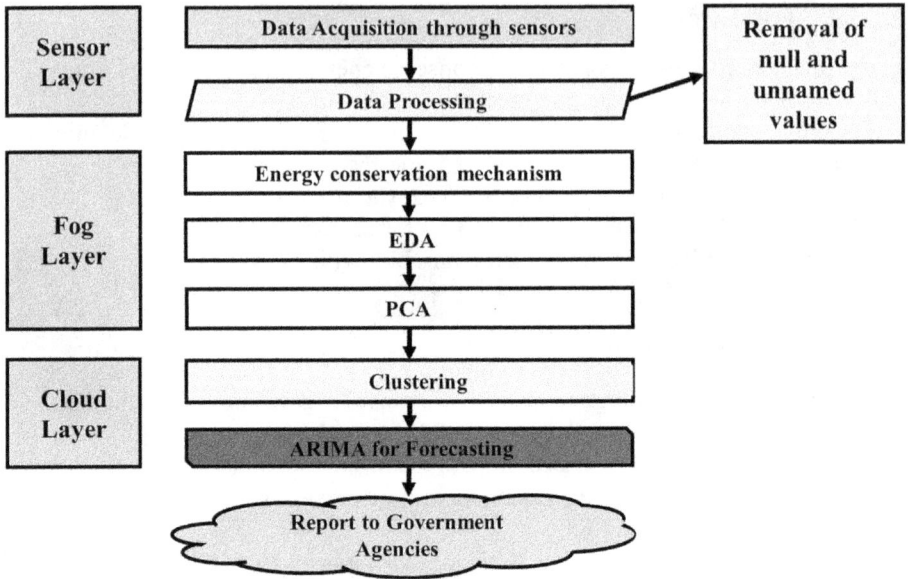

FIGURE 12.1   Flow of the Proposed Model.

Source: Khang (2021)

12.4   PERFORMANCE ANALYSIS AND RESULTS

The results obtained using the improved forest fire prediction system using a differential approach are presented in this section. All four classes are predicted using the proposed mechanism.

First, the results in terms of classification accuracy were elaborated. The classification accuracy was obtained using Equation 12.1:

$$\text{Clasifications}_{Acc} = \frac{\text{TrueP} + \text{TrueN}}{\text{TrueP} + \text{TrueN} + \text{FalseP} + \text{FalseN}} \tag{12.1}$$

TrueP indicates true-positive values, and TrueN indicates true-negative values. FalseP indicates false-positive values, and FalseN indicates false-negative values (Table 12.4).

The training dataset values were normalized between 0 and 1 to reduce the complexity of the operation. The visualization corresponding to the classification accuracy differs from the existing work without ARIMA by 5–6%, which is significant and worth studying.

The visualization results corresponding to the traffic prediction are shown in Figure 12.2.

TABLE 12.4

Classification Accuracy Result with Varying Dataset Sizes.

Dataset size	Classification accuracy (%) using forest fire prediction without ARIMA	Classification accuracy (%) using forest fire prediction with ARIMA
1000	85	95
2000	83	94.2
3000	82	94
4000	79	93.5
5000	78	93

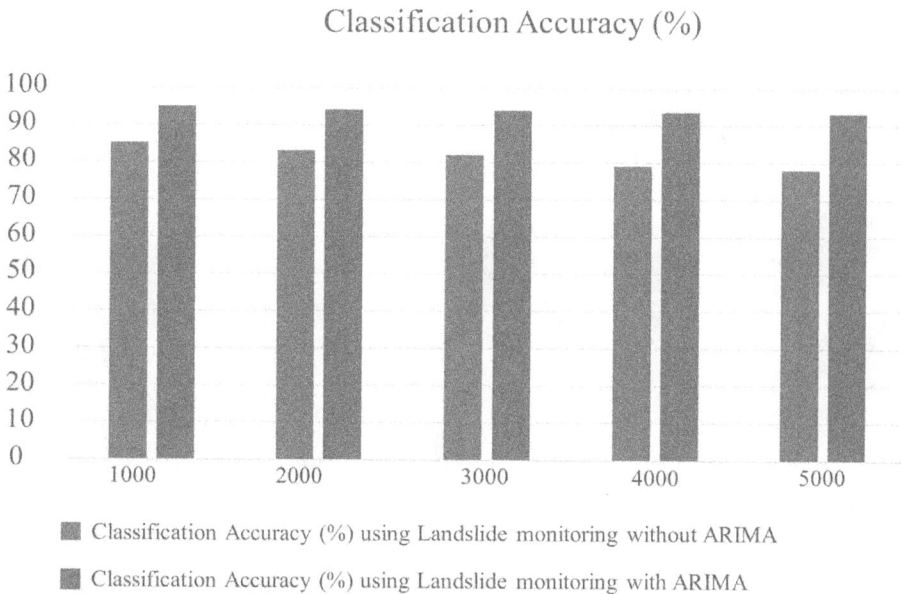

FIGURE 12.2 Visualization Result Corresponding to Classification Accuracy.

The result, in terms of sensitivity, is considered next. This metric indicates the percentage of correctly classified instances positively into any class. The sensitivity result is given by Equation 12.2:

$$Sensitivity = \frac{TrueP}{TrueP + FalseN} \tag{12.2}$$

The sensitivity results are listed in Table 12.5.

The visualization results corresponding to the sensitivity are shown in Figure 12.3.

TABLE 12.5
Sensitivity Results.

Dataset size	Sensitivity (%) using forest fire prediction without ARIMA	Sensitivity (%) using forest fire prediction with ARIMA
1000	72	75
2000	70	73.6
3000	65	73.2
4000	64	72
5000	63	71

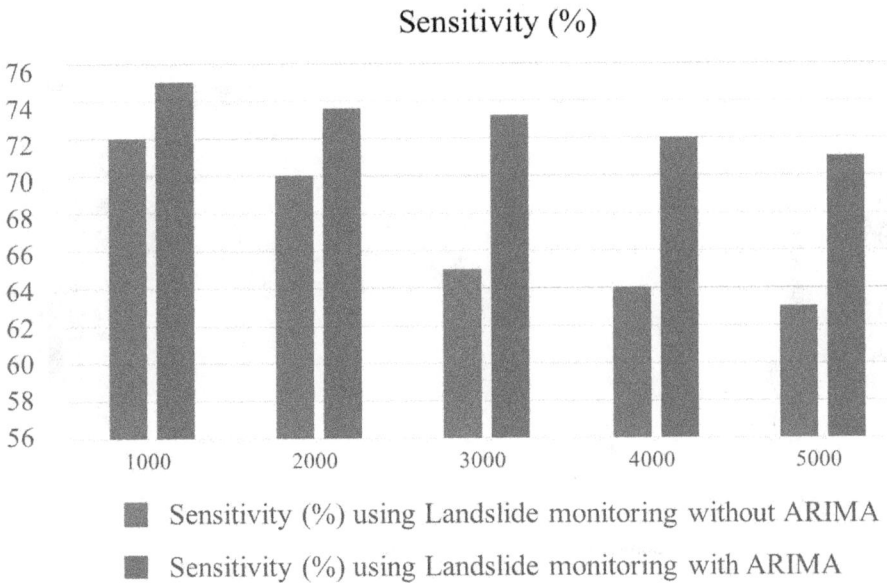

FIGURE 12.3   Sensitivity by Varying Dataset Size.

The last result is in the form of specificity, which is the result in terms of correctly negatively classified instances from the dataset. The specificity is given by Equation 12.3:

$$Specificity = \frac{TrueN}{TrueN + FalseP} \qquad (12.3)$$

The results corresponding to specificity are given by Table 12.6.

TABLE 12.6

Results of Specificity Corresponding to the Visualization Results Given in Figure 12.4.

Dataset size	Specificity (%) using forest fire prediction without ARIMA	Specificity (%) using forest fire prediction with ARIMA
1000	28	25
2000	30	27
3000	35	27
4000	36	28
5000	37	29

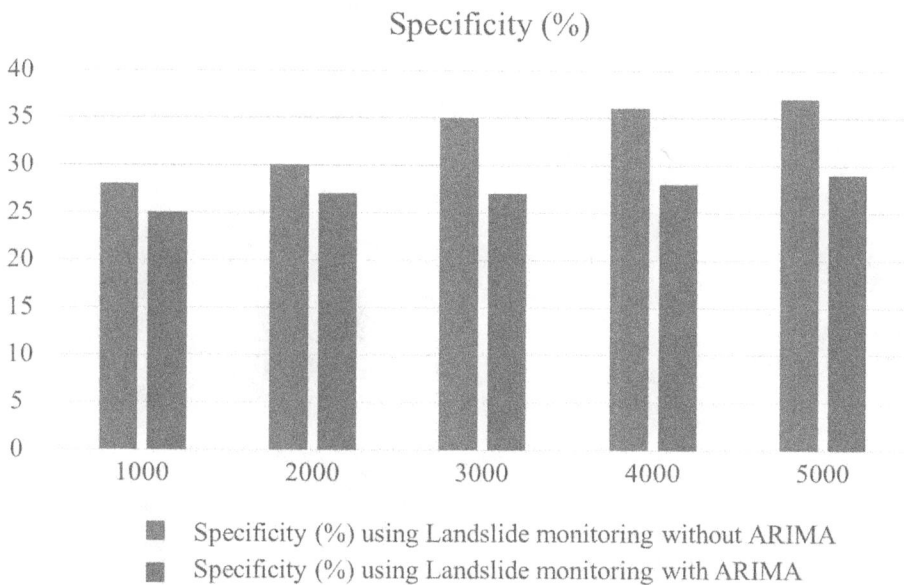

FIGURE 12.4   Specificity Results with and without ARIMA.

12.5   IMPLEMENTATION

The program was executed by clicking on the main form. The user must fill in the threshold value to check whether the temperature exceeds that limit, as shown in Figure 12.5.

If the limit exceeds this limit, an alarm is set up. This alarm is indicated with the help of a red light in the form shown in Figure 12.6.

In this case, the initial value of the threshold was taken as 25. After inputting the threshold value, the user must click on the activated sensor options (Hahanov et al., 2022).

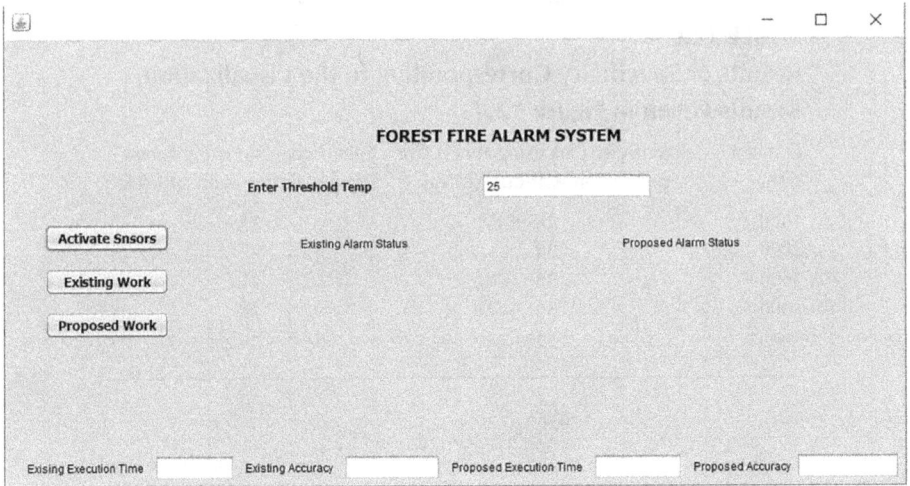

FIGURE 12.5   Starting Screen.

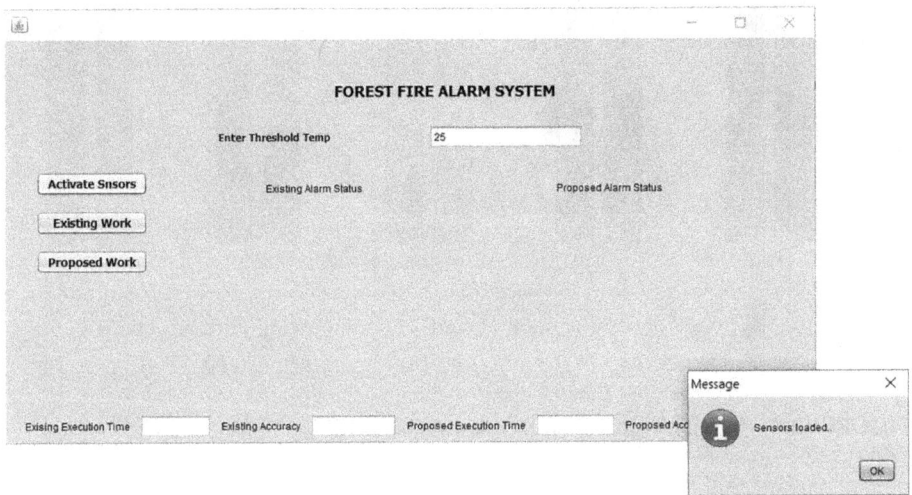

FIGURE 12.6   Screen with Threshold Values.

By clicking on the activated sensor option, the heat and smoke sensors are activated. This step is necessary to check the abnormal temperature conditions in the open environment, as shown in Figure 12.7.

If the read values are greater than the threshold, an alarm in the form of a red light will be generated. Otherwise, a green light is visible over the form shown in Figure 12.8. Clearly, the figure values exceed the threshold, indicating that fire is detected.

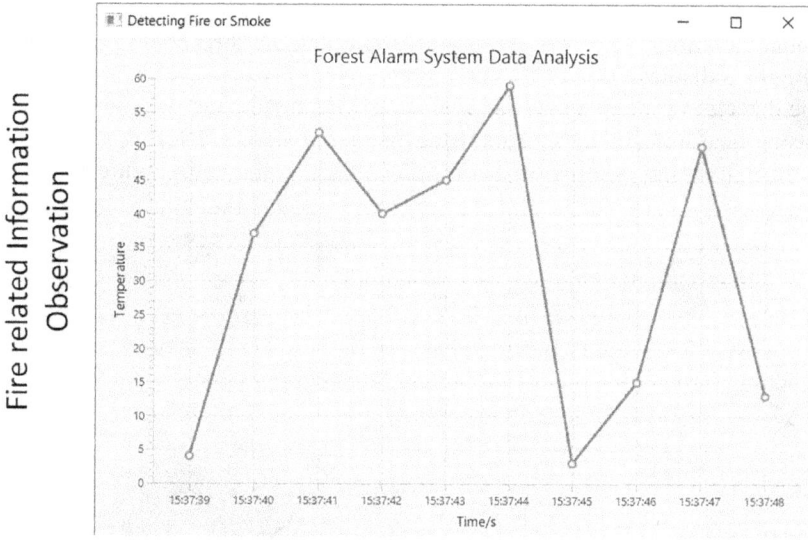

FIGURE 12.7 Fire-Related Information Observation.

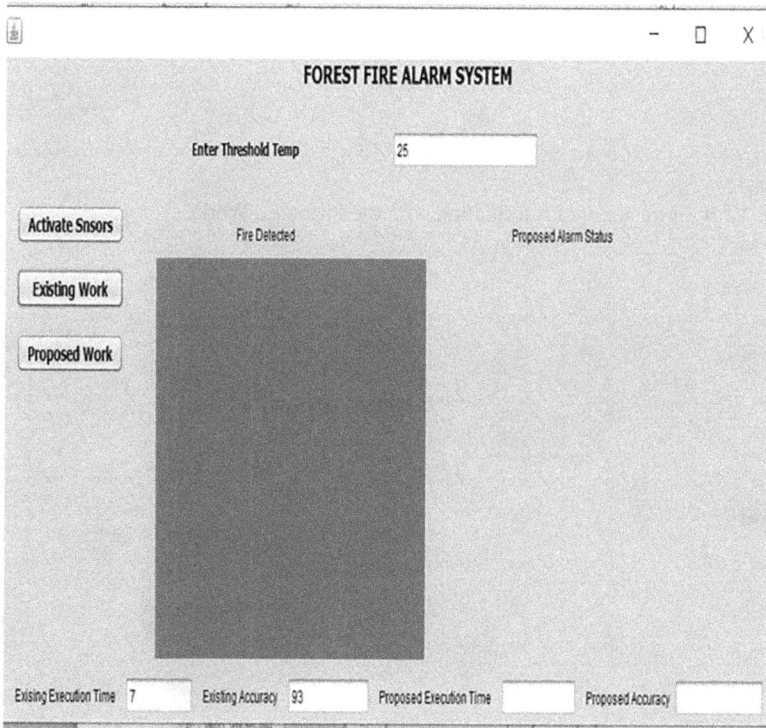

FIGURE 12.8 Results from the Existing Work.

The red bar on the form indicates that heat and smoke were detected because of threshold violations, as shown in Figure 12.9. When the threshold is not violated, a green light is displayed.

The detection process takes the input from GSON file and then performs detection using heat and smoke sensors using fuzzy C means clustering approach. The mechanism uses the existing threshold value of 25, as shown in Figure 12.10.

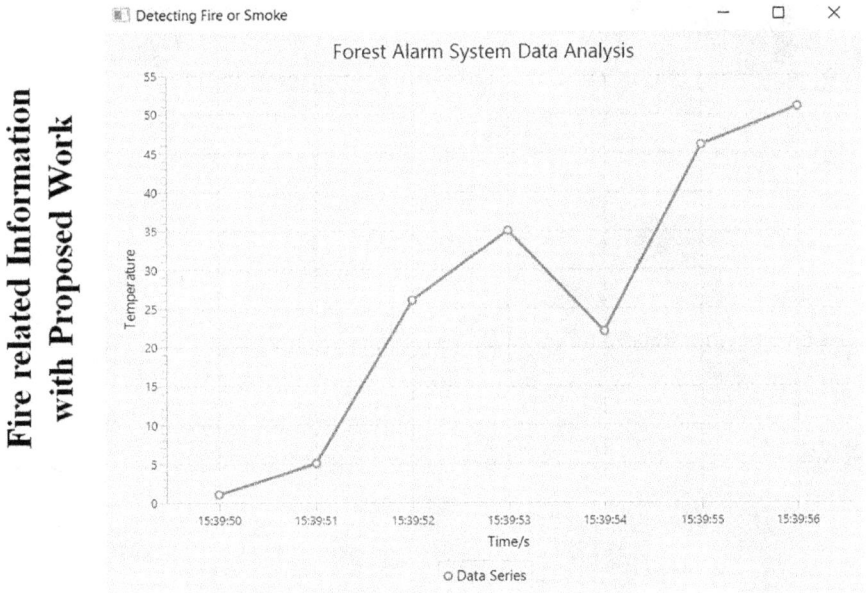

FIGURE 12.9   Fire-Related Information with the Proposed Work.

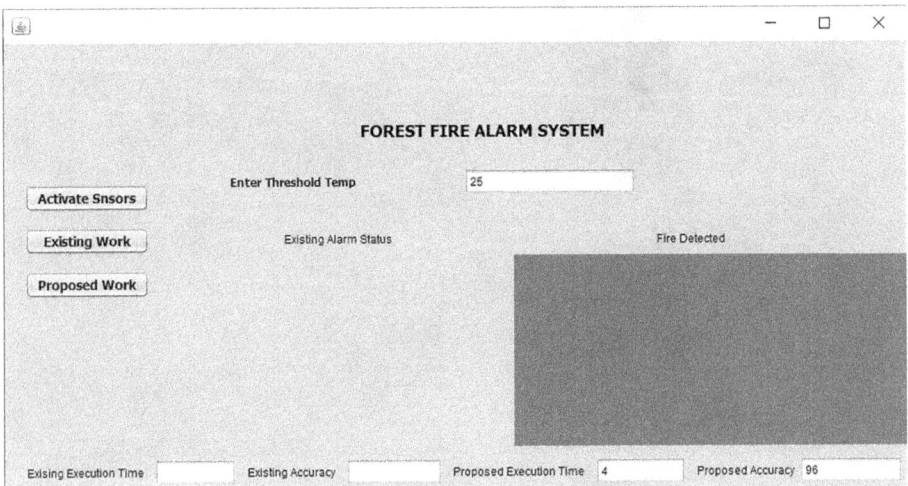

FIGURE 12.10   Results from the Proposed Work.

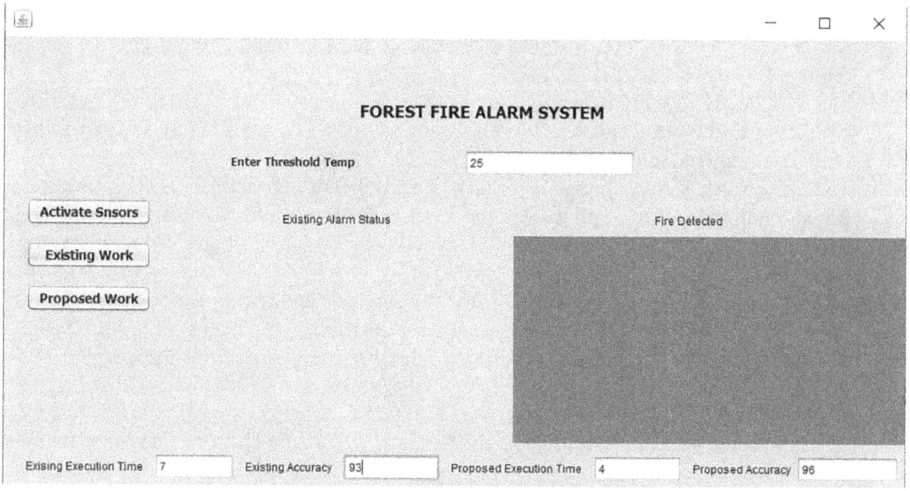

FIGURE 12.11 Comparison between Existing and Proposed Work.

The red bar on the form indicates that heat and smoke were detected because of threshold violations, as shown in Figure 12.11. When the threshold is not violated, a green light is displayed.

On the other hand, the existing approach produced an execution time of 7 ms and classification accuracy of 93%. The results of the proposed approach are certainly better than those of the existing approach.

12.6 CONCLUSION

This chapter presented the fog-based model for the prediction of forest fires. The dataset for forest fire prediction was derived from the benchmark website. The dataset pre-processing mechanism within the acquisition layer handles all the abnormalities and finalizes the classification accuracy stored within the cloud layer.

The data acquisition layer is fed into the fog layer. The fog layer contains the mechanism of energy conservation, which is achieved through a reduction mechanism using PCA. The exploratory data analysis mechanism reduces the size based on the correlation calculated using the PCA. The obtained result for the fog layer is entered within the cloud layer (Geetha et al., 2023).

The result from the cloud layer can be extracted by administrators with an account within the cloud layer. The result of the classification accuracy is in the range of 95%, which is almost 7% better than that of the existing model, which proves the worth of the study.

REFERENCES

Alam M., (2016). "Use of ICT in higher education," *The International Journal of Indian Psychology*, 3(4), 68, 162–171. www.ijip.in/Archive/v3i4/18.01.208.20160304.pdf

Amit S. N. K. B., Yoshimitsu A., (2017). "Disaster detection from aerial imagery with convolutional neural network," In *Proceedings—International Electronics Symposium* on Knowledge Creation and Intelligent Computing, IES-KCIC 2017 2017-January (December), pp. 239–245. Institute of Electrical and Electronics Engineers Inc. https://doi.org/10.1109/KCIC.2017.8228593

Anderson S. R., (2015). "Mobile technology in complex emergencies: A study of digital data collection in the Norwegian NGO relief sector," *MSc University of Stavanger*. https://core.ac.uk/download/pdf/52119484.pdf

Ayalew L., Hiromitsu Y., Norimitsu U., (2004). "Forest fire susceptibility mapping using GIS-based weighted linear combination, the case in Tsugawa area of Agano River, Niigata Prefecture, Japan," *Forest Fires*, Vol. 1, No. 1, pp. 73–81. Springer Verlag. https://doi.org/10.1007/s10346-003-0006-9

Azmoon B., Aynaz B., Zhen L., Ye S., (2021). "Image-data-driven temperature stability analysis for preventing forest fires using deep learning," *IEEE Access 9*, pp. 150623–150636. Institute of Electrical and Electronics Engineers Inc. https://doi.org/10.1109/ACCESS.2021.3123501

Barmpoutis P., Periklis P., Kosmas D., Nikos G., (2020). "A review on early forest fire detection systems using optical remote sensing," *Sensors*, 20(22), 6442. https://doi.org/10.3390/s20226442

Benzekri L., Gauthier Y., "Clinical markers of vitiligo activity," Am Acad Dermatol. 2017. 76(5):856–862. PMID: 28245942. doi: 10.1016/j.jaad.2016.12.040

Beraja Christian K. Wolf. (Sept 2021). *NBER Working Paper Series Demand Composition and the Strength of Recoveries*. Working Paper 29304 http://www.nber.org/papers/w29304

Bhambri P., Rani S., Gupta G., Khang A., (2022). *Cloud and Fog Computing Platforms for Internet of Things*. CRC Press. https://doi.org/10.1201/9781003213888

Bolourchi P., Uysal S., (2013). Forest fire detection in wireless sensor network using fuzzy logic. In *IEEE Fifth International Conference on Computational Intelligence, Communication Systems and Networks*, pp. 83–87. IEEE: Madrid. https://ieeexplore.ieee.org/abstract/document/ 6571347/

Bornmann L., Leydesdorff L., (2014). "Scientometrics in a changing research landscape," *EMBO Reports*, 15(12), 1228–1232. www.embopress.org/doi/full/10.15252/embr.201439608

Dai L., Mingcang Z., Zhanyong H., Yong H., Zezhong Z., Guoqing Z., Chao W., et al., (2021). "Forest fire risk classification based on ensemble machine learning," In *International Geoscience and Remote Sensing Symposium (IGARSS) 2021 July*, pp. 3924– 3927. Institute of Electrical and Electronics Engineers Inc. https://doi.org/10.1109/IGARSS47720.2021.9553034.

Dhar S., Rai A. K., Nayak P., (2017). "Estimation of seismic hazard in Odisha by remote sensing and GIS techniques," *Natural Hazards*, 86(72), 695–709. https://link.springer.com/article/10.1007/s11069-016-2712-3

ESCAP, (2015). *Building E-Resilience: Enhancing the Role of ICTs for Disaster Risk Management (DRM)*. https://repository.unescap.org/handle/20.500.12870/1132

Geetha C., Neduncheliyan S., Khang A., (Eds.). (2023). "Dual access control for cloud based data storage and sharing," *Smart Cities: IoT Technologies, Big Data Solutions, Cloud Platforms, and Cybersecurity Techniques*. CRC Press. https://doi.org/10.1201/9781003376064-17

Giuntini F. T., Beder D. M., Ueyama J., (2017). "Exploiting self-organization and fault tolerance in wireless sensor networks: A case study on wildfire detection application," *International Journal of Distributed Sensor Networks*, 13. https://journals.sagepub.com/doi/pdf/10.1177/1550147717704120

Guha-Sapir D., Hoyois P., Below R., (2016). "Annual disaster statistical review 2015: The numbers and trends," *The Centre for Research on the Epidemiology of Disasters*. http://lib.riskreductionafrica.org/bitstream/handle/123456789/1010/annual%20disaster%20statistical%20review%202008.pdf?sequence=1

Hahanov V., Khang A., Litvinova E., Chumachenko S., Hajimahmud V. A., Alyar V. A., (2022). "The key assistant of smart city—sensors and tools," In *AI-Centric Smart City Ecosystems: Technologies, Design and Implementation*, Vol. 17, No. 10 (1st ed.). CRC Press. https://doi.org/10.1201/9781003252542-17

Hartomo K. D., Sri Y., Joko M., (2017). "Spatial model design of forest fire vulnerability early detection with exponential smoothing method using google API," In *Proceedings—2017 International Conference on Soft Computing, Intelligent System and Information Technology: Building Intelligence through IoT and Big Data, ICSIIT 2017 2018-January (July)*. Institute of Electrical and Electronics Engineers. www.sciencedirect.com/science/article/pii/S1364815212002460

Hibino J., Shaw R., (2014). "Role of community radio in post disaster recovery: comparative analysis of Japan and Indonesia," *Disaster Recovery: Used or Misused Development Opportunity*, 385–410. https://doi.org/10.1007/978-4-431-54255-1_20

IFRC, (2013). *World disaster report-focus on technology and the future of humanitarian organization*. International Federation of Red Cross and Red Crescent Societies (*International Committee of Red Cross*). www.ifrc.org/PageFiles/134658/WDR%202013%20complete.pdf

Jan M. A., Nanda P., He X., Liu R. P., (2018) "A sybil attack detection scheme for a forest wildfire monitoring application," *Future Generation Computer Systems*, 80, 613–626. www.sciencedirect.com/science/article/pii/S0167739X16301522

Jana N. C., Singh R. B., (May 18, 2022). *Climate, Environment and Disaster in Developing Countries*, p. 536. https://link.springer.com/content/pdf/ 10.1007/978-981-16-6966-8.pdf.

Jha R. K., Tewari A., Shrivastava P., (2015). "Performance analysis of disaster management using WSN technology," *Procedia Computer Science*, 49. www.sciencedirect.com/science/article/pii/S1877050915007498

Johnson K. A., Busdieker-Jesse N., McClain W. E., Lancaster P. A., (2019). "Feeding strategies and shade type for growing cattle grazing endophyte-infected tall fescue," *Livestock Science*, 230, 103829. https://doi.org/10.1016/j.livsci.2019.103829

Juyal A., Sachin S., (2021). "A study of forest fire susceptibility mapping using machine learning approach," In *Proceedings of the 3rd International Conference on Intelligent Communication Technologies and Virtual Mobile Networks, ICICV 2021, February*. Institute of Electrical and Electronics Engineers. https://doi.org/10.1201/9781003145011

Kabenge M., Elaru J., Wang H., Li F., (2017). "Characterizing flood hazard risk in data-scarce areas, using a remote sensing and GIS-based flood hazard index," *Natural Hazards*, 89(3), 1369–1387. https://link.springer.com/article/10.1007/s11069-017-3024-y

Kang et al., (23, Jan 2017). "Microbiota transfer therapy alters gut ecosystem and improves gastrointestinal and autism symptoms: an open-label study," *Microbiome*, 5(1), 10. https://doi.org/10.1186/s40168-016-0225-7

Kansal A., Singh Y., Kumar N., Mohindru V., (2015). "Detection of forest fires using machine learning technique: a perspective," In *IEEE Third International Conference on Image Information Processing*, pp. 241–245. IEEE. https://ieeexplore.ieee.org/abstract/document/7414773/

Kaur H., Sood S. K., Bhatia M., (2019). "Cloud-assisted green IoT-enabled comprehensive framework for wildfire monitoring," *Cluster Computing*, 23(2), 1149–1162. https://doi.org/10.1007/S10586-019-02981-7.

Khang A., (2021). "Material4Studies," *Material of Computer Science, Artificial Intelligence, Data Science, IoT, Blockchain, Cloud, Metaverse, Cybersecurity for Studies*. www.researchgate.net/publication/370156102_Material4Studies

Khang A., Ragimova N. A., Hajimahmud V. A., Alyar V. A., (2022). "Advanced technologies and data management in the smart healthcare system," In *AI-Centric Smart City Ecosystems: Technologies, Design and Implementation*, Vol. 16, No. 10 (1st ed.). CRC Press. https://doi.org/10.1201/9781003252542-16

Khang A., Rana G., Tailor R. K., Hajimahmud V. A., (Eds.). (2023a). *Data-Centric AI Solutions and Emerging Technologies in the Healthcare Ecosystem*. CRC Press. https://doi.org/10.1201/9781003356189

Khang A., Hahanov V., Litvinova E., Chumachenko S., Triwiyanto C., Hajimahmud V. A., Nazila A. R., Abuzarova Vusala A. R., Anh P. T. N., (2023b). "The analytics of hospitality of hospitals in healthcare ecosystem," In *Data-Centric AI Solutions and Emerging Technologies in the Healthcare Ecosystem*, p. 4. CRC Press. https://doi.org/10.1201/9781003356189-4

Khanh H. H., Khang A., (2021). "The role of artificial intelligence in blockchain applications," *Reinventing Manufacturing and Business Processes through Artificial Intelligence*, 2(20–40) (CRC Press). https://doi.org/10.1201/9781003145011-2

Komac M., (2006). "A forest fire susceptibility model using the analytical hierarchy process method and multivariate statistics in Perialpine Slovenia," *Geomorphology*, 74(1–4), 17–28. https://doi.org/10.1016/j.geomorph.2005.07.005

Lee S., (2005). "Application of logistic regression model and its validation for forest fire susceptibility mapping using GIS and remote sensing data," *The International Journal of Remote Sensing*, 26(7), 1477–1491 (Taylor and Francis Ltd.). https://doi.org/10.1080/01431160412331331012

Liang X., Ho M. C. W., Zhang Y., Li Y., Wu M. N., Holy T. E., Taghert P. H. (2019). "Morning and evening circadian pacemakers independently drive premotor centers via a specific dopamine relay," *Neuron*, 102(4), 843–857.e4. http://dx.doi.org/10.1016/j.neuron.2019.03.028

Loomba R. S., Anderson R. H., (2018). "Are we allowing impact factor to have too much impact: The need to reassess the process of academic advancement in pediatric cardiology?," *Congenital Heart Disease*, 13(2), 1–4. https://onlinelibrary.wiley.com/doi/abs/10.1111/chd.12593

Marjanović M., Miloš K., Branislav B., Vít V., (2011). "Forest fire susceptibility assessment using SVM machine learning algorithm," *Engineering Geology*, 123(3), 225–234. https://doi.org/10.1016/j.enggeo.2011.09.006

Mitchell G., (2013). *Selecting the Best Theory to Implement Planned Change*. PMID: 23705547 https://doi.org/10.7748/nm2013.04.20.1.32.e1013

Mobin M., Seyed M. M., Mohammad K., Madjid T., (Jan 2018). "A hybrid desirability function approach for tuning parameters in evolutionary optimization algorithms," *Measurement*. 114, 417–427. https://doi.org/10.1016/j.measurement.2017.10.009

Monacelli G., Galluccio M. C., Abbafati M., (2005). "Drought assessment and forecasting," *Drought within the Context of the Region VI*. http://danida.vnu.edu.vn/cpis/files/Refs/Drought/DROUGHT%20ASSESSMENT%20AND%20FORECASTING.pdf

Mouradian C., Naboulsi D., Yangui S., Glitho R. H., Morrow M. J., Polakos P. A., (2017). "A comprehensive survey on fog computing: State-of-the-art and research challenges," *IEEE Communications Surveys and Tutorials*, 20(1), 416–464. https://ieeexplore.ieee.org/abstract/document/8100873/

Rajasekaran Sathish M. D., Kylie Kvinlaug M. D., Jonathan T. Finnoff. (13 Dec, 2012). *Exertional Leg Pain in the Athlete*. https://doi.org/10.1016/j.pmrj.2012.10.002

Rani S., Bhambri P., Kataria A., Khang A., Sivaraman A. K., (2023). *Big Data, Cloud Computing and IoT: Tools and Applications* (1st ed.). Chapman and Hall/CRC. https://doi.org/10.1201/9781003298335

Rani S., Chauhan M., Kataria A., Khang A., (Eds.). (2021). "IoT equipped intelligent distributed framework for smart healthcare systems," In *Networking and Internet Architecture*, Vol. 2, p. 30. CRC Press. https://doi.org/10.48550/arXiv.2110.04997

Rau J. Y., Jyun P. J., Ruey J. R., (2013). "Semiautomatic object-oriented forest fire recognition scheme from multisensor optical imagery and dem," *IEEE Transactions on Geoscience and Remote Sensing*, 52(2), 1336–1349. https://doi.org/10.1109/tgrs.2013.2250293

Reddy P. N. N., Basarkod P. I., Manvi S. S., (2011). "Wireless sensor network based fire monitoring and extinguishing system in real time environment," *IEEE*. www.researchgate.

net/profile/Prabhugoud-Basarkod/publication/268420619_Wireless_Sensor_Network_ based_Fire_Monitoring_and_Extinguishing_System_in_Real_Time_Environment/ links/558d3b4f08aee43bf6aec16b/Wireless-Sensor-Network-based-Fire-Monitoring-and-Extinguishing-System-in-Real-Time-Environment.pdf

Rhodes et al., (2020). Cell Reports 30, Cohesin Disrupts Polycomb-Dependent Chromosome Interactions in Embryonic Stem Cells. VOLUME 30, ISSUE 3 820–835. TheAuthor(s). https://doi.org/10.1016/j.celrep.2019.12.057

Saoudi M., Bounceur A., Euler R., Kechadi T., (2016). "Data mining techniques applied to wireless sensor networks for early forest fire detection," In *Proceeding International Conference Internet of Things and Cloud Computing*, pp. 1–7. ACM. https://dl.acm.org/ doi/abs/10.1145/2896387.2900323

Sarwar M. I., Muhammad M., (2022). *Vulnerability and Exposures to Forest Fires in the Chittagong Hill Region, Bangladesh: A Case Study of Rangamati Town for Building Resilience*, pp. 391–399. Springer. https://doi.org/10.1007/978-981-16-6966-8_21

Shinde S. et al., (2017). *Correlation of Third Molar Calcification with Retromolar Space.* https://www.academia.edu/44120553/Shinde_S_et_al_2017_Correlation_of_third_ molar_calcification_with_Retromolar_space

Sinha A., Kumar P., Rana N. P., Islam R., Dwivedi Y. K., (2017). "Impact of Internet of Things (IoT) in disaster management: A task-technology fit perspective," *Annals of Operations Research*, pp. 1–36. https://link.springer.com/article/10.1007/s10479-017-2658-1

Sun Q., Zhang L., Ding X. L., Hu J., Li Z. W., Zhu J. J., (2015). "Temperature deformation prior to Zhouqu, China forest fire from InSAR time series analysis," *Remote Sensing of Environment*, 156, 45–57 (Elsevier Inc.). https://doi.org/10.1016/j.rse.2014.09.029

Thein T. L. L., Myint M. S., Murata K. T., Tungpimolrut K., (2020). "Real-time monitoring and early warning system for forest fire preventing in Myanmar," In *2020 IEEE 9th Global Conference on Consumer Electronics, GCCE 2020, October*, pp. 303–304. Institute of Electrical and Electronics Engineers Inc. https://ieeexplore.ieee.org/abstract/ document/9291809/

Trivedi K., Srivastava A. K., (2014). "An energy efficient framework for detection and monitoring of forest fire using mobile agent in wireless sensor networks," In *IEEE International Conference Computational Intelligence and Computing Research*, pp. 1–4. IEEE. https://ieeexplore.ieee.org/abstract/document/7238433/

Vinck P., et al., (2013). *World Disasters Report 2013: Humanitarian Technology. International Federation of Red Cross and Red Crescent Societies (IFRC).* http://worlddisastersreport. org/en/

Yoon I., Noh D. K., Lee D., Teguh R., Honma T., Shin H., (2012). "Reliable wildfire monitoring with sparsely deployed wireless sensor networks," In *IEEE 26th International ConferenceAdvanced Information Networking and Applications*, pp. 460–466. IEEE. https:// ieeexplore.ieee.org/abstract/document/6184906/

Zhao J., Liu Y., Cheng Y., Qiang Y., Zhang X., (2014). "Multisensory data fusion for wildfire warning," In *IEEE 10th International Conference Mobile Ad-hoc and Sensor Networks*, pp. 46–53. IEEE. https://ieeexplore.ieee.org/abstract/document/7051749/

13 Hybrid Particle Swarm Optimization with Random Forest Algorithm Used in Job Scheduling to Improve Business and Production

Vanitha R., Nidhya M. S., Surrya Prakash Dillibabu, and Harishchander Anandaram

13.1 INTRODUCTION

A common and significant study area in cloud and grid computing is the job scheduling system. It serves similar purposes in grid and cloud computing. The optimal resources in a cloud or grid to select for critical care unit (CCU) jobs must be considered by the job scheduling mechanism. In addition, some static and dynamic parameter limits of the CCU workloads are considered.

The majority of research on grid computing is immediately transferable to cloud computing. Currently, grid computing work scheduling is the subject of extensive research. Centralized and decentralized schedulers are used in the cloud or grid.

Because of the complexity of the decentralized scheduler's implementation, the majority of related studies have focused on centralized schedulers (Santoro et al., 2019). The source provides a succinct overview of a modeling technique, as shown in Figure 13.1.

As we can see from references, for cloud computing services with a dynamic scheduling system, a method of Quality of Service (QoS) performance analysis has been proposed (Khanh & Khang, 2021).

In recent years, an increasing number of academics have started to examine the QoS of work scheduling systems (Dash et al., 2019). However, the majority of research articles seldom refer to the cloud computing environment's differential service-oriented QoS-assured job scheduling mechanism.

DOI: 10.1201/9781003400110-13

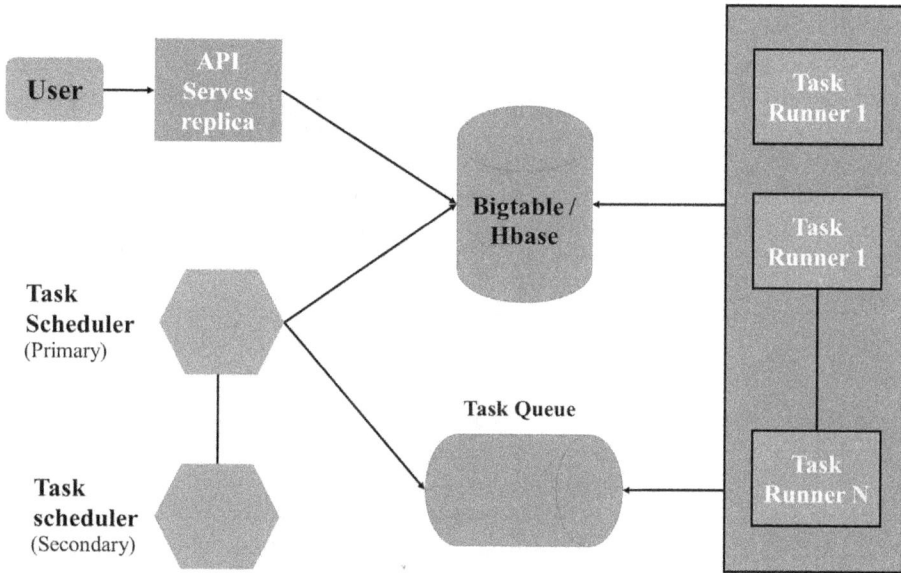

FIGURE 13.1 Distributed Job Scheduler.

Source: Khang (2021)

13.1.1 JOB SCHEDULING

Job management is key concept in cloud computing systems. Problems with task scheduling are significant because they affect the overall effectiveness of the cloud computing system (Rani et al., 2023).

Job scheduling is a technique that links user tasks to the choice and utilization of relevant resources. It is convenient and flexible for scheduling jobs.

Depending on the operational requirements, needs, and priorities, jobs and job streams can be scheduled to run at any time. Without the requirement of support workers, job streams and procedures can be scheduled daily, weekly, monthly, and yearly in advance and executed as needed.

13.1.2 DATA CLEANING

Data cleaning is the process of deleting inaccurate, flawed, poorly organized, duplicate, or insufficient data from a dataset.

Regardless of whether local or international web-based information systems are used, data cleaning has become increasingly crucial. This is because sources typically include redundant information in various formats (Borner et al., 2019). To provide access to correct and consistent data, it is necessary to combine various data representations and to remove redundant data.

13.1.3 MACHINE LEARNING

Modern inventions, such as machine learning, have enhanced our daily lives as well as a number of corporate and professional tasks. Statistical approaches are used to develop intelligent computer systems that are capable of learning from existing databases. This is a subset of artificial intelligence.

Big data warehouses are used to analyze and extract information from large volumes of data (Chan et al., 2020). The ability of a machine to automatically learn from experience and advance without being specially programmed is referred to as machine learning.

13.1.4 BIG DATA ANALYSIS

Businesses can use big data analytics to make use of their data and uncover their unrealized potential. Thus, decisions made by the business are made more carefully, operations are successfully conducted, earnings rise, and clients are satisfied.

Big data and advanced analytics are valuable to businesses in several ways, including saving cash. It helps with business process optimization, which leads to cost savings, increased productivity, and improved customer satisfaction (Gunasekaran et al., 2018). It is possible to manage hiring and HR better.

13.1.5 PARTICLE SWARM OPTIMIZATION

A straightforward bio-inspired approach for identifying the best solution in an available space is called particle swarm optimization (PSO). It varies from other approaches to optimization in that it only needs the objective function and does not need the gradient or any other differential form of the objective. Moreover, there are few hyper-parameters (Zhu et al., 2020).

13.1.6 RANDOM FOREST

A potent supervised machine learning method for classification and regression issues is the random forest algorithm. We are aware that a forest is made up of many different types of trees, and the more trees there are, the more robust the forest is (Rana et al., 2021).

Data scientists employ random forests in various fields, including banking, stock trading, medicine, and e-commerce, in their professional jobs. It is used to predict factors such as patient history, consumer behavior, and safety that help different businesses run smoothly (He et al., 2022), as shown in Figure 13.2.

All information relevant to a business and its operations is referred to as business data. Any statistical information, uncooked analytical information, consumer feedback information, sales figures, or other information could be included.

Data are crucial to businesses. Good data gathering and analysis allows one to make the right decisions quickly (Balyan et al., 2022). Data analysis can offer insights that can assist in deciding what is best for your business when deciding where to market next, how to price your goods, or how to optimize a certain advertising campaign.

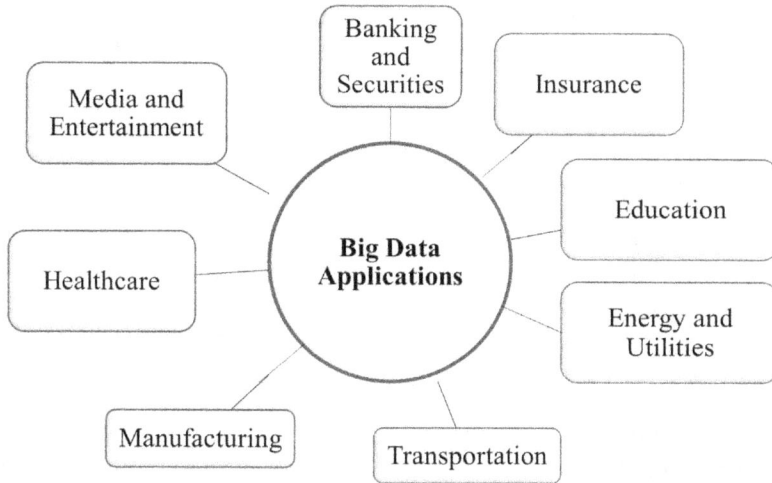

FIGURE 13.2   Real-Time Big Data Applications.

Source: Khang (2021)

The remainder of this chapter is organized into five parts. Section 1.2 explains the present method and its drawbacks of job scheduling in business and production. Section 1.3 explains the proposed techniques for assigning jobs using hybrid PSO techniques. Section 1.4 presents the experimental results and their graph analysis, and section 1.5 provides a conclusion.

13.2   LITERATURE REVIEW

Linear programming (LP) was utilized by Abdel-Basset et al. (2019). It is a mathematical technique for determining how to obtain the optimal result, for example, the model's highest profit or lowest cost. A particular instance of mathematical optimization is LP.

Dynamic programming (DP) was used by Guo et al. (2019) as an implicit enumerative search technique. In DP, a complex problem is deconstructed into multiple smaller, independent subproblems before attempting to solve it. Every minor subproblem is solved, and the solutions are saved for later use.

Chan et al. (2020) presented a DP method to the lot scheduling problem. On a regular basis, they considered the timing of production of multiple items inside the same facility. The infrastructure of the system allows the production and processing of only one item at once. Setup time and cost are related to the item produced. Every item's demand rate is known and remains the same over an endless planning horizon. All requests must be fulfilled.

Anna et al. (2018) created the branch-and-bound algorithms for the makespan minimization problem with the permutation flow shop.

Tanaka and Tierney (2018) used the branch-and-bound method to schedule jobs with sequence-dependent setup times on a single processor to reduce the overall number of tool adjustments.

This approach has limited computational flexibility and is only appropriate for small-scale applications. Deng et al. (2019) examined two distinct flow-shop scheduling issues. These occur when one machine in a two-machine situation exhibits sequence-dependent setup time.

The goal was to find a schedule to reduce the makespan. An effective DP formulation was created for each of these problems after the best permutation schedules have been determined. Dai et al. (2019) developed genetic algorithm-based algorithms that draw inspiration from Darwin's idea of natural selection and survival.

According to Rius-Sorolla et al. (2020), formulating scheduling issues as mathematical programming models is a natural approach to solving them. The outcomes were unsatisfactory because many of these strategies could not produce workable answers to issues. As a result, the practical utility of these strategies is limited.

Tomazella and Nagano (2020) reduced the cost of machine configuration, Istatic scheduling issues with m machines, and resolved n jobs mathematically using the branch-and-bound technique. They hypothesized that the setup times were dependent on the order of events. The processing times were not considered.

Only small issues were solved using this technique. This provided insight into the composition of ideal solutions. They can be applied to create effective heuristic rules. Heuristic rules are typically more suitable for solving large-scale scheduling problems when the computational work for optimizing procedures increases rapidly with problem size.

The adaptive heuristic method known as artificial immune system (AIS) is influenced by the human immune system. The two key components of the immune response mechanism are as follows:

1) Clonal selection principle
2) Affinity maturation.

Silva et al. (2020) employed a heuristic for hybrid flow shops based on AIS. Houben and Van de Poel (2019) reduced the overall flow time in permutation flow shops by using two ant colony optimization algorithms. An initial seed sequence was produced using these algorithms. At the conclusion of each iteration, a local search was conducted.

In a flow shop scheduling setting, an artificial neural network (ANN) was utilized to minimize the dual criterion of the makespan and total flow time. They demonstrated that the ANN produces a solution quality that is either comparable to or superior to that of classical heuristics. They concluded that as the number of training exemplars increased, the ANN gradually increased the quality of the solutions.

According to Iqbal et al. (2020), big data have a huge impact on creating useful smart cities and sustaining contemporary societies. Businesses use big data to enhance operations, develop tailored marketing campaigns, and perform a variety of other duties that could ultimately boost revenue and profitability (Khang et al., 2022a).

Wang et al. (2020) developed paths for information business that may be used to maximize the value of data by data mining, analysis, and sharing in the enormous data set. This resulted in significant economic benefits. Meanwhile, it can provide strategies for making decisions regarding social and economic progress.

Data are used as a resource by a new service economic model known as the big data service architecture, which loads and extracts the data received from multiple data sources. This service architecture provides a range of individualized data processing, data analysis, and visualization services to service consumers (Bhambri et al., 2022).

Modern operations management (OM) requires big data analytics. Current big data-related analytics methodologies were examined by Choi et al. (2018), who also noted their main features, shortcomings, and strengths.

Various big data analytics approaches have been examined to address the corresponding computational and data challenges. Big data analytics has practical applications in well-known companies using case studies.

13.3   SYSTEM DESIGN

This section discusses the proposed system job scheduling for business and production using particle swarm optimization (PSO) and random forest (RF).

13.3.1   PARTICLE SWARM OPTIMIZATION FOR DATA IN FEATURE SELECTION

The ideal application of PSO is to identify the maximum or minimum value of a function in a multidimensional vector space. Ant colonies, schooling fish, flocking birds, bee swarming, and other natural phenomena are examples of swarm intelligence. Particle swarm optimization is a stochastic optimization technique based on the movement and behavior of swarms.

Particle swarm optimization uses social interaction to resolve disputes. It uses a swarm of particles (called agents) that roam the search space to search for an optimum solution.

The PSO-based feature selection of data in the business and production fields is shown in Figure 13.3. Particle swarm optimization is an effective and efficient global search technique. Owing to its superior representation, it is a suitable method for dealing with feature selection issues because it has the ability to explore large spaces, is less computationally expensive, is easier to create, and requires fewer parameters.

13.3.2   TRAINING DATA

The network security lab, knowledge discovery in databases cup dataset contains 30,000 random samples of network records. A total of 19,200 normal data points (64%) and 10,800 intrusive ones (36%) made up the chosen dataset. The features of this dataset were then projected into the primary space to determine their importance to the classifier.

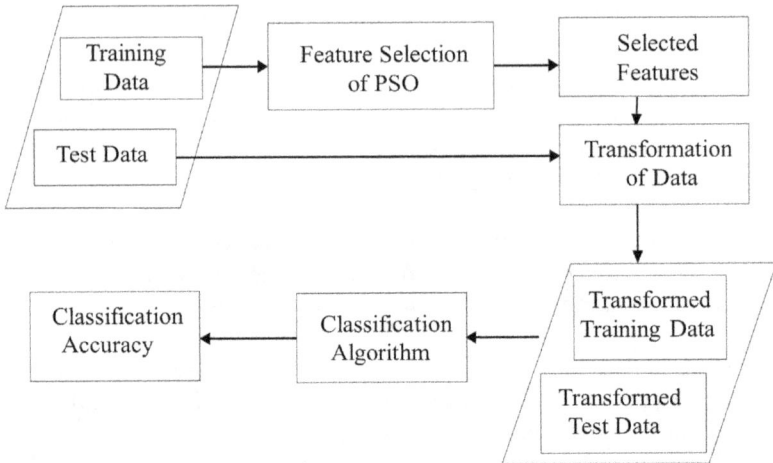

FIGURE 13.3 Feature Selection of Data Using PSO.

The performance of the PSO network was examined using the testing dataset. At this point, the weights of the neural network were frozen; 30% (2250) of the dataset (7500) was employed in this study to test the performance of the trained PSO.

Another training process method is cross-validation. This method stops training when the error starts to increase, and it regulates the error in a separate batch of data. For cross-validation, a dataset size of 20% for normal generalization and 40% for strong generalization is advised. In addition, 1500 datasets were employed in this cross-validation study.

13.3.3 Testing Data

The weights of the PSO are frozen after the training procedure is completed, and its performance is evaluated. There are two processes to test the PSO: verification and generalization steps.

13.3.4 Verification Step

For the training dataset, the goal of the verification step is to evaluate the learning capacity of the network. If a network has been trained well, its output will be comparable to the actual output. In this study, 1500 data points, or 30% of the training dataset (7500), were employed.

13.3.5 Particle Swarm Optimization Algorithm

A swarm of particles is used by PSO to find updates from iteration to iteration. Each particle in the swarm gets closer to its prior best position (P_{best}) and the overall best position of the swarm to obtain the ideal response (G_{best}), as shown in Equations 13.1 and 13.2:

$$P_{best} (i,t) = \text{argmin } [f(P_i(K))], k = 1, \ldots, t \qquad (13.1)$$

$$G_{best} (t) = \text{argmin } [f(P_i(K))], i = 1 \ldots N_p \qquad (13.2)$$

where i stands for the particle index, and N_p is the overall particle count. Updates to particle position P and velocity V are calculated using Equations 13.3 and 13.4:

$$V_i (t + 1) = wV_i (t) + c1r1 (P_{best}(i, t) - P_i (t)) + c2r2(G_{best}(t) - P_i (t)) \qquad (13.3)$$

$$P_i (t + 1) = P_i (t) + V_i (t + 1) \qquad (13.4)$$

where w is the inertia weight, V is the speed, $r1$ and $r2$ are random variables with uniform distributions in the range [0, 1], and $c1$ and $c2$ are "acceleration coefficients."
The enhanced PSO pseudocode is described as follows:

- **Step 1**. Enter the raw material and the processing time data for the batik production process, then initialize the PSO settings.
- **Step 2**. Execute the program, and save it with the highest fitness value (G_{best}).
- **Step 3**. Update the velocity particles.
- **Step 4**. Update the particle's location.
- **Step 5**. Slightly reduce the inertia weight () from 0.9 to 0.4.
- **Step 6**. Verify whether the position of the particle has been violated: if pos(j) > mp, then pos(j) = mp. If the particle position pos(j) = 1, otherwise.
- **Step 7**. If the velocity of particle vel(j) > mv, then vel(j) = mv, as a test for particle velocity violation. Alternatively, if the particle's velocity is vel(j)—mv, then its pos(j) is equal to −mv.
- **Step 8**. Linearly reduce the weight of inertia () from 0.9 to 0.4. Until a criterion is attained, repeat steps 2–8.

The hybrid PSO technique is described in Code block 13.1.
Code block 13.1. Algorithm for hybrid PSO technique.

```
G_best: The best position is found so far
1. start
2. Initialize the swarm randomly;
3. for i =1 to N do
4. v_i = a random vector within [LB, UB]^D;
5. v_i = a random vector within [LB, UB]^D;
6. P_best i = X_i;
7. end for
8. to find G_best;
9. t = 1;
10. while t ≤ T do
11. for i = 1 to N do
12. r1, r2 ← two independent vectors randomly generated from [0,1]^D;
```

13. **if** $f(X_i^t) < f(P\dfrac{t-1}{best_i})$ **then,**

14. $f(P\dfrac{t-1}{best_i}) = f(X_i^t)$

15. **end if**

16. **end for**

17. **to find** $g\dfrac{t}{best}$;

18. $t = t + 1;$

19. **end while**

20. **end**

13.3.6 RF IN CLASSIFICATION ALGORITHM

Three key hyper-parameters must be determined before training random forest algorithms. The random forest classifier may then be utilized to address classification or regression issues, for instance, identifying an email as "spam" or "not spam."

The steps of the method are as follows:

1. Choose a random sampling of the data or training set.
2. Each training dataset is used to build a decision tree using this technique.
3. The choice tree is averaged during voting.
4. The prediction outcome that garnered the greatest support is chosen as the final prediction outcome.

Perform the following steps to implement a random forest:

1. Separate the training and testing groups in the data.
2. Include a column in the data that, for instance, indicates 0 for all rows in the training group and 1 for all rows in the testing group.
3. Combine the two groups once more to create a new dataset, and use the new column as the target variable of the random forest model.
4. Build a model using random forest, as shown in Figure 13.4.

Random forests are used in the workplace by data scientists in a range of industries, including finance, stock trading, healthcare, and online shopping. They are used to forecast factors, such as patient history, consumer behavior, and safety, which are crucial for the effective operation of many industries.

Random forests can handle categorical, continuous, and binary data. Generally, random forest is a rapid, simple, flexible, and dependable model with some limitations. The random forest approach enhances the performance of a model by combining different classifiers as a part of ensemble learning.

Compared with other algorithms, random forest requires less training time. In addition, it accurately predicts results, and even with a vast dataset, it functions

Random Forest Classifier

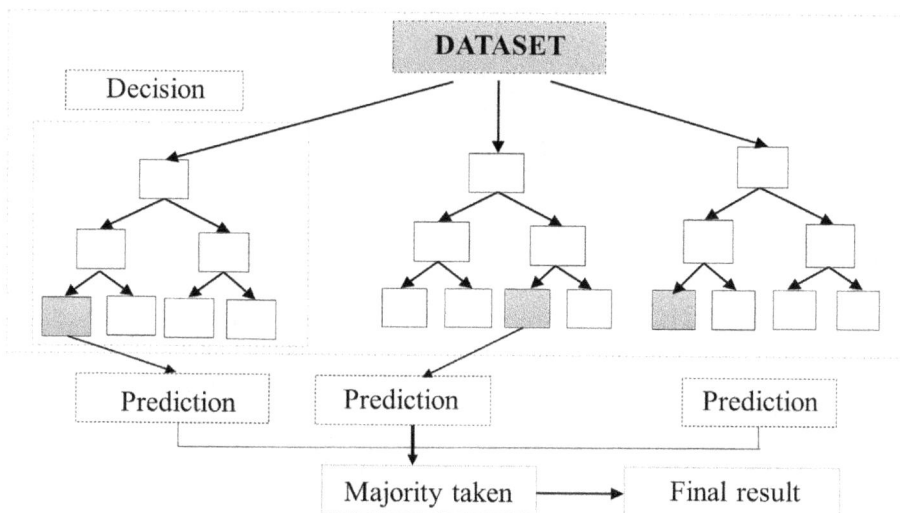

FIGURE 13.4　Classification Algorithm Using RF.

effectively. Even when a large amount of data are missing, the accuracy can still be maintained.

Random forest consists of several decision trees, each of which becomes fully developed without the need for quick processing. Random forest produces more accurate results and is less prone to overfitting if it contains more trees as an algorithm in Code block 13.2.

Overall estimation is performed using the random forest algorithm, which also has a benefit of automatically choosing features. Therefore, the primary issues listed as follows must be resolved.

Code block 13.2. Algorithm 1: for random forest.

```
for i ← 1 to B do
Draw a bootstrap sample of size N from the training data;
while node size! = minimum node size do
randomly select a subset of m predictor variables from total p;
for j ← 1 to m do
if jth predictor optimizes splitting criterion then
split the internal node into two child nodes;
break;
end
end
end while
end
return the ensemble tree of all B sub-trees generated in the outer for loop;
Alt-text: Code block 13.2 displays the code of the random forest algorithm.
```

For the minor class, a bootstrap sampling is created with a replacement technique, and then the k major class samples are found, which are the nearest-neighbor samples. Spark can be used to accomplish this in parallel.

A random forest classifier is produced for each set of samples. Only the training dataset, which is provided on the same node, is processed by each classifier. The model is then used to concurrently predict the testing dataset.

Finally, a Spark Driver is used to obtain the prediction findings. The bootstrap sampling of the random forest algorithm is shown in Code block 13.3.

Code block 13.3. Algorithm 2: bootstrap sampling.

```
Dataset T parameters {δ, k}
Sample batch
Bootstrap (T, δ, K)
Minor ← broadcast (T.minor)
Dist ← T.major.map (l => dist (l, minor))
Candidate ← dist.filter (l => l.distance< δ)
Batch ← sample (candidate, K)
Return batch
Alt-text: Code block 1.3 depicts the bootstrap sampling with the replace-
ment technique, and then finds the k major class samples, which are the
nearest-neighbor samples.
```

Various decision trees are used in a random forest system. A decision tree is composed of a decision where there is a node, a leaf node, and a root node. A majority vote is required to determine the outcome. In this case, the output of the majority of the decision trees determines the final output of the rainforest system.

13.4 RESULTS AND DISCUSSION

This section examines the performance of the current system, which includes accuracy ratings of 83.65%, sensitivity ratings of 85.59%, specificity ratings of 87.65%, and precision ratings of 87.65%.

The proposed system, with a novel method of disease prediction, is more effective than the current system. Furthermore, the accuracy of this prediction procedure was enhanced.

Confusion matrix: A confusion matrix provides examples of correctly categorized TN values that belong to a different class as well as flawlessly classified TP values, FP values that belong to one class but not another, FN values that belong to one class but not another, and FP values that belong to one class but not the other.

Accuracy, precision, sensitivity, and specificity scores are the frequently often used performance measures for categorization based on these criteria. The confusion matrix is shown in Figure 13.5.

Real Label

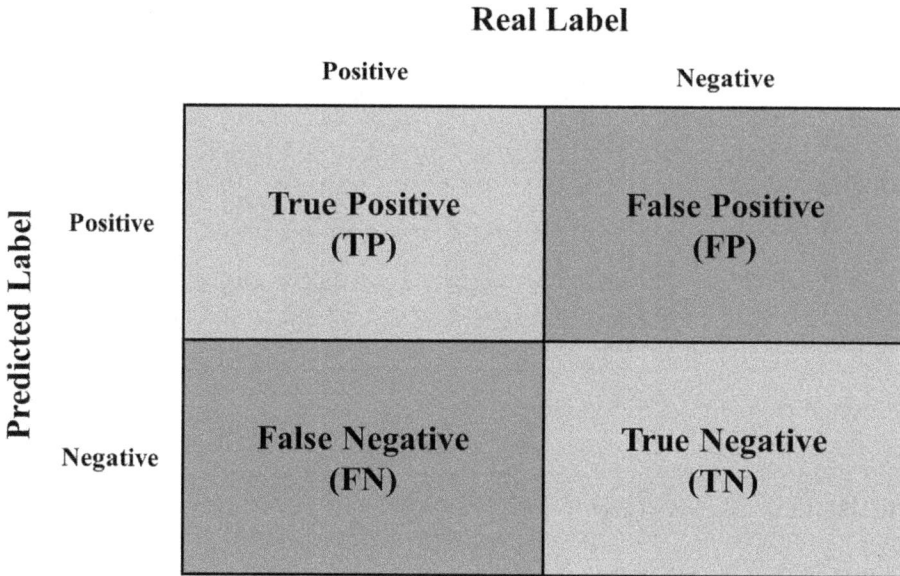

FIGURE 13.5 Confusion Matrix Prediction.

Source: Khang (2021)

TABLE 13.1
Accuracy Values for the Proposed and Existing Systems.

Algorithm	Result
PSO	77.21
Support vector machine (SVM)	78.36
RF	80.45
Hybrid particle swarm optimization (HPSO) + RF	83.65

Accuracy: Accuracy is calculated by dividing the total number of correct predictions by the total number of observations in the dataset.

The highest accuracy is 1.0, while the lowest is 0.0. To calculate accuracy, use Equation 13.5:

$$Accuracy = (TP + TN)/(TP + TN + FP + FN) \qquad (13.5)$$

The accuracy values of the various algorithms are listed in Table 13.1 and shown in Figure 13.6.

The accuracy graph is described in Figure 13.6.

Precision: Precision is determined by dividing the true positives by anything predicted as positive (Equation 13.6).

$$Precision = TP/(TP + FP) \qquad (13.6)$$

Accuracy

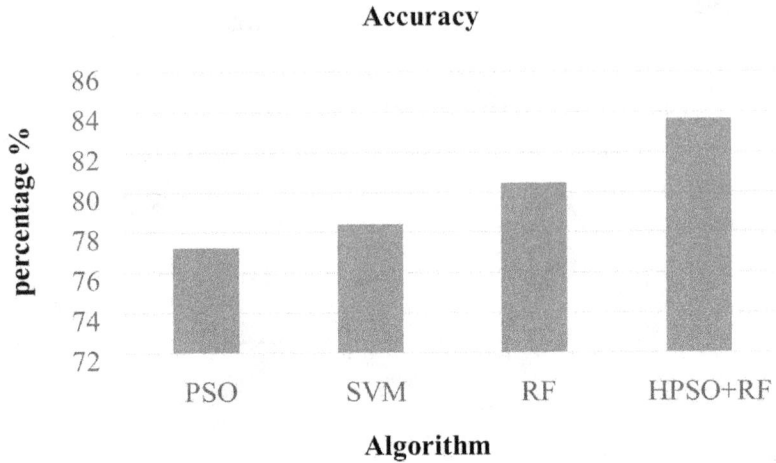

FIGURE 13.6   Accuracy Graph.

TABLE 13.2
Precision Values for the
Proposed and Existing Systems.

Algorithm	Result
PSO	76.12
SVM	78.21
RF	81.65
HPSO + RF	87.65

The precision values of the various algorithms are listed in Table 13.2.

The precision graph is shown in Figure 13.7.

Specificity: Calculating specificity involves dividing the total number of negatives by the number of accurate negative predictions (TN) (N), as shown in Equation 13.7:

$$\text{Specificity} = TN/(TN + FP) \tag{13.7}$$

The specificity values of the various algorithms are listed in Table 13.3.

The specificity graph is described in Figure 13.8.

Sensitivity: The TP rate is the proportion of positive values among all truly positive events (TPR), as shown in Equation 13.8:

$$\text{Sensitivity} = TP/(TP + FN) \tag{13.8}$$

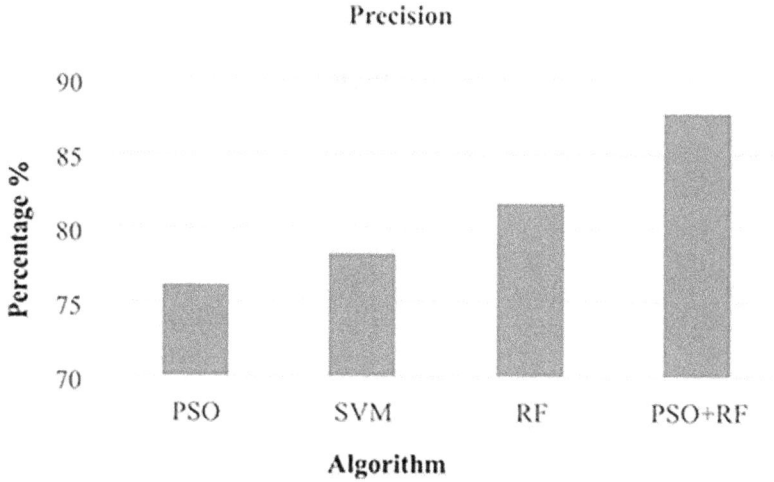

FIGURE 13.7 Precision Graph.

TABLE 13.3
Specificity Values for the Proposed and Existing Systems.

Algorithm	Result
PSO	76.12
SVM	78.21
RF	81.65
HPSO + RF	87.65

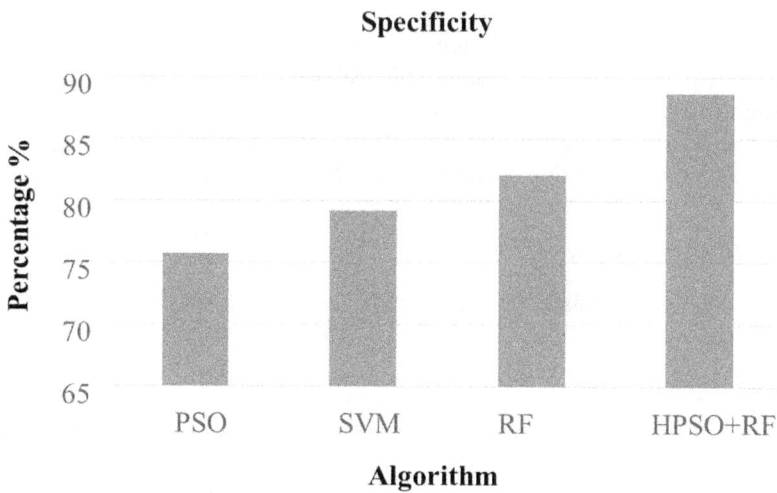

FIGURE 13.8 Specificity Graph.

TABLE 13.4

Sensitivity Values for the Proposed and Existing Systems.

Algorithm	Result
PSO	71.69
SVM	72.55
RF	78.20
HPSO + RF	85.59

Sensitivity

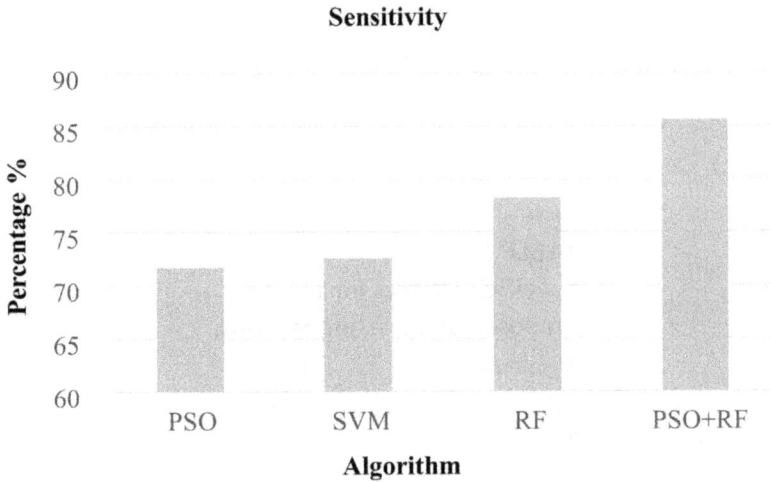

FIGURE 13.9   Sensitivity Graph.

The sensitivity values of the various algorithms are listed in Table 13.4.

The sensitivity graph is described in Figure 13.9.

The above graph for sensitivity explains the proposed PSO + RF machine learning techniques achieves better performance in task scheduling for employees while compare to any other existing system (Khang et al., 2023a).

13.5   CONCLUSION

We proposed and tested an approach for intrusion detection using SVMs on a set of selected data. It provides high data accuracy and demonstrates a compatible level of performance.

Everything in the Internet of Things (IoT) sends data packets, and as the number of linked devices grows exponentially, this communication requires reliable connectivity, storage, and security (Khang et al., 2022b).

An organization must manage, watch over, and secure enormous amounts of data and connections from scattered devices when using IoT.

A healthy, democratic society requires privacy, which is more than just a concern for individualistic personal data. As technology advances, so do personal care and healthcare.

We investigated the newly proposed privacy-preserving SVM classification system and discovered that its primary protocol is a secure technique for determining whether the sign of encrypted numbers has a soundness issue or a security leak. Then, we propose a new technique that can be applied to assist privacy-preserving SVM classification and is secure and accurate for finding the signs of encrypted integers (Khang et al., 2023b).

The experiments show that the MR-SVM model can effectively solve the problem of rapid increases in the SVM cost associated with data volume increases. However, the experiment used two nodes, and the data transfer speed was limited. Multiple nodes can be established for cybersecurity in future studies.

REFERENCES

Abdel-Basset M., Gunasekaran M., Mohamed M., Smarandache F., (2019). "A novel method for solving the fully neutrosophic linear programming problems," *Neural Computing and Applications*, 31, 1595–1605. https://link.springer.com/article/10.1007/s00521-018-3404-6

Anna et al. (2018). *World Health Organization, Regional Office for Europe, European Observatory on Health Systems and Policies. Organization and Financing of Public Health Services In Europe: Country Reports*. World Health Organization. Regional Office for Europe. https://apps.who.int/iris/handle/10665/326190

Balyan A. K., Ahuja S., Lilhore U. K., Sharma S. K., Manoharan P., Algarni A. D., Raahemifar K., (2022). "A hybrid intrusion detection model using EGA-PSO and improved random forest method," *Sensors*, 22(16), 5986. www.mdpi.com/1424-8220/22/16/5986

Bhambri P., Rani S., Gupta G., Khang A., (2022). *Cloud and Fog Computing Platforms for Internet of Things*. CRC Press. https://doi.org/ 10.1201/9781003213888

Borner K., Bueckle A., Ginda M., (2019). "Data visualization literacy: Definitions, conceptual frameworks, exercises, and assessments," *Proceedings of the National Academy of Sciences*, 116(6), 1857–1864. www.pnas.org/doi/abs/10.1073/pnas.1807180116

Chan W., Saharia C., Hinton G., Norouzi M., Jaitly N., (Nov 2020). "Imputer: Sequence modelling via imputation and dynamic programming," *In International Conference on Machine Learning*, pp. 1403–1413. PMLR. http://proceedings.mlr.press/v119/chan20b.html

Choi T. M., Wallace S. W., Wang Y., (2018). "Big data analytics in operations management," *Production and Operations Management*, 27(10), 1868–1883. https://onlinelibrary.wiley.com/doi/abs/10.1111/poms.12838

Dai M., Tang D., Giret A., Salido M. A., (2019). "Multi-objective optimization for energy-efficient flexible job shop scheduling problem with transportation constraints," *Robotics and Computer-Integrated Manufacturing*, 59, 143–157. www.sciencedirect.com/science/article/pii/S0736584518305222

Dash S., Shakyawar S. K., Sharma M., Kaushik S., (2019). "Big data in healthcare: Management, analysis and future prospects," *Journal of Big Data*, 6(1), 1–25. https://link.springer.com/article/10.1186/s40537-019-0217-0

Deng W., Xu J., Zhao H., (2019). "An improved ant colony optimization algorithm based on hybrid strategies for scheduling problem," *IEEE Access*, 7, 20281–20292. https://ieeexplore.ieee.org/abstract/document/8635465/

Gunasekaran A., Yusuf Y. Y., Adeleye E. O., Papadopoulos T., (2018). "Agile manufacturing practices: The role of big data and business analytics with multiple case studies," *International Journal of Production Research*, 56(1–2), 385–397. www.tandfonline.com/doi/abs/ 10.1080/00207543.2017.1395488

Guo Y., Ma J., Xiong C., Li X., Zhou F., Hao W., (2019). "Joint optimization of vehicle trajectories and intersection controllers with connected automated vehicles: Combined dynamic programming and shooting heuristic approach," *Transportation Research Part C: Emerging Technologies*, 98, 54–72. www.sciencedirect.com/science/article/pii/ S0968090X18303279

He S., Wu J., Wang D., He X., (2022). "Predictive modeling of groundwater nitrate pollution and evaluating its main impact factors using random forest," *Chemosphere*, 290, 133388. www.sciencedirect.com/science/article/pii/S0045653521038625

Houben M., Van de Poel B., (2019). "1-Aminocyclopropane-1-carboxylic acid oxidase (ACO): The enzyme that makes the plant hormone ethylene," *Frontiers in Plant Science*, 10, 695. www.frontiersin.org/articles/10.3389/fpls.2019.00695/full?fbclid=IwAR3xHgO_ qvNdRMwPWPvfVWvr4U8SYuHI6h4Yr2ZdneAl4PuXhL7lDs8p8oM

Iqbal R., Doctor F., More B., Mahmud S., Yousuf U., (2020). "Big data analytics: Computational intelligence techniques and application areas," *Technological Forecasting and Social Change*, 153, 119253. www.sciencedirect.com/science/article/pii/ S0040162517318498

Khang A., (2021). "Material4Studies," *Material of Computer Science, Artificial Intelligence, Data Science, IoT, Blockchain, Cloud, Metaverse, Cybersecurity for Studies*. www. researchgate.net/publication/370156102_Material4Studies

Khang A., Gupta S. K., Hajimahmud V. A., Babasaheb J., Morris G., (2023a). *AI-Centric Modelling and Analytics: Concepts, Designs, Technologies, and Applications* (1st ed.). CRC Press. https://doi.org/10.1201/9781003400110

Khang A., Gupta S. K., Rani S., Karras D. A., (Eds.). (2023b). *Smart Cities: IoT Technologies, Big Data Solutions, Cloud Platforms, and Cybersecurity Techniques*. CRC Press. https:// doi.org/10.1201/9781003376064

Khang A., Rani S., Sivaraman A. K., (2022a). *AI-Centric Smart City Ecosystems: Technologies, Design and Implementation* (1st ed.). CRC Press. https://doi.org/10.1201/9781003252542

Khang A., Hahanov V., Abbas G. L., Hajimahmud V. A., (2022b). "Cyber-physical-social system and incident management," In *AI-Centric Smart City Ecosystems: Technologies, Design and Implementation*, Vol. 2, No. 15 (1st ed.). CRC Press. https://doi. org/10.1201/9781003252542-2

Khanh H. H., Khang A., (2021). "The role of artificial intelligence in blockchain applications," *Reinventing Manufacturing and Business Processes through Artificial Intelligence*, 2(20–40) (CRC Press). https://doi.org/10.1201/9781003145011-2

Rana G., Khang A., Sharma R., Goel A. K., Dubey A. K., (Eds.). (2021). *Reinventing Manufacturing and Business Processes through Artificial Intelligence*, CRC Press. https://doi. org/10.1201/9781003145011

Rani S., Bhambri P., Kataria A., Khang A., Sivaraman A. K., (2023). *Big Data, Cloud Computing and IoT: Tools and Applications* (1st ed.). Chapman and Hall/CRC. https://doi. org/10.1201/9781003298335

Rius-Sorolla G., Maheut J., Estellés-Miguel S., Garcia-Sabater J. P., (2020). "Coordination mechanisms with mathematical programming models for decentralized decision-making: A literature review," *Central European Journal of Operations Research*, 28, 61–104. https://link.springer.com/article/10.1007/s10100-018-0594-z

Santoro G., Fiano F., Bertoldi B., Ciampi F., (2019). "Big data for business management in the retail industry," *Management Decision*, 57(8), 1980–1992. www.emerald.com/insight/ content/doi/10.1108/MD-07–2018–0829/full/html

Silva G. C., Carvalho E. E., Caminhas W. M., (2020). "An artificial immune systems approach to case-based reasoning applied to fault detection and diagnosis," *Expert Systems with Applications*, 140, 112906. www.sciencedirect.com/science/article/pii/S0957417419306244

Tanaka S., Tierney K., (2018). "Solving real-world sized container pre-marshalling problems with an iterative deepening branch-and-bound algorithm," *European Journal of Operational Research*, 264(1), 165–180. www.sciencedirect.com/science/article/pii/S0377221717305106

Tomazella C. P., Nagano M. S., (2020). "A comprehensive review of branch-and-bound algorithms: Guidelines and directions for further research on the flowshop scheduling problem," *Expert Systems with Applications*, 158, 113556. www.sciencedirect.com/science/article/pii/S0957417420303808

Wang J., Yang Y., Wang T., Sherratt R. S., Zhang J., (2020). "Big data service architecture: A survey," *Journal of Internet Technology*, 21(2), 393–405. https://jit.ndhu.edu.tw/article/view/2261

Zhu B., Feng Y., Gong D., Jiang S., Zhao L., Cui N., (2020). "Hybrid particle swarm optimization with extreme learning machine for daily reference evapotranspiration prediction from limited climatic data," *Computers and Electronics in Agriculture*, 173, 105430. www.sciencedirect.com/science/article/pii/S0168169919320472

14 Robotic Process Automation Applications in Data Management

Vivek Sharma, Alex Khang,
Pragati Hiwarkar, and Babasaheb Jadhav

14.1 INTRODUCTION

Although robotic process automation (RPA) tools, such as UiPath or Blue Prism, are readily available in the market, their systematic applications, especially for data management, are seldom discussed. Moreover, industries face the ever-increasing burden of optimizing front-end and back-end operations.

To handle front-end operations, RPA requires natural language, and it can readily be used for activities such as validating data, customer facing, customer onboarding, and billing management. It can also be used for front-end processes, such as management and automated customer management.

Owing to RPA, employees can set up, launch, and administer virtual workers instead of middleware. Thus, the time required for testing and engagement is reduced.

Robotic process automation tools help automate rule-based tasks; therefore, they are the go-to options for dealing with and managing high volumes of unstructured data. However, it is still a task to effectively analyze existing business processes and identify the right process for automation.

A simple use case of RPA are RPA bots, which can compare resumes with the description for a particular job and shortlist those resumes. With the help of RPA, offer letters are customized according to the selected candidate. It also helps to check and track time-to-time company reviews.

14.1.1 ROBOTIC PROCESS AUTOMATION TOOLS AND TECHNIQUES

Robotic process automation is a miser's dream come true, as it helps save money for the organization by automating any mundane and repetitive task that any person can do with a mouse and keyboard.

Robotic process automation also helps in upgrading legacy systems that APIs and web services fail to access. Robotic process automation bots are useful in many business processes, such as

 DOI: 10.1201/9781003400110-14

- Accessing emails and attachments
- Filling forms
- Editing databases and reading from them
- Doing calculations
- Gathering statistics from social media
- Documenting data extraction

14.1.2　RELATION OF ROBOTIC PROCESS AUTOMATION WITH ARTIFICIAL INTELLIGENCE PROCESSES

Figure 14.1 shows how RPA is related to artificial intelligence (AI) processes. Robotic process automation tools are useful for the automation of business processes using various technologies. The initial application of RPA was the automation of business process outsourcing (BPO) processes using workflow configurations and screen scrapping (Rana et al., 2021).

Robotic process automation tools also help in cognitive learning, thus combining them with AI. Therefore, these tools are also termed tools for intelligent automation rather than automation, as shown in Figure 14.1.

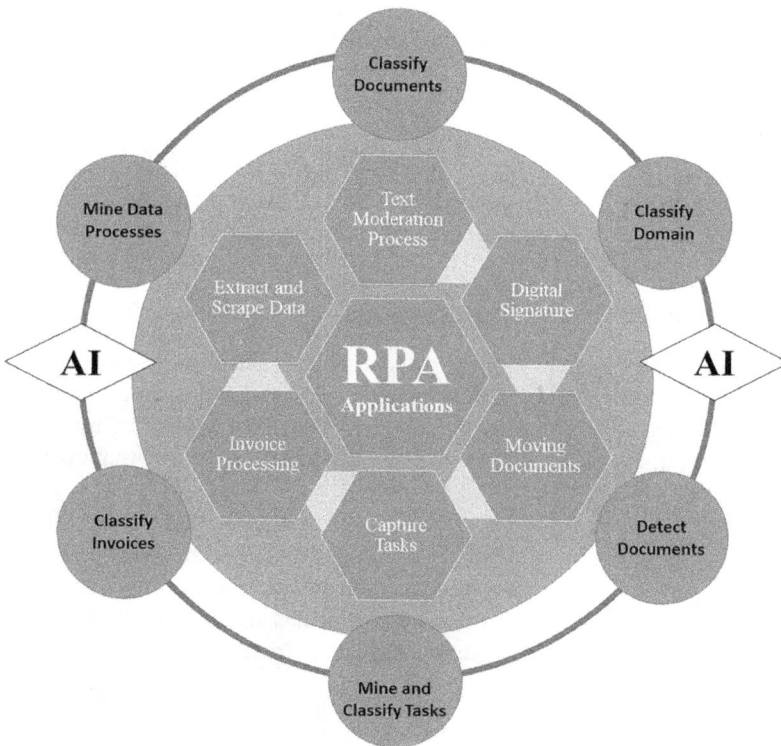

FIGURE 14.1　Relation of RPA Applications with AI-Embedded Processes.

Source: Khang (2021)

14.2 ROBOTIC PROCESS AUTOMATION IN DATA MANAGEMENT

Robotic process automation is an easy-to-use minimum-code technology that even non-voice developers in business departments can utilize to automate their business processes.

Robotic process automation can be easily understood as software robots. They are programs that perform repetitive tasks that are beneficial to firms looking to automate data management tasks. These tasks include data cleansing, data normalization, data wrangling, and management of metadata.

14.2.1 ROBOTIC PROCESS AUTOMATION APPLICATION IN DATA CLEANSING

Robotic process automation is applied in data cleansing to detect corrupt data and update records with accurate and useful information.

Uniqueness of the data is preserved by removing duplicate files. In data cleansing, incorrect or corrupted data are removed, and grammatical errors and incorrect number files are corrected.

In data cleansing, the usefulness of data is also considered so that data are standardized and unnecessary content is cleared. Thus, data cleansing provides pre-robotics solutions by standardizing the information from multiple sources in bot-understandable templates.

If data cleansing is not dealt with diligently, it can cost companies financial losses and delays. A pre-robotics solution in terms of data cleansing provides an efficient and agile approach to handling unstructured information. This is done through information mining following specific rules in business and computerized processes with minimum human intervention (Khanh & Khang, 2021).

14.2.2 ROBOTIC PROCESS AUTOMATION APPLICATION IN DATA NORMALIZATION

The process of data normalization is often cumbersome and time-consuming, and can carry many errors. Complex data systems constitute a large part of the current workplace. The verification of mappable (i.e., able to be represented by a map) data from source to target is an important aspect of data onboarding so that uniformity in data is maintained (Christian & Peter, 2021).

To manage data efficiently and effectively, calculated columns are added. Data are restored according to custom criteria. Matching data formatting with an analytics engine using normalization ensures a better interaction with the dataset.

Susceptibility to human errors can be minimized if this dataset is handled correctly; otherwise, it can result in significant downstream rework and an increase in staff hours.

14.2.3 HOW IS ROBOTIC PROCESS AUTOMATION USED IN DATA WRANGLING?

Data wrangling is the immediate process after data collection. Data wrangling begins with data discovery and ends in publishing.

The following are the steps of data wrangling that can be repeated until the data are ready to be analyzed:

- **Discovery**: This step forms the basis of data wrangling, as you try to know more about data and their pattern so that any discrepancies can be taken care of, and the data can be made clean and usable.
- **Structuring**: After discovering the data, the data structure must be understood so that it can be manipulated easily. This structure involves understanding and deciding on the best-suited format for raw data for manipulation and storage.
- **Cleaning**: After deciding on storage, an important part of data wrangling is cleaning. During cleaning, unwanted and redundant information is filtered so that the final dataset contains minimum errors. Approaches for data cleaning include handling outliers, removing null values, converting the given data into numerical form, and removing undesired rows or columns.
- **Enriching**: After cleaning comes enriching the dataset, where more relevant data are added if required. This data addition boosts the analysis. Data augmentation is also a part of data enrichment, where more subsets or variables are created from existing datasets.
- **Validating**: In validation, the consistency in the datasets is checked by addressing discrepancies so that the variables are properly allocated and correct distributions are adhered to. Automated processes are used to check discrepancies and readiness of data for usage.
- **Publishing**: The final step is publishing, where the discrepancy-free, enriched, clean, and structured data are ready to be published and analyzed by the organization.

14.2.4 ROBOTIC PROCESS AUTOMATION APPLICATION MANAGEMENT OF METADATA

The business approach to future-proofing includes the use of sustainable models and strategic plans for the longevity of the company. This helps in adapting, optimizing, and planning future events. 'Future-proof' business applications are the need of the hour.

The business practice of future-proofing helps ensure a company's longevity, as it helps improve sustainable models and strategic plans. The runtime rendering engine or application browser can be designed to consume metadata and provide the application user interface dynamically.

14.2.5 VARIATIONS OF ROBOTIC PROCESS AUTOMATION

Some RPA approaches depend on evolving technologies and are therefore constantly being redefined according to the needs of the market. Robotic process automation can be classified as follows:

- **Attended RPA**: This automation is alternatively called robotic desktop automation. This is a fundamental form of RPA used for call center tasks, in which representatives perform tasks such as information lookup while customers interact.

- **Unattended RPA**: After the attended RPA, unattended RPA, which is used in human-free automation, was developed. An example is the triggering of a bot after an event, such as receiving an invoice from a customer. Unattended RPA is generally used for back-office purposes.
- **Cognitive RPA or intelligent process automation**: Alternatively, intelligent process automation is also called cognitive RPA. Intelligent process automation uses AI for system intelligence, which increases with time. Thus, human interventions are minimized, as RPA software performs better decision-making and learning. Understanding variations plays an important role, as different RPA systems are meant for different purposes.

14.3 WHAT MAKES ROBOTIC PROCESS AUTOMATION SO SPECIAL?

People often get confused between RPA and business process automation (BPA), as well as business process management (BPM) and BPO (Tom & Apress Publication, 2020). Before stating the differences, we discuss the definitions of all the three terms stated.

14.3.1 BUSINESS PROCESS AUTOMATION

Business processes are not just record keeping or data handling and manipulation activities, but they have a wider scope. Business process automation is defined as the automation of difficult business processes and functions using advanced technologies.

The focus of BPA is always on running and not counting the businesses, thereby making it effective and efficient. Business process automation is always an event-driven, focused fundamental process.

A business process is an activity or a set of activities with a target to fulfill the goals of an organization, such as producing a product, building an employee resource pool, and creating a new customer base.

Business processes occur in many domains, such as HR and marketing, operations, and management, to carry out recurring, time-intensive, collaborative tasks that require compliance and audit trials. Examples are employee onboarding, IT service desk support, and marketing automation.

14.3.2 BUSINESS PROCESS MANAGEMENT

Business process management is a domain, in which business processes are discovered, modeled, measured, analyzed, improved or optimized, and automated. The methods used to manage a company's work come under business processes. These processes can be structured or unstructured and generally use technology for growth purposes.

Business process management is a time-consuming and extensive process, as it is detail oriented. Detailing is required for both documents and training. Thus, regulated industries, such as healthcare and banking, generally prefer BPM.

The regulations in these industries are also seen as drawbacks, as they hamper growth, and agility, are procedure-oriented, and lack innovation. Examples are monitoring, optimization, and re-engineering.

14.3.3 Business Process Outsourcing

Business process outsourcing is the practice of using a third-party service provider to perform various functions related to a company. The pioneering industry that first used BPO was manufacturing, in which firms outpoured tasks for the efficient management of supply chains.

The BPO industry serves many sectors, such as healthcare, asset management, energy, pharmaceuticals, and e-commerce. Business process outsourcing provides companies with the advantage to use new and innovative methods to improve customer experience and gain competitive advantage (Khang et al., 2023c).

14.3.4 Advantages of Robotic Process Automation

The advantages of RPA are as follows:

- **Better productivity**: RPA helps to automate repetitive tasks, thereby improving the efficiency of each employee and increasing the productivity of the company.
- **Rapid growth and quick gains**: The main advantages of RPA are its quick implementation and rapid generation of results by thinking of, designing, developing, and deploying in just the time of the week.
- **Relatively less initial cost**: RPA bot licenses are affordable compared to other software tools, and bots can easily outperform humans, also known as full-time equivalents, thereby reducing costs and requiring low initial investments.
- **Low cost of processing**: The processing cost becomes less significant, as the cost of the bot is nearly 50% of the employee cost.
- **Better accuracy and quality**: Bots execute the assigned work with 100% accuracy; hence, no rework is required.
- **Better compliance**: Logged RPA activities can be reviewed at a given time. Thus, it provides good control over any operation and greater oversight over both the desktop and server RPA.

14.4 ROBOTIC PROCESS AUTOMATION LEARNING WITH CLOUD

Programming skills are least required for RPA, but technological concepts are applied in detail while implementing RPA.

On-premise technology is generally used in any IT system approach, in which the company uses its own purchased hardware and software in its data center (Rani et al., 2021). The cloud also has the following approaches:

14.4.1 Public Cloud

In this model, a third-party provider manages the on-demand computing services and infrastructure. Many organizations also access this infrastructure using the Internet, which is available to the public.

Even cloud-based services, such as IaaS, PaaS, and SaaS, can be accessed by users using multiple options, such as pay-per-use fees or pay monthly. Public cloud service providers monitor these services. Therefore, there is no requirement for users to hold these services at their own data centers.

A public cloud is helpful for companies that do not wish to expand their physical infrastructure. A virtual desktop license can be a good alternative to a physical desktop machine, as the virtual machine can be used and deactivated in a very short period (minutes) and is portable.

Data storage is another application of the public cloud, as the stored data are easily accessible from anywhere and have backup as per the needs. The infrequently used data can also be stored inexpensively on the cloud. One of the concerns of using the public cloud is security, which must be considered (Bhambri et al., 2022).

14.4.2 Private Cloud

A private cloud can also be termed an exclusive service of cloud facilities and computing given to an organization where other organizations are automatically not involved. Thus, the company owns the data center. Private clouds are more secure and controllable than public clouds.

14.4.3 Hybrid Cloud

Hybrid cloud is a combination of public and private clouds, with features from both. For example, a public cloud can handle fewer mission-critical functions. In RPA, cloud services are used differently and have different implications.

First, a platform must deal with complex distributed applications, which may be difficult when custom programs are developed by the company on cloud services.

In addition, RPA platforms generally fail to scale up from the on-premise software to the cloud. Developing cloud-native systems is a difficult and uphill task for an RPA because deep connections are required across many applications and environments.

Sometimes, cloud services, such as AWS, may load an on-premise system with RPA. Although this loading can be beneficial, it is not cloud-native because of the degradation and maintenance required by the software (Rani et al., 2023).

14.5 GENERALLY USED ROBOTIC PROCESS AUTOMATION TOOLS

The following are the five generally used RPA tools:

- Blue Prism
- UiPath

TABLE 14.1

Five Popular RPA Tools Usually Applied in Corporations in the Industry 4.0 Era.

Vendor	Kofax Kapow	Blue Prism	UiPath	Automation Anywhere	NICE
Factors-based technology	.NET, JAVA	C#	Microsoft SharePoint Framework Elastic search, Kibana	Microsoft	VB Scripting and C#
Reliability	Moderate	High	Moderate	High	Moderate
Cognitive capacity	Medium	Low	Low	Medium	Low
Reusability	Yes	Yes	Yes	Yes	Yes
Accuracy	High for web automation and file handling	Available for web, desktop, and Citrix automation	Good in Citrix environment designed for BPO automation	Rational accuracy cross medium	Good accuracy for tasks that require little or no subjective judgment
User-friendliness	High (no coding required)	High	High	Medium	Medium
Operational scalability	Easily scalable with a stateless, multi-thread architecture	High speed of execution	Frequently crash in medium projects	Large-scale robot deployment is limited	Fast execution seamlessly scalable

- Automation Anywhere
- Kofax Kapow
- Neptune Intelligence Computer Engineering (NICE)

These tools can give a comprehensive, understandable result if they are combined. A comparison of these tools is presented in Table 14.1.

14.5.1 BLUE PRISM

It is a software company that builds RPA software. This software is useful for automation of complicated end-to-end procedures in business.

The Blue Prism RPA tool has the following essential components:

- Process diagram
- Application modeler
- Object studio
- Process studio

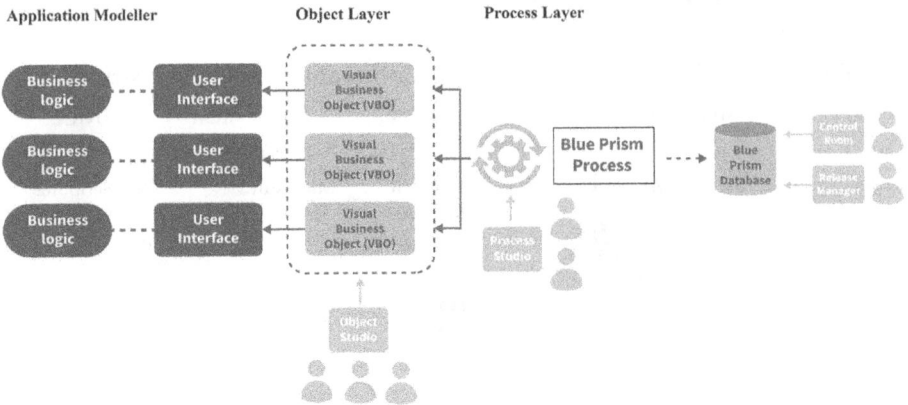

FIGURE 14.2 Diagram of Blue Prism Layers.

Source: Khang (2021)

- Control room
- Release manager
- System manager
- Work queue
- Exception handling

The essential features of Blue Prism include scalability, flexibility, resilience, security, and compliance. It contains three main layers: object layer, application modeler, and process layer (Figure 14.2).

The implementation of Blue Prism requires minimum IT skills, such as coding and programming. It can be implemented within IT infrastructure and, most importantly, does not require process change for implementation, providing a good return on investment (Khang, 2024a).

14.5.2 UiPath

UiPath helps automate repetitive and tedious tasks through its platform (UiPath, 2023). It believes in engaging humans in creative and inspiring activities, as shown in Figure 14.3.

14.5.2.1 UiPath Studio

UiPath Studio is used to design robotic processes. It uses a visual interface and flowchart-based modeling. It encourages collaboration within the same workflow and is relatively fast. It has a recording facility and an error display system that makes modeling much simpler and easier even for laypeople.

Studio harbors both simple and complex solutions for the integration of applications. It can easily automate third-party applications. It is a useful tool for administrative IT tasks and business IT processes. It forms the core of the automation

UiPath Studio	UiPath Orchestrator	UiPath Robots
Digitize your process	**Manage and Secure**	**Deploy and scale**
Easily create automations in a visual editor across various software applications	Control, monitor, and analyze the performance of your virtual workforce	Your attended *and* unattended robots

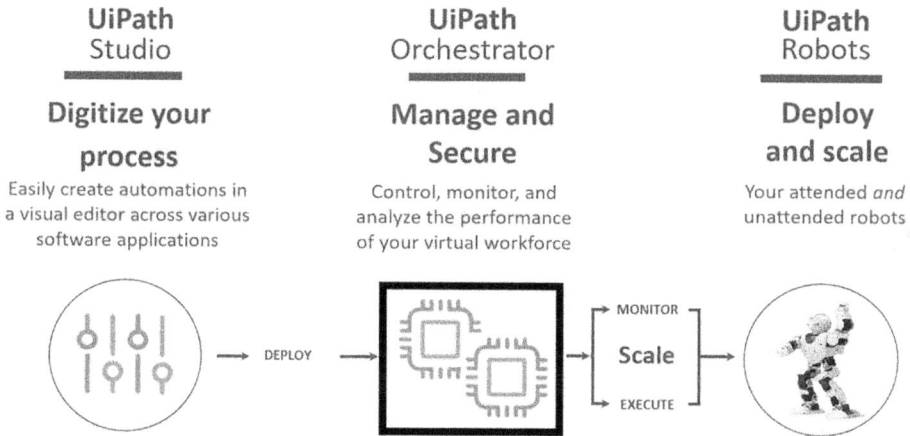

FIGURE 14.3   Illustration of the UiPath Platform.

Source: Khang (2021)

performed by UiPath. It uses activities as a basic component of its workflow, which is further executed by the UiPath Robot and then published by the UiPath Orchestrator.

UiPath Studio facilitates the arrangement of steps in a task or fully records the sequence of events or steps. UiPath Studio is therefore very simple to handle, design, and implement for a person with very little technical knowledge. The steps can appear alternatively as data flow diagrams, which can be easily understood.

The enterprise environment provides hints for the flow of work used for the development of the robots. The studio can be alternatively seen as a workflow designing the complete control of the execution order and activities, which can be as simple as clicking a button, writing into, or reading a file.

14.5.2.2   UiPath Orchestrator

UiPath is a server-based application that orchestrates the execution of repetitive business processes. It uses all robots within the network with the help of a browser-based interface, which is used for the orchestration of robots.

The creation, monitoring, and deployment of resources in a given environment can be managed using Orchestration. The main capabilities of Orchestration are as follows:

- Connections between robots can be performed and maintained using Orchestration.
- Queues are effectively handled using Orchestration.
- Orchestration ensures that the robot receives the correct packages.
- Orchestration is useful in identifying robots.
- Orchestration uses SQL or Elasticsearch to sort and index logs.

14.5.2.3 UiPath Robot

The steps developed or recorded by UiPath Studio can be executed step by step using UiPath Robot. It is a window service that is alternatively known as an execution or a runtime agent. Therefore, it is referred to as a robot.

Robots connected to Orchestrator can execute processes. Local licensing of robots can also be used as an alternative approach for program execution. Some robots are licensed, while others are not, but their use completely depends on the manner in which they are deployed.

UiPath software robots can be classified as attended or unattended.

- **Attended robot**: It operates in coordination with humans to complete daily tasks allocated at the workstation. User events trigger the attended robot, and Orchestrator cannot be used for the same purpose. These robots cannot run under a locked screen.
- **Unattended robot**: This type of robot is independent of the user, as it does not require any attention in a virtual environment. This robot can automate several processes. The configuration set in the Orchestrator is responsible for the execution of this program. Along with the capabilities of an attended robot, this robot can also perform scheduling, remote execution, and monitoring tasks. It can also provide support for work queues. One of the advantages of the unattended robot is that it easily captures employee performance data, departmental task daily volume data, and other related data.
- **Free robots**: These robots work similarly to unattended robots, but they have no application in the production environment and can be used only for development testing purposes.

14.5.3 AUTOMATION ANYWHERE

Automation helps make business processes more efficient and effective by reducing or replacing the human workforce. Automation Anywhere is one of the most used RPA tools that automate complex business tasks using powerful features. It uses cloud technology along with AI and is a web-based platform. It gives end-to-end automation (Tailor et al., 2022a).

Different bots, such as IQ bots, meta bots, and task bots, can be developed using components of Automation Anywhere. It easily adapts to new environments and is useful for deployment in the cloud, data centers, or desktops. It requires zero programming skill.

The recorders can be used for manual action recording, and the task editor can be used to edit the action as follows:

- It provides intelligent automation for business and IT operations.
- It provides bank grade-level security.
- It can be hosted on premises or in the cloud.
- It has low-code or no-code platform for automating processes.
- It automates complicated and complex tasks.

BOT CREATORS
Desktop Application

BOT RUNNERS
Classify Documents

CONTROL ROOM

Automation
Anywhere

Source Control

User Management

Data Comparison

Data Validation

Dashboard

and more…

Developer 1

Developer 2

Developer 3

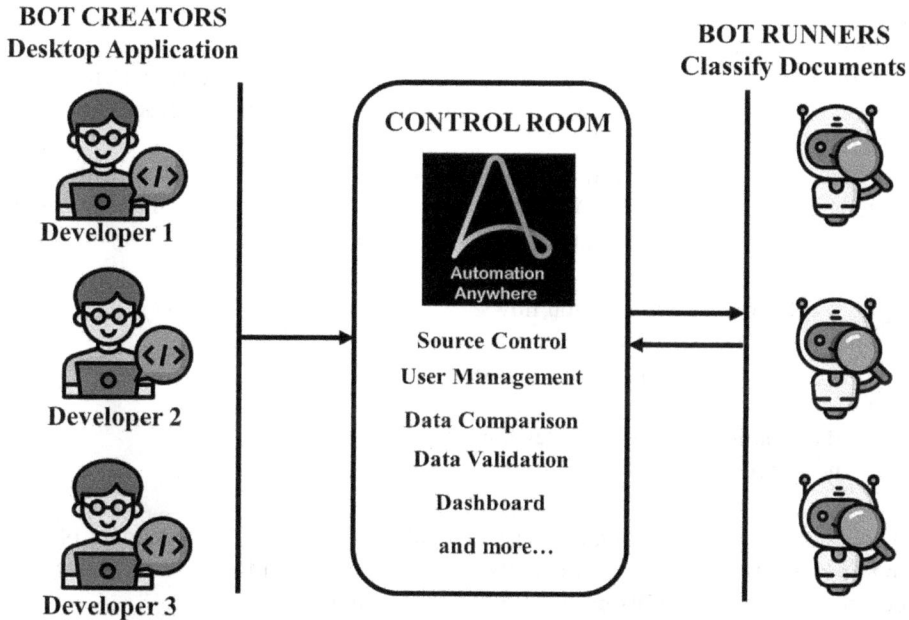

FIGURE 14.4   Architecture of the Automation Anywhere Platform, Where the Control Room Is the Heart of the Automation Anywhere Platform.

Source: Khang (2021)

- Its intelligent automation uses AI and cognitive services with an IQ bot.
- It provides real-time reports and analytics with bot insights.

The most important part of the Automation Anywhere architecture is the control room, which is a web-based server dealing with bots created by bot creators. It controls the entire execution and management of clients, roles, security, etc.

All automated tasks or scripts are uploaded in the control room. The management and scheduling of the execution of tasks in the host server are performed by the control room and bot runner, as shown in Figure 14.4.

- Source control
- License management
- Automation deployment
- Dashboard
- User management

14.5.4   KOFAX KAPOW

Kofax Kapow RPA is an information integration and process automation software manufactured by Kofax. Information can be acquired, enhanced, and delivered using Kofax Kapow. The source of this information can be business applications or web-based sources, without custom coding.

Kofax also provides a facility for collaboration, as it cost-effectively pulls useful market information and automates repetitive information-driven processes (Luke et al., 2023c).

Data from a legacy-driven content management system can also be easily automated by Kofax Kapow (2023). Thus, person-hours can be reduced, and time can be properly managed using this automation tool in the following ways:

- It enhances data integration flows.
- It fosters collaboration.
- It provides bidirectional integration.
- It accesses market intelligence.
- It dramatically lowers the development costs.
- It connects different applications and systems.

It integrates with partners and improves real-time searching independent of coding or consulting. Its platform can also be easily extended to desktops and mobile devices for automation. Its prominent features include:

- Process management
- RPA
- E-signature
- Mobility
- Customer communication services to multiple customers

The widest variety of internal and external data types and sources is integrated. Integrations are easily built and published as web apps for users to explore and take action on data.

Data are integrated with business systems, processes, databases, data warehouses, and business intelligence tools. Large volumes of data are acquired, enhanced, and delivered on a secure enterprise platform. Kofax Kapow facilitates the creation of prototypes within hours, thus facilitating faster market capitalization and competitive leverage in real time. It also exploits an AI platform and intelligent software robots to automate business processes, as shown in Figure 14.5.

14.5.5 NEPTUNE INTELLIGENCE COMPUTER ENGINEERING

Neptune Intelligence Computer Engineering (NICE) is a public enterprise in Israel. It uses big data with operations that deliver both server and desktop automations. It is useful in automating cross-application processes (NICE RPA, 2023). It can work together with third-party or home systems. It combines customer relationship management (CRM), billing, MS Office, virtualization, and networking.

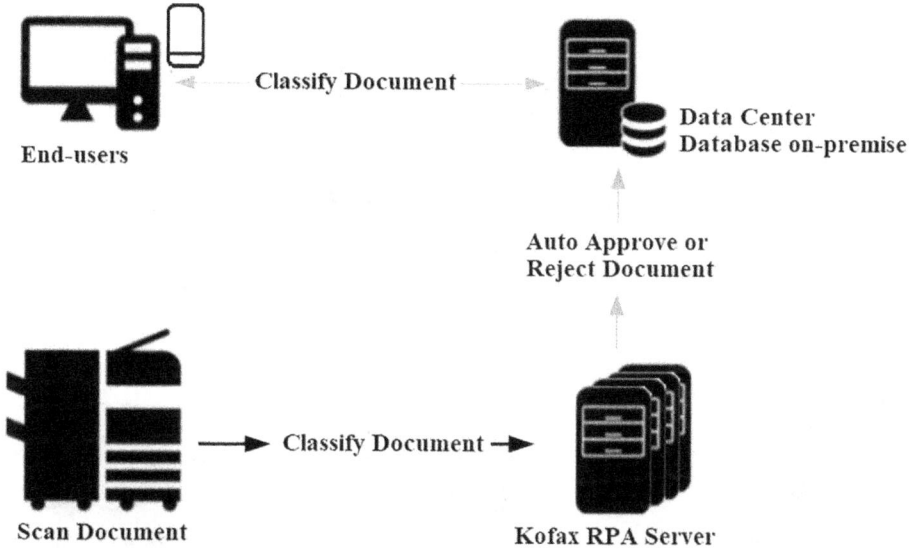

FIGURE 14.5   Approve and Reject Document Processes.

Source: Khang (2021)

These RPA tools that extend past generation bots can be classified into three broad categories depending on the purpose they serve (Nandan & Arun Kumar, 2020):

- The set of inputs is understood and coded for the RPA implementation. The bots signify the first-generation RPA tools.
- **Self-learning solutions**: These tools understand the process of human employee activity and start performing the same task using the platform.
- **Cognitive or intelligent automation bots**: These bots learn and handle both structured and structured data. These are modified forms of self-learning bots that use advanced functionalities, such as natural language processing (NLP), image recognition, and machine learning.

A real-world example is extracting data or contents from a specific website (the so-called website scrapping) on the Internet, such as

- Employee onboarding
- Credit card applications
- Scheduling systems
- Claims administration
- Call center operations
- Compliance reporting
- Customer order processing

The NICE RPA tool can be used, as shown in Figure 14.6.

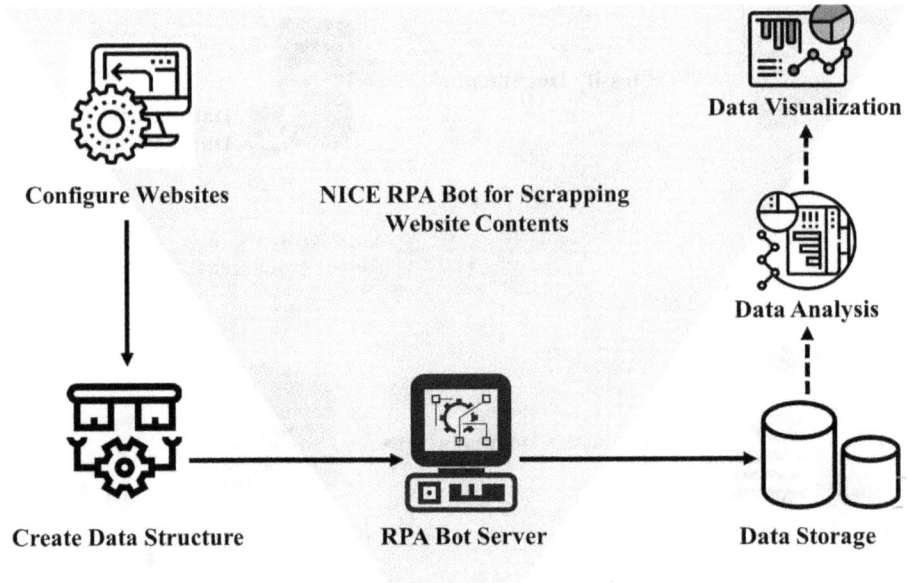

FIGURE 14.6 RPA Workflow for Scrapping Contents from a Website.

Source: Khang (2021)

14.5.6 Key Differences between the Best Robotic Process Automation Tools

The three best automation tools are UiPath, Automate Anywhere, and Blue Prism. Although these tools are extensively used in the industry, there are some important differences that must be considered. The differentiating points for UiPath are as follows:

- **Drag and drop of the workflow**: UiPath converts drag-and-drop processes into visual workflows by offering users the ability to build visual path processes.
- **Advanced scraping options for screen and data**: Data scraping can be used to build tables at runtime and search queries and other repetitive structures.
- **Image and textual Automation**: These techniques are used when UI automation is not feasible for virtual machines. The main drawback of these techniques is that they are costly.

The same differentiating points for Blue Prism (n.d) can be considered as follows:

- **Intelligent automation platform**: Intelligent automation is provided by Blue Prism, as it uses AI along with machine learning for repetitive tasks and creates value for organizations.
- **Design studio**: The design studio offered by Blue Prism allows users to reuse events and actions and create processes across organizations.

- **Digital workforce**: Autonomous software robots and intelligent technologies imitate and learn business processes parallel to humans. Blue Prism enables the automation of these processes by offering a digital workforce.

The differentiating points for Automation Anywhere (n.d.) can be considered as follows:

- **Bot Store**: Automation is accelerated when Automation Anywhere uses Bot Store. It is the largest RPA marketplace, as it offers a rich pool of pre-built intelligent automation solutions. The time consumed by these solutions is relatively low.
- **RPA workspace**: Automation engine is used to automate all repetitive tasks and processes using Automation 360. It transforms unrelated processes into automation and unifies their operations.
- **IQ Bot**: It uses AI technologies, such as machine learning, NLP, computer vision, and fuzzy logic, for extracting and validating information from business processes (Khang et al., 2023b).

14.6   ROBOTIC PROCESS AUTOMATION IN BUSINESS DATA MANAGEMENT

Different strategic objectives can help build and implement an efficient RPA. Trivial and manual tasks, such as entering sales data and orders, SAP automation, Citrix Automation, and automating user interface and recordings, can be replaced by RPA.

Smaller routine tasks can be integrated into a major task to create good business models. The categories of RPA can be understood according to the complexity of the processes they execute, given by:

- **Routine task**: Copied and combined data from different application systems are used.
- **Structured task**: Data from various application systems are used and evaluated using some pre-defined rules. The decisions made by these tasks are rule-based.
- **Unstructured tasks**: In these tasks, experience and cognition play a vital role along with existing data and rules.

Robotic process automation works similarly to other information systems. Robotic process automation systems have a spectrum of variations ranging from single systems to fully integrated subsystems. These systems operate in the cloud, on premise, or on hybrid systems.

Robotic process automation implementations still pose a challenge, as they typically result from standard software systems. Software robots are intelligent machines with the following components:

- **Sensors**: They leverage and manipulate data from the environment.
- **Actuators**: They communicate with the environment, and their outputs can be used as inputs for other systems.
- **Intelligent Center**: It generates output data by using sensor input data.

14.7 ROBOTIC PROCESS AUTOMATION USE CASES IN BUSINESS DATA MANAGEMENT

As research and development in the RPA domain progresses, new use cases can be explored to implement the same. Some are back-end processes, such as customer onboarding, billing management, validation of data, and customer facing, and others are front-end processes, such as customer service management, contract initiation, and management.

Front-end processes generally use AI technologies, such as machine learning, deep learning, and other technologies, to obtain the input and classify unstructured data.

These data are then provided to enterprise applications for application creation. These AI-enabled methods are known as cognitive RPA.

- **How variable the input source is**: The input source plays a very significant role in cognitive automation. It also means that for good cognitive automation, the variation must be short.
- **How readable the source is**: The machine-readable format of the input process is of primary importance in the RPA. It is converted into a readable format if it is not machine-readable.
- **What technology can be leveraged**: The cost of intelligent automation technology should be bearable with the technology being mature enough to be used by the firm without compromising its security policies.
- **How much data are available**: Data availability is an essential component of an AI engine. Data must be available in ample amounts always so that AI engines can leverage the efficiencies of machine learning models (Khang et al, 2023a).
- **How to standardize and re-engineer**: Message delivery to the customers should always be in a standard format. It should not use cognitive automation; instead, common formats can be adapted. In addition, algorithms can be used on unstructured data. Thus re-engineering and standardization can help make RPA systems more efficient with minimum use of AI.
- **How to model risk**: Underwriting is an effective process for cognitive automation, but it is not helpful in efficient decision-making. Therefore, proper assessment methods must be followed before taking machine-based decisions that involve human judgment.

Robotic process automation differs from cognitive automation in many aspects, some of which are listed as follows:

- Approach used to work with human workers
- Data types
- Project timelines
- Approaches used for programming

According to Tailor et al. (2022), RPA is a technology that takes the robot out of the human, whereas in cognitive automation, we put humans into robots. Business intelligence and cognitive RPA can be used together for effective business handling and insightful business understanding.

Business intelligence (BI) analysts can always leverage insights from the cognitive automation of activities that involve enormous efforts, repetition, and standardization. Thus, it can be said that BI and cognitive RPA together are more effective than when taken individually.

Robotic process automation effectively collects data without program integration and supports different formats. It maintains data in desired, understandable, and easy-to-use format. Thus, RPA is an easy-to-use software for the development of products and its implementation for single and multiple processes. Its limitation is that, although manual processes can be optimized, they cannot be eliminated in all cases.

Some of the optimizations supported by RPA are listed as follows:

- Data from competitive companies or their websites can be effectively collected and compiled.
- Information about customer behavior can be collected, and necessary inferences can be drawn.
- Market activities can be effectively automated.
- Prices can be modified, and products can be effectively priced so that maximum customer footfall is generated.
- Supply chains can be properly managed using effective communication.
- Processes can be monitored after deployment to achieve effective performance.
- Innovative product launches can be handled effectively to minimize the relative risk of using RPA.

14.7.1 How Customer Support Management Use Robotic Process Automation

Customer retention and satisfaction play an important factor in any retail business. To do so effectively, retailers need to manage many activities, such as streamlining the checkout process, customer complaint monitoring, and generating effective feedback.

For this purpose, data must be effectively and efficiently used to meet the ever-increasing demands of customers. Some use cases of RPA in the customer service domain are discussed as follows:

- Handling time optimization in call centers
- Improving first call resolution rate
- Effective usage of chatbots, RPA, and AI
- Review of effective tracking and management of complaints

14.7.2 Natural Language Use in Business Data Processing with Robotic Process Automation

Natural language processing processes human language and yields necessary analysis, manipulations, and interpretations for a better understanding. Knowledge understanding is facilitated by developers to perform important activities, such as text summarization, Named Entity Recognition Algorithms (NER), topic segmentation, relationship extraction, semantic analysis, and speech recognition.

Companies utilize NLP for its various processes, such as improving the accuracy of documentation and identifying information from large databases. Natural language processing can always be a good and potential complement to the RPA life cycle for designing specific executions and routines and identifying automation candidates using semantic analysis.

Natural language processing and RPA have many functions that are effective in business data processing. Some of these functions include the following:

14.7.2.1 Data Extraction from Unstructured Sources

A structured dataset can be quantified, formatted effectively, is easy to manage, and is understandable by algorithms and software alike. Structured data examples include transaction receipts and invoices.

The CRM (customer relationship data) in finances is generally generated on structured data, such as customer feedback and emails. In contrast, pdf invoices, printed receipts, social media posts, sensor data, and mobile data all fall under the unstructured data category (Santiago & Alejandro, 217).

Unstructured data are difficult to parse and process using automation tools because they lack a simple structure; hence, automation is a challenge (Ibrahim et al., 2016). Thus, NLP, along with automation tools, can facilitate data extraction, validation, and conversion to a structured form.

Robotic process automation bots with NLP capabilities can read the data source to understand its context and work along with different enterprise applications that generate unstructured data. An example is income invoice processing using bots.

14.7.2.2 Generating Concise Reports

Bots that use NLP can be used to gather information from data. They can also be used to provide data to pre-assigned documents, such as database tables and spreadsheets, and to manipulate the data. They can generate essential reports in the stipulated time frame and the required format, thereby helping to optimize the use of time and then further forward them to the designated staff.

14.7.2.3 Good Customer Experience through Efficient Services

Bots, along with automation, can assist customer service representatives in case of heavy traffic at their place or for their jobs as they can work for $24/7$ handling customer queries and providing their solutions.

14.7.2.4 Customer Sentiment Analysis

Intelligent bots use NLP to manage and improve brand image (Lukasz et al., 2020). They handle customer feedback and queries posted on different media platforms, emails, and feedback forms so that well-informed and need-based streamlined services can be delivered according to customer priority.

14.7.2.5 Information Segregation and Classification

Not-so-clean data become difficult to handle and cannot provide business insights. Intelligent bots using NLP can serve the purpose of text classification in the following ways:

- Automatically detect spam in the email and transfer it to the respective folder.
- Vital information on a candidate's resume, such as education, skills, and work experience, can be automatically extracted using NLP-enabled bots.
- Intelligent bots can interpret customer responses or feedback and automatically suggest corrective actions (Shanee et al., 2020). For example, if the customer sends a message, 'The product is awesome but too expensive', the information that could be inferred would be the customer is interested, but the cost is a concern. In this case, emails can be sent to customers at discounted prices.
- Intelligent bots can effectively transfer tickets raised by customers regarding any issue for a transaction or service to respective departments or persons. This would help improve customer engagement and satisfaction.

Natural language processing plays an important role in business operations (Shruti et al., 2017) in the following ways:

- **Financial domain and banking institutions**: Market data analysis can be performed using NLP, thereby helping banks and other financial institutions in optimal strategy design for better output. In addition, NLP can be used for fraud detection and identifying illegal transactions.
- **Insurance domain**: Claim settlement and false claim identification are the major concerns of insurance companies that can be complemented by NLP and RPA. Natural language processing can analyze the doubtful communications of fraudulent customers, thereby flagging financial fraud and, in some cases, preventing it. Automatic feature extraction using RPA and NLP can help insurance companies outsmart their competitors and develop effective strategies and product planning (Jianwei & Haixia, 2022).
- **Retail domain**: Customer behavior can be automatically analyzed to provide insights into the sentiments and behaviors of customers. This can help retailers make better and quicker decisions about products, pricing, drives, etc. Streamlined and need-based information can be fed to customers so that their needs are identified and automatically satisfied, thereby increasing their foot count.
- **Healthcare domain**: Symptomatic and diagnostic treatments can be provided to patients on a timely basis using automated analysis of the communications they performed through emails, help lines, chatbots, and interactions. These activities can be performed by using NLP along with automation tools (Khang et al., 2024b).

Natural language processing is thus distinctively applied to automated business processes, as it checks for semantic errors and creates legible texts for process compliance (Rui et al., 2016).

14.8 CONCLUSION

The main work in this chapter discusses different RPA tools and techniques and how they are applied by companies for data management. It also focuses on illustrations and novel use cases, in which RPA finds its takers, its future, and comparison with cognitive process automation.

In addition to the discussion of the decision process for choosing between different RPA tools and techniques, the cost of automation is also discussed.

Cognitive automation and RPA can be combined in many ways, and different use cases demonstrate the same result. This combination can be performed for structured and unstructured data.

REFERENCES

Automation Anywhere, (n.d.). *Automation Anywhere Platform*. www.automationanywhere.com/

Bhambri P., Rani S., Gupta G., Khang A., (2022). *Cloud and Fog Computing Platforms for Internet of Things*. CRC Press. https://doi.org/ 10.1201/9781003213888

Blue Prism, SS & C Blue Prism | Intelligent Robotic Process Automation—RPA. (n.d.) www. blueprism.com/

Christian C., Peter F., (Eds.). (2021). *Walter de Gruyter. Robotic Process Automation* (1st ed.). GmbH Publication. https://books.google.com.vn/books?id=HFw8EAAAQBAJ

Ibrahim A. H., Tan Z., Nicolai Baek T., Xiaodong D., (2016). "User acceptance of social robots, international conference on advances in computer-human interaction, 2016," *Physical Review*, 47, 777–780. www.researchgate.net/profile/Ibrahim-Hameed/publication/298060667_User_Acceptance_of_Social_Robots/links/5ce1115492851c4ea-bacf1ce/User-Acceptance-of-Social-Robots.pdf

Jianwei M., Haixia J., (2022). "Application of financial robots based on RPA technology in small and medium-sized enterprises," *International Conference on Knowledge Engineering and Communication Systems (ICKES)*, 2022, 1–7. https://ieeexplore.ieee.org/abstract/document/10060387/

Khang A., (Eds.). (2024a). *(AIoCF) AI-Oriented Competency Framework for Talent Management in the Digital Economy: Models, Technologies, Applications, and Implementation*. CRC Press. https://doi.org/10.1201/9781003440901

Khang A., Abdullayev V. A., Vladimir H., Vrushank S., (2024b). *Advanced IoT Technologies and Applications in the Industry 4.0 Digital Economy* (1st ed.). CRC Press. https://doi.org/10.1201/9781003434269

Khang A., Gupta S. K., Shah V., Misra A., (Eds.). (2023a). *AI-Aided IoT Technologies and Applications in the Smart Business and Production*. CRC Press. https://doi.org/10.1201/9781003392224

Khang A., Gupta S. K., Hajimahmud V. A., Babasaheb J., Morris G., (2023b). *AI-Centric Modelling and Analytics: Concepts, Designs, Technologies, and Applications* (1st ed.). CRC Press. https://doi.org/10.1201/9781003400110

Khang A., Rana G., Tailor R. K., Hajimahmud V. A., (Eds.). (2023c). *Data-Centric AI Solutions and Emerging Technologies in the Healthcare Ecosystem*. CRC Press. https://doi.org/10.1201/9781003356189

Khang A., (2021). "Material4Studies," *Material of Computer Science, Artificial Intelligence, Data Science, IoT, Blockchain, Cloud, Metaverse, Cybersecurity for Studies.* https://www.researchgate.net/publication/370156102_Material4Studies

Khanh H. H., Khang A., (2021). "The role of artificial intelligence in blockchain applications," *Reinventing Manufacturing and Business Processes through Artificial Intelligence,* 2(20–40) (CRC Press). https://doi.org/10.1201/9781003145011-2

Kofax Kapow, (2023). *RPA Software—What is Robotic Process Automation.* www.kofax.com/products/rpa

Lukasz A., Wroclaw University of Science and Technology (Poland), AVAYA (USA), (2020). *Applied NLP—Data, Models, Pipelines, and Business Decisions: Lessons from the Field.* www.hilarispublisher.com/open-access/applied-nlp--data-models-pipelines-business-decisions-lessons-from-the-field.pdf

Luke J., Khang A., Vadivelraju C., Antony R. P., Kumar S., (Eds.). (2023c). "Smart city concepts, models, technologies and applications," In *Smart Cities: IoT Technologies, Big Data Solutions, Cloud Platforms, and Cybersecurity Techniques* (1st ed.). CRC Press. https://doi.org/10.1201/9781003376064-1

Nandan M., Arun Kumar A., (2020). *Robotic Process Automation Projects* (1st ed.). Packt Publication. https://link.springer.com/article/ 10.1007/s12525-019-00365-8

NICE RPA, (2023). *NICE Robotic Process Automation & Artificial Intelligence.* www.nicerpa.com/

Rana G., Khang A., Sharma R., Goel A. K., Dubey A. K., (Eds.). (2021). *Reinventing Manufacturing and Business Processes through Artificial Intelligence.* CRC Press. https://doi.org/10.1201/9781003145011

Rani S., Bhambri P., Kataria A., Khang A., Sivaraman A. K., (2023). *Big Data, Cloud Computing and IoT: Tools and Applications* (1st ed.). Chapman and Hall/CRC. https://doi.org/10.1201/9781003298335

Rani S., Chauhan M., Kataria A., Khang A., (Eds.). (2021). "IoT Equipped Intelligent Distributed Framework for Smart Healthcare Systems," In *Networking and Internet Architecture,* Vol. 2, p. 30. CRC Press. https://doi.org/10.48550/arXiv.2110.04997

Rui V. O., do Nascimento V. P., Dal Sasso Freitas C. M., (2016). "A systematic literature review on natural language processing in business process identification and modeling," *International Conference on Advances in Computer-Human Interaction.* Copyright (c) IARIA, 2016. www.lume.ufrgs.br/handle/10183/150923

Santiago A., Alejandro R., (2017). "Automation of a business process using robotic process automation (RPA): A case study, workshop on engineering applications, Communications in Computer and Information Science book series (CCIS, volume 742)," *Physical Review,* 47, 777–780 (Springer). https://link.springer.com/chapter/10.1007/978-3-319-66963-2_7

Shanee H., Alon B., Oron-Gilad T., (2020). *Using Customers' Online Reviews to Identify and Classify Human-Robot Interaction Failures in Domestic Robots, Companion of the 2020 ACM/IEEE International Conference on Human-Robot Interaction, 2020.* https://dl.acm.org/doi/abs/10.1145/3371382.3378323

Shruti J., Suma S., Sarika C. G., (2017). "Survey on socially intelligent robots by using NLP," *International Journal of Computer Applications (0975–8887),* 171(1). www.ijcaonline.org

Tailor R. K., Ranu P., Khang A., (2022). "Robot process automation in blockchain," In Khang A., Chowdhury S., Sharma S., Eds., *The Data-Driven Blockchain Ecosystem: Fundamentals, Applications, and Emerging Technologies,* Vol. 8, No. 13, pp. 149–164 (1st ed.). CRC Press. https://doi.org/10.1201/9781003269281-8

Tom T., Apress Publication, (2020). *The Robotic Process Automation Handbook. A Guide to Implementing RPA Systems.* Monrovia, CA, USA. https://link.springer.com/content/pdf/10.1007/978-1-4842-5729-6.pdf

UiPath, (2023). AI-powered UiPath Business Automation Platform™. www.uipath.com/

15 Artificial Intelligence-Enabled Bibliometric Analysis in Tourism and Hospitality Using Biblioshiny and VOSviewer Software

Dilip Kumar, Abhinav Kumar Shandilya, and Sajjan Choudhuri

15.1 INTRODUCTION

In the era of globalization, customization, and consumerism, there is an indiscriminate race to gain the trust and faith of customers/guests to generate and survive in the tourism and hospitality industry (T&HI).

The most influential factor in tourism and hospitality (T&H) is loyalty, which includes satisfaction and trust (Florencio et al., 2020). Guest satisfaction has long been a crucial subject in tourism research. According to Hu et al. (2009) and Velázquez et al. (2015), high levels of customer satisfaction also encourage people to suggest hotels, pay more, and stay there again.

Service providers personalize customer experiences to promote uniqueness and loyalty. Tourist accommodation experience is a fundamental component of the overall visiting experience (Kau & Lim, 2005).

Because of travel restrictions, border closures, government restrictions, event cancellations, quarantine procedures, and spread-related concerns, COVID-19 has temporarily halted T&HI operations (Gössling et al., 2020).

The second decade of 2020 ended with a significantly more upsetting and unpleasant incidence of the new diseases (Divvela et al., 2023), in addition to the more than 30 novel viruses that the world has encountered during the preceding 30 years (Nkengasong, 2020).

The spread of COVID-19 has brought the world to its knees, and people have lost their lives and resources. Globally, the coronavirus has been identified in over

DOI: 10.1201/9781003400110-15

590 million people and has caused the death of over 6 million people (www.worldometers.info/coronavirus/).

The travel and hospitality sectors have recognized the potential usefulness of big data analytics. It has developed the capacity to transform big data into beneficial real-world information and insights, such as enhancing T&H management practices, which are considered significant viable benefits (Liu et al., 2017).

Tourism is still one of the most lively industries in the economy and has evolved significantly over time. It is well recognized that adopting information and communication technology may benefit general businesses and hotel establishments (Buhalis & Law, 2008). This industry had to quickly adjust to a changing technological environment to stay competitive.

The relevance of the topic is highlighted by the presence of only a few thorough literature evaluations of ample data in the T&H sector (Lyu et al., 2022).

Businesses in the hotel sector can use artificial intelligence (AI) to better utilize on-site services and procedures and enhance client (guest) experiences. It is crucial to stay in touch with visitors and attend to their demands to maintain the overall quality.

Because AI makes it possible to provide individualized experiences, the idea of a "smart hotel" has recently gained significant attention from academic (Buhalis & Leung, 2018) and corporate (Perini, n.d.) communities (Krce et al., 2012). Therefore, machine learning (ML)/AI has become necessary to survive and propel toward the objectives.

Recent research indicates that ML fundamentally alters commercial landscapes. Its vast possibilities have been shown to disrupt labor and drive organizations to reevaluate their marketing tactics (Juaneda-Ayensa et al., 2016).

The present study uses Biblioshiny (Bibliometrix analysis using R) to examine the scientific development of AI and ML in T&HI, and VOSviewer to investigate the publication and citation trend (objective 1), scientific production of countries (objective 2), scientific production of authors (objective 3), scientific production of institutions (objective 4), scientific collaboration of countries (objective 5), scientific production by sources and dissemination of sources (objective 6), classification and analysis of keywords (objective 7), citation analysis (objective 8), and segregation of keywords based on centrality and density.

15.2 ARTIFICIAL INTELLIGENCE/MACHINE LEARNING

The push to incorporate ML, a subset of artificial AI, into the T&H sector has grown since the dawn of the 21st century. This topic has recently gathered attention with numerous advancements in AI-based decision assistance approaches.

A branch of AI called ML uses computers to analyze data and build models that can be used to solve problems. Machine learning differs from conventional programming in several aspects. Traditionally, rules are not explicitly learned from the data but are coded in a computer language.

In contrast to traditional programming, ML creates predictive models that can be applied to forecasts using upcoming data. Owing to the intricacy of the code, much work might be required to design a rule-based program for certain problems.

In such situations, ML can be employed if sufficient data are available for the problem under consideration (Geisel, 2018). Managers can alter the business models of their firms by applying AI and ML.

For example, using AI and ML improves risk management, client identification, client retention, and monitoring (Jain & Pandey, 2017). Customer emotions are influenced by effort expectancy and performance (Roy et al., 2020).

Academics and businesspeople have described customers' experiences as a complex construct, including social, emotional, behavioral, cognitive, and sensory (Zollo et al., 2020). The T&H sector has always been seen as a service sector that places greater emphasis on experience (Ii & Disney, 2002).

Predicting algorithms will eventually be crucial in determining client experience (Hoyer et al., 2020). Sánchez-Franco et al. (2019) claimed that organizations in the hospitality sector have been using concepts such as "guest experience" to identify opportunities to improve hospitality services.

Companies' adoption of a crude algorithm model enables them to apply a supervised ML method to process a sizable amount of data.

As a result, the manager can categorize hotel reviews with more accuracy and lower computational costs. Customer support is delivered more efficiently and rapidly through ML.

Chatbots can help reduce waiting times by responding to all inquiries immediately and by using ML. (Rushdi et al., 2011).

Since 2008, Airbnb has allowed travelers to travel in a more distinctive, customized fashion, and it has quickly risen to the top of the list of websites used to book vacation accommodations (Lutz & Newlands, 2018).

Sharma et al. (2021) used explainable ML techniques to understand the features of customer experience that are important for pricing a variety of accommodations, which can assist businesses in deciding the components of the customer experience for future progress.

15.3 METHODOLOGY

The present study used a bibliometric analysis method, beginning with a search of the Scopus database, to better understand, quantify, and analyze the available scientific literature on AI in the T&HI.

The study was conducted using a three-phase technique that included an implementation plan, data collection, and bibliometrics. The bibliometric analysis was made up of two events.

The first focused on the domain and used numerous metrics, such as Bradford's law and Lotka's law, to provide unbiased and quantifiable information to understand trends.

The second, which focused on knowledge structures, assessed social, intellectual, and conceptual structures using factorial analysis and scientific mapping utilizing bibliometric techniques, such as cooperation, co-citation, and co-word (Cuccurullo et al., 2016).

Science mapping enables statistical investigation and creation of a global image of scientific knowledge. It primarily uses the three knowledge structures to convey the dynamics of scientific inquiry (Van, 2007) and structural elements and to identify representations of intellectual links (Small, 1999).

The social structure demonstrates the relationships among authors, institutions, and countries. The conceptual structure indicates what science is debating, including critical issues and trends, whereas the intellectual framework illustrates how a particular author's contributions impact a specific scientific field. Together, these structures provide a comprehensive view of knowledge (Aria & Cuccurullo, 2017).

Therefore, the main research questions of this bibliometric study are as follows:

- RQ1: What are the main keywords of AI in the hospitality/tourism/aviation/catering industry?
- RQ2: How do authors determine the impact of AI on the hospitality/tourism/aviation/catering industry?
- RQ3: How do authors, institutions, and countries collate their AI findings in the hospitality/tourism/aviation/catering industry?
- RQ4: How do we explore the understudied and overstudied research keywords in the last 35 years? The data collection and search criteria were as follows.

A data search was performed on July 5, 2022, through one of the most reliable data sources (i.e., Scopus) from the virtual private network (VPN) of Chandigarh University. Keywords, such as "Machine learning" OR "Artificial intelligence," "AI" OR "Robotics," AND ("Hospitality" OR "Tourism" OR "Aviation" OR "Catering"), and language=English were searched in the Scopus database since its inception.

A research article, conference proceedings, and published and unpublished articles were extracted. A total of 1348 metadata were extracted, but some of the data had missing, incorrect, or inappropriate information, which was removed during data cleaning. After removing 196 data (153 lecture notes; 43 no author names) from the metadata, 1152 data were found fit for final analysis; 12.44% of the annual growth rate of the present publication context was found during the study.

In the current study, articles accessible through the Scopus database between 1987 and July 5, 2022, were considered and exported as an a.csv file containing citations, bibliographic data, abstract and keyword data, funding, details, and other information.

15.4 RELATED WORK

15.4.1 ANALYSIS AND VISUALIZATION OF DATA

For the bibliometric analysis, we used open-source statistical software (i.e., Bibliometric R package 4.2.0) through the Biblioshiny version in the present study.

Biblioshiny produces an abundance of outputs, including a summary of the key facts, annual citations, annual scientific output, three field plots, sources, the most pertinent and locally cited sources, Bradford's law, source impact, and source dynamics.

In addition, it generates authors based on their importance, local citation frequency, productivity over time, Lotka's law, and influence. Affiliations, countries, documents, clustering, conceptual structures, intellectual structures, and social structures can also be analyzed and viewed (Aria & Cuccurullo, 2017).

For publication trends and citation analysis, Microsoft Excel 2016 (www.microsoft.com) was used to present the histogram, chart, graph, and trend line and to better understand the increasing/decreasing importance of AI since its inception.

The VOSviewer software was used to visualize the network in the present study for mapping and clustering, as it uses a unified structure (Waltman et al., 2010). Since 2006, more than 500 publications have used it for graphical representation (www.vosviewer.com/publications).

The VOSviewer software is a handy tool for building and viewing large bibliometric maps, with focus on a visual representation that includes journals, institutions, or authors based on co-citation, bibliographic coupling, citation, or co-authorship relationships. It uses circles, lines, and colors to depict the relationships between authors, documents, countries, etc.

Bibliometrix (Biblioshiny) and the VOSviewer software are used to visualize and extract the desired information in tables and images required for further analysis.

15.4.2 MAIN INFORMATION ABOUT THE STUDY

This study explores the importance of ML in the hospitality/tourism/aviation/catering industry using a quantitative approach to metadata extracted from the Scopus database.

Data regarding AI from 1987 to July 5, 2022, found 629 sources (journals, books, etc.) and 1152 documents with an annual growth rate of 12.44%, which can be considered very well and explains the area's growing demand.

The average age of the documents was 4.68, the average number of citations per year was 11.2, and the total number of references was 1568.

Two types of keywords were identified (i.e., 3748 KeyWords Plus and 3127 authors' keywords). A total of 3030 authors were identified in the metadata, and 179 single-author documents were found in the study.

While checking the authors' collaboration, 195 single-author documents were identified, 3.11 co-authored per document were identified, and 3.472% of international co-authorships were found. The information contained various materials, including articles, books, book chapters, and conference papers.

The maximum contribution was found in a conference paper (i.e., 565), whereas the minimum contribution was from a short survey (i.e., one shown in Table 15.1).

15.4.3 EMERGENCE OF SOURCES AND CITATION ANALYSIS

The presence of documents on AI has been in the hospitality and tourism industry since 1987 and has explored 1348 documents, but 196 data were removed from the dataset during sorting and cleaning. In the recent past (five years), there has been an upward trend in publications, especially after the COVID-19 outbreak.

The maximum number of publications can be seen in 2021, followed by 185 in 2021, and by July 5, 2022, 121 documents have already been published. The maximum

TABLE 15.1
Information About the Metadata.

Description	Results	Description	Results
Main information about the data		**Authors' collaboration**	
Time span	1987:2022	Single-author documents	195
Sources (journals, books, etc.)	629	Co-authored per document	3.11
Documents	1152	International co-authorships (%)	3.472
Annual growth rate (%)	12.44	**Document types**	
Document average age	4.68	Article	494
Average citations per document	11.2	Book	8
References	1568	Book chapter	30
Document contents		Conference paper	565
KeyWords Plus (ID)	3748	Editorial	7
Author's keywords (DE)	3127	Erratum	1
Authors		Letter	3
Authors	3030	Note	2
Authors of single-author documents	179	Retracted	7
		Review	34
		Short survey	1

(Source: Scopus database; software: Biblioshiny)

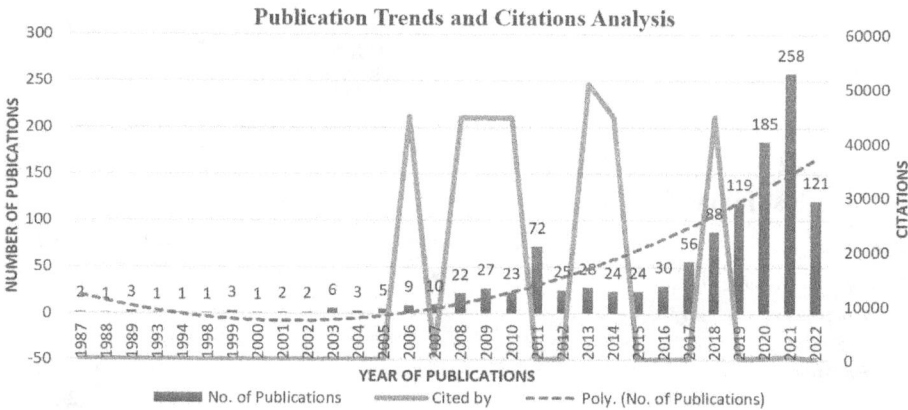

FIGURE 15.1   Publications, Citations, and Publication Trends.

Source: Scopus database; software: Microsoft Excel

number of citations related to the present study can be seen in 2013 (approximately 5000). The trend line also appears in a bell-curved shape, as shown in Figure 15.1.

In total, 1152 documents were published from 629 sources. Bradford's distribution law states that if journals are split into three categories, each group will contain one-third of the articles. The distribution of journals in each group was arranged in the proportion of 1:n:n2 (Bradford, 1934).

In the present study, it is divided into three clusters (i.e., cluster 1 contains the top 24 journals (3.81%) with 385 publications (33.42%), cluster 2 contains the following top 225 journals (38.95%) with 387 publications (33.59%), and cluster 3 contains 480 journals (76.31%) with 380 publications (32.99%). These documents have 322,004 total citations, as shown in Table 15.2.

15.4.3.1 Authors

A total of 3030 authors have been identified in writing research articles related to AI/ML/robotics in the hospitality/tourism/aviation/catering industry because 1987.2.63 authors per article were found, whereas the average number of articles per author was found to be 0.38.

The number of co-authors per document was 3.11. There were 179 (16.92%) authors of single-author documents. The percentage of international co-authorship was 3.472%.

Articles written by the authors ranged from 1 to 14, where one article was written by 2725 authors (0.899%), two articles were published by 209 (0.069%), and three articles were published by 46 authors (0.015%), which constitute the maximum proportion of authors, as shown in Table 15.3.

Three authors (0.001%) published 10 articles, and two authors (0.001%) published 12 articles. In contrast, only one author has published 14 articles on AI in the hospitality/catering/tourism/aviation industry, as shown in Figure 15.2.

15.4.3.2 Countries

Since 1987, 61 countries have been identified as contributing to the field of AI in the hospitality/tourism/aviation/catering industry.

TABLE 15.2
Top 10 Most Impactful Sources.

Source	h-index	g-index	m-index	TC	NP	PY_start
International Journal of Contemporary Hospitality Management	10	17	0.476	531	17	2002
Tourism Management	10	11	0.5	975	11	2003
Annals of Tourism Research	7	7	1.75	437	7	2019
Expert Systems with Applications	7	7	0.5	996	7	2009
International Journal of Hospitality Management	7	9	0.194	433	9	1987
Tourism Review	7	13	0.538	215	13	2010
Advances in Intelligent Systems and Computing	5	5	0.625	48	16	2015
Current Issues in Tourism	5	6	0.333	76	6	2008
Journal of Tourism Futures	5	6	1.667	75	6	2020
Sustainability (Switzerland)	5	7	1	64	14	2018

(Source: Scopus database; software: Biblioshiny)

(TC, total citations; NP, number of publications; PY_start, year of first publications)

TABLE 15.3
Author's Productivity Using Lotka's Law.

Documents written	No. of authors	Proportion of authors
1	2725	0.899
2	209	0.069
3	46	0.015
4	16	0.005
5	12	0.004
6	8	0.003
7	4	0.001
8	1	0
9	3	0.001
10	3	0.001
12	2	0.001
14	1	0

(Source: Scopus database; software: Biblioshiny)

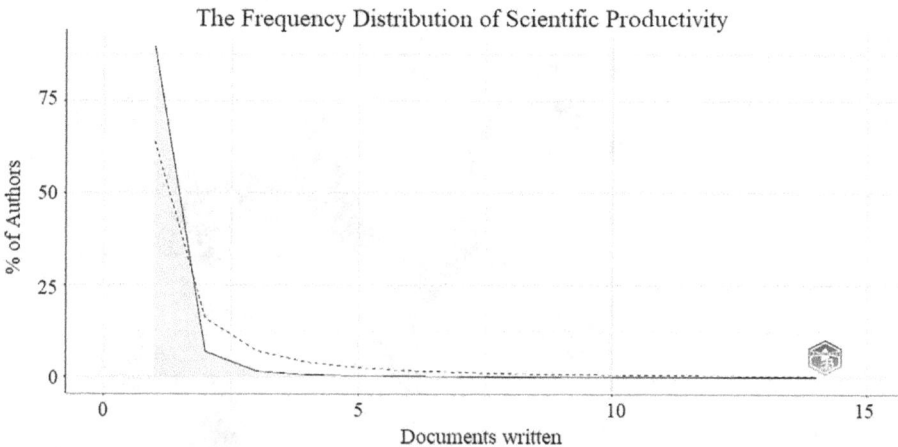

FIGURE 15.2 Author's Productivity Using Lotka's Law.

Source: Scopus database; software: Biblioshiny

China was found to be the most significant contributor, with 234 publications (20.31%), the USA was ranked second with 54 publications (4.69%), and Spain ranked third with only 42 publications (3.65%) in the mentioned area.

Thailand, Germany, and Canada ranked 20th, 19th, and 18th with only nine publications each in the field of AI. Figure 15.3 shows the countries' contributions to AI and depicts their future planning and its importance and utilities soon.

China always remains ahead in every research area, and the same can be seen in the present study. In the present study, there were fewer multiple-country publications (M.C.P.) than single-country publications (S.C.P.), as shown in Figure 15.4.

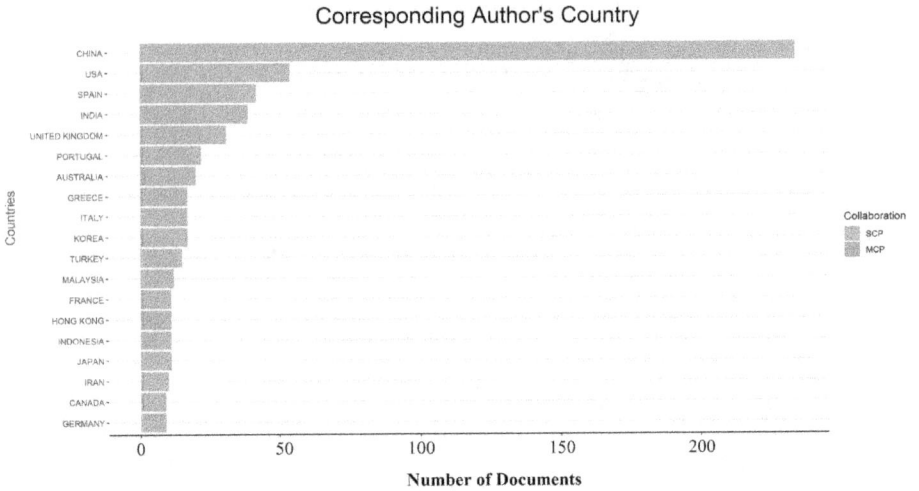

FIGURE 15.3 Collaboration of Countries.

Source: Scopus database; software: Biblioshiny

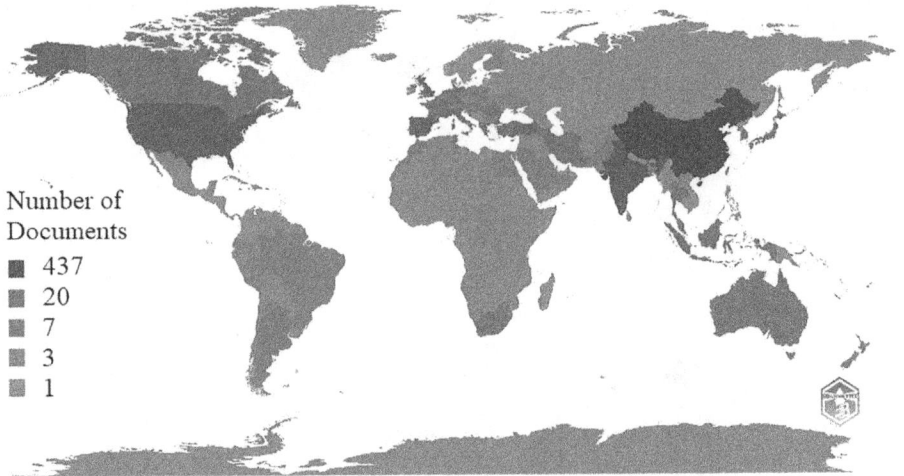

FIGURE 15.4 Scientific Production of Countries.

Source: Scopus database; software: Biblioshiny

15.4.3.3 Citations

The metadata were extracted from the Scopus database on July 5, 2022; only 1152 clean documents were extracted. A total of 12,906 citations were received, and the average citations/document was 11.2.258, which is the maximum number of documents published in 2021, whereas the maximum citations were found in 2020 (i.e., 1852 published 185 documents). The maximum average of 139 citations was found in 2005, but only five documents were published.

In 2019, 118 documents were published, and the average number of citations was 15.25 (more than 50 documents published in a particular year). The lowest citations were in 2000 and 1988, with only one average citation.

A document with the maximum number of citations is "Modeling the world from Internet photo collections" by Snavely et al. (2008), published in the *International Journal of Computer Vision* in 2008, having citations of 1567, as shown in Table 15.4.

In contrast, "A comparison of three different approaches to tourist arrival forecasting" (Cho, 2003), published in *Tourism Management* in 2003, received a minimum of 200 citations, although it was published five years later than the most cited documents, as shown in Table 15.5.

TABLE 15.4
Top 10 Most Globally Cited Documents.

TI	Authors	SO	PY	TC	TCpY
Modeling the world from Internet photo collections	Snavely et al. (2008)	*International Journal of Computer Vision*	2008	1567	104.47
Sentiment classification of online reviews to travel destinations by supervised machine learning approaches	Ye et al. (2009)	*Expert Systems with Applications*	2009	440	31.43
Learning concept hierarchies from text corpora using formal concept analysis	Hotho et al. (2005)	*Journal of Artificial Intelligence Research*	2005	412	22.89
Intelligent tourism recommender systems: A survey	Borràs et al. (2014)	*Expert Systems with Applications*	2014	303	33.67
User-generated content: The use of blogs for tourism organizations and tourism consumers	Akehurst (2009)	*Service Business*	2008	268	19.14
Effects of COVID-19 on hotel marketing and management: A perspective article	Jiang & Wen (2020)	*International Journal of Contemporary Hospitality Management*	2020	265	88.33
Structural, semantic interconnections: A knowledge-based approach to word sense disambiguation	Navigli & Velardi (2004)	*IEEE Transactions on Pattern Analysis and Machine Intelligence*	2005	263	14.61
From high-touch to high-tech: COVID-19 drives robotics adoption	Zeng et al. (2020)	*Tourism Geographies*	2020	211	70.33
SenticNet: A publicly available semantic resource for opinion mining	Cambria et al. (2010)	*AAAI Fall Symposium*	2010	200	15.38
A comparison of three different approaches to tourist arrival forecasting	Cho (2003)	*Tourism Management*	2003	196	9.80

(Source: Scopus database; software: Microsoft Excel)

TI, title; SO, source; PY, publication year; TC, total citations; TCpY, total citations per year.

TABLE 15.5
Top 10 Local Author Impact.

Authors	h-index	g-index	m-index	TC	NP	PY_start
Law R	7	8	0.35	706	8	2003
Gursoy D	6	7	1.5	315	7	2019
Ivanov S	6	6	1.2	134	6	2018
Webster C	6	6	1.2	149	6	2018
Chi Oh	4	4	1.333	127	4	2020
Li G	4	5	1.333	67	5	2020
Wang S	4	6	0.286	44	7	2009
Buhalis D	3	5	0.75	403	5	2019
Camacho D	3	3	0.167	42	3	2005
Chang Y-C	3	3	0.214	44	3	2009

(Source: Scopus database; software: Biblioshiny)
TC, total citations; NP, number of publications; PY_start, start of publication year

The increasing number of publications and citations suggests that this area is growing, and the future of the T&H sector is shifting toward AI/ML/digitalization.

15.4.4 Keyword Analysis

In the present study, we have written 1152 articles with 3127 authors' keywords and 3748 KeyWords Plus. Zhang et al. (2016) found that keywords are more effective than the authors', especially in scientific knowledge structure, but less impactful in explaining the article's content, as shown in Figure 15.5.

"Artificial" appeared 509 times, "tourism" 137 times, and "decision-making" 57 times (top three keywords), whereas "learning systems," "big data," and "forecasting" emerged fourth, fifth, and sixth in ranking, having 43, 42, and 35 appearances, as shown in Figure 15.6.

15.4.5 Analysis Based on Structures of Knowledge

To understand the research questions in the present study, three structures of knowledge are analyzed with the help of bibliometric analysis using Biblioshiny.

15.4.5.1 Conceptual Structure of Knowledge

Aria & Cuccurullo (2020) discovered that conceptual structure investigates the key ideas and issues, and illustrates the connections between research topics. The conceptual structure of the current study is mapped using two methods, namely, co-words network and factor analysis.

The multiple correspondence analysis (MCA) technique is used to analyze the factor of the author's keywords to reduce the dimensionality of the data. This technique is beneficial in analyzing various relationships of the categorical dependent variables, also known as an extension of correspondence analysis (Abdi & Valentin, 2007).

electric commerce   tourism

learning systems   online teaching

decision making

big data   sustainable development

industry technology   data science

deep learning

data mining   data analytics

machine learning   AI engineering

artificial intelligence   blockchain   cryptocurrency

information technology   healthcare ecosystem

internet of things   medical

FIGURE 15.5   Word Cloud KeyWords Plus.

Source: Scopus database; software: Biblioshiny

electric commerce   tourism   **Robotics**

IIoT   learning systems   online teaching

decision making   COVID-19   Google

big data   sustainable development   RPA

Cloud IoT industry technology   data science

deep learning

data mining   data analytics

Azue   machine learning   AI engineering

artificial intelligence   blockchain   cryptocurrency

AWS   NLP

information technology   healthcare ecosystem

internet of things   medical

FIGURE 15.6   Word Cloud Author Keywords.

Source: Scopus database; software: Biblioshiny

Figure 15.7 represents the conceptual structure map using MCA, having five clusters of the author's keywords presented in different colors (blue, red, green, purple, and orange) to explain a specific topic. Clusters represent the co-occurrence of keywords in the published article.

The VOSviewer software was used to map all the keywords and found that AI, tourism, smart tourism, forecasting, geographical information system, innovation,

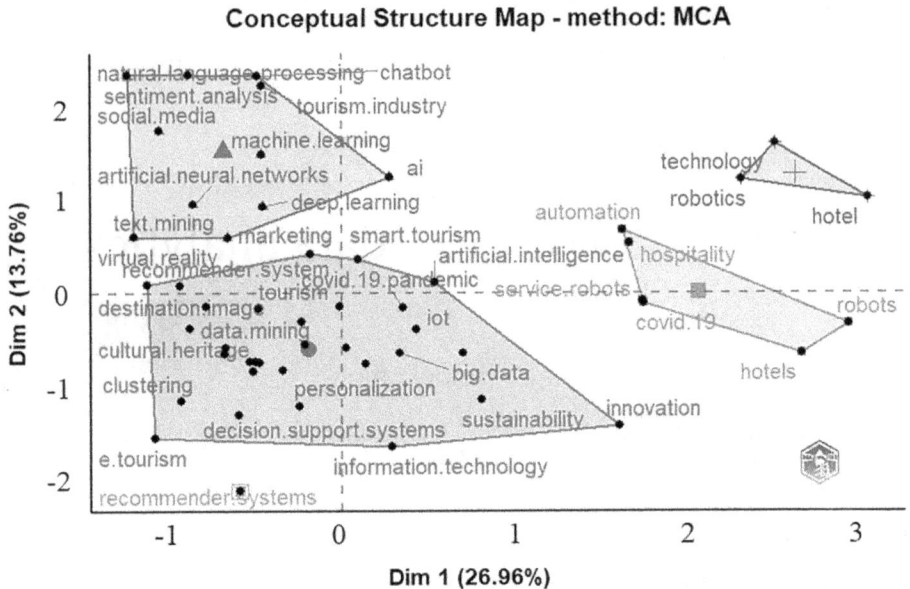

FIGURE 15.7   Conceptual Structure Map.

Source: Scopus database; software: Biblioshiny

tourism development, commerce, and decision support system have a strong relationship and show strong co-occurrence, as shown in Figure 15.8.

The most recent research words are "automation," "smart tourism," "leisure industry," "tourism management," "innovation," "automation," "hospitality industry," "robotics," and "service robots," as shown in Figure 15.9.

It can be easily seen that the cluster related to AI has 14 items (i.e., AI technology, big data, commerce, Internet, Internet of Things (IoT), leisure industry, marketing, science spot, software engineering, surveys, tourism, tourism attractions, and tourism destinations) (Rani et al., 2023), as shown in Figure 15.10.

15.4.5.2   Intellectual Structure of Knowledge

The intellectual structure of the analysis was used to distinguish the thought process of the study. This study determined the influence and contributions of various authors. According to Small (1999), co-citation analysis is considered the most popular citation analysis.

When a third document cites two documents, it is considered a co-citation. In the present study, 1152 documents had 38,391 total bibliographic references.

Figure 15.11 shows the bifurcation of these authors into five clusters (red, blue, green, purple, and yellow) based on AI, tourism, and the hospitality industry.

During the bibliometric analysis through VOSviewer, 53 authors were identified after restricting the minimum number of cited references (seven).

The top five authors and their documents are "Artificial intelligence in service" (Huang & Rust, 2018), "Developing and validating a service robot integration

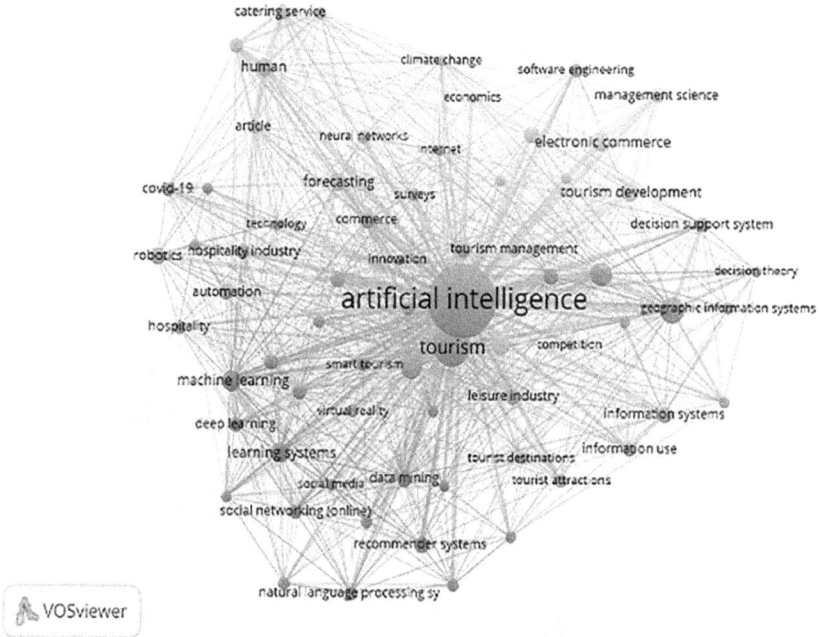

FIGURE 15.8   Network Visualization of Co-Occurrence of Keywords.

Source: Scopus database; software: VOSviewer

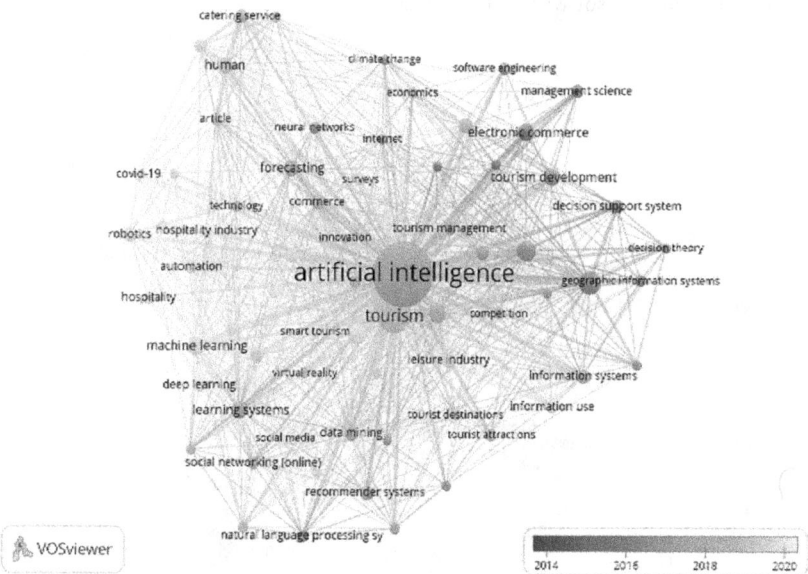

FIGURE 15.9   Overlay Visualization of Authors' Keyword Co-Occurrence.

Source: Scopus database; software: VOSviewer

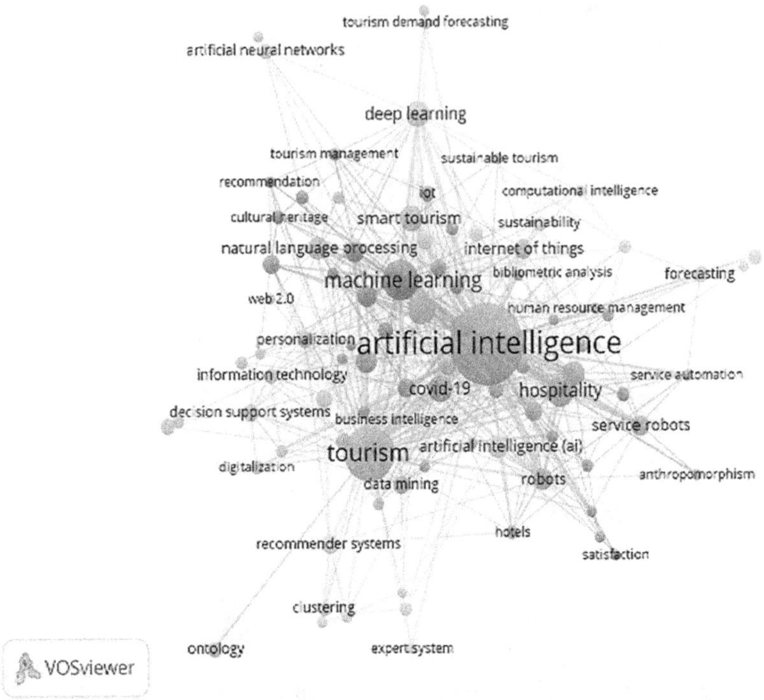

FIGURE 15.10 Network Visualization of Index Keyword Co-Occurrence.

Source: Scopus database; software: VOSviewer

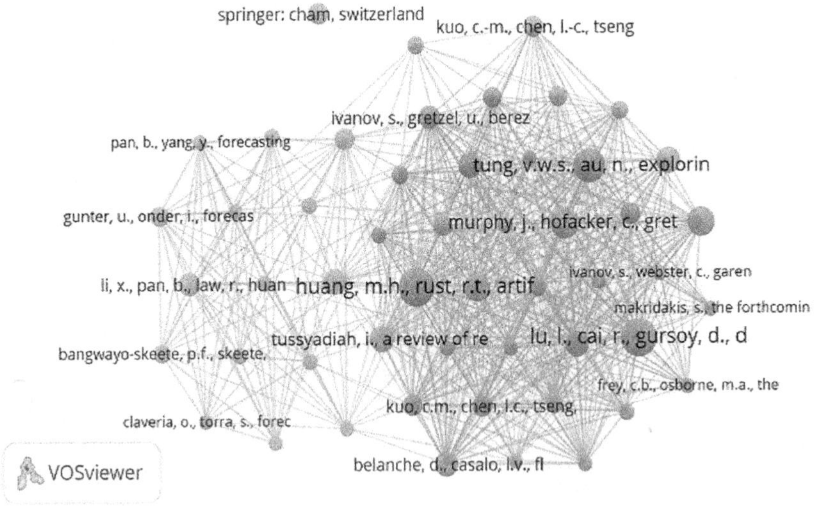

FIGURE 15.11 Bibliographic Coupling of Author's Analysis.

Source: Scopus database; software: VOSviewer

willingness scale" (Lu et al., 2019), "Exploring customer experiences with robot-ics in hospitality" (Tung & Au, 2018), "Dawning of the age of robots in hospi-tality and tourism: Challenges for teaching and research" (Murphy et al., 2017), and "The potential for T&H experience research in human-robot interactions" (Tung & Law, 2017).

The most influential journals are *Tourism Management, Annals of Tourism Research, International Journal of Contemporary Hospitality Management, Journal of Travel Research,* and *International Journal of Hospitality Management,* having 1250, 553, 478, 417, 1, and 340 citations. It can be easily seen that highly impacted journals are co-cited by others' journals, as shown in Figure 15.12.

15.4.5.3 Social Structure of Knowledge

Evaluating collaborative networks, such as co-authorship networks, including coun-tries, authors, and organizations, can be measured through social network analysis (Fagan et al., 2018).

Social structure analysis is utilized to understand the relationship between authors, countries, and organizations and to find the most dominant authors, countries, and research organizations in the context of the studied area (Moed et al., 2004).

Figure 15.13 represents the co-authorship network visualization of authors. During the visualization process, an author's minimum number of documents and citations was restricted to five, in which out of 3007 authors, only 30 met the desired threshold value.

While visualizing, the most extensive set of connected items was found, only 23, but all 30 items were used in visualization and classified into 10 clusters (red, green, cobalt blue hue, yellow, violet, crimson blue, orange, brown, purple, and pink). The closest circle shows the authors with the closest research collaboration, as shown in Figure 15.13.

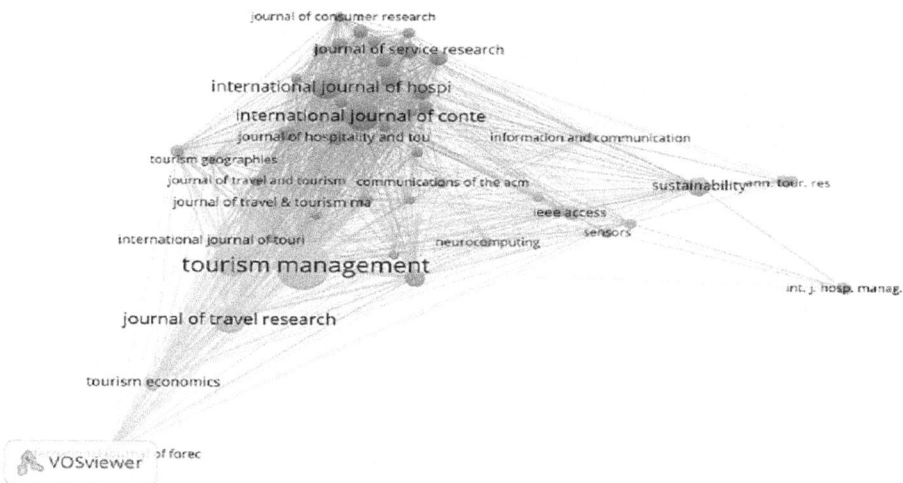

FIGURE 15.12   Bibliographic Coupling of Analysis of Sources.

Source: Scopus database; software: VOSviewer

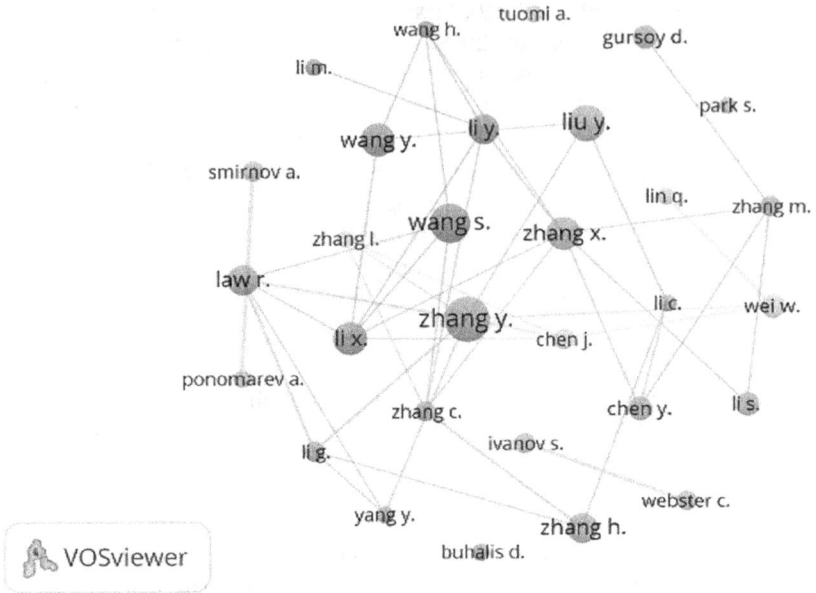

FIGURE 15.13   Co-Authorship and Author's Analysis.

Source: Scopus database; software: VOSviewer

The thickest is the connecting line, and the most substantial is collaboration. China's biggest circle represents the maximum number of publications, followed by the USA, the UK, and India.

The collaboration of the countries, represented by thick lines, is China (21 links)—India, Macau, Hong Kong, Taiwan, Australia, South Korea, South Africa, USA, and the UK; the USA (28 links)—South Africa, China, South Korea, Bulgaria, India, the UK, Italy, Canada, and Australia; the UK (25 links)—India, USA, China, Malaysia, Turkey, Hong Kong, Italy, and Greece; Spain (13 links)—France, Portugal, Japan, China, South Africa, and Italy. Ten clusters were identified. The USA was identified as the country with the maximum number of collaborations. The UK, China, and Spain have followed the USA in collaboration, as shown in Figure 15.14.

Only 46 of the 2163 organizations that were found to be active in the context of the current investigation satisfied the required standards.

Only five items were seen as the most extensive set of connected items; however, all 46 items were considered, and 36 clusters were developed for visualization. There is robust, diversified collaboration between Asian, African, and American organizations.

"Shandong Xiehe University" (China) has nine documents, the "School of Tourism and Hospitality, University of Johannesburg (South Africa)" has four documents, and the "School of Hospitality Business Management, Carson College of Business, Washington State University (United States)" has four documents, which were found as the most productive organizations (Figure 15.15).

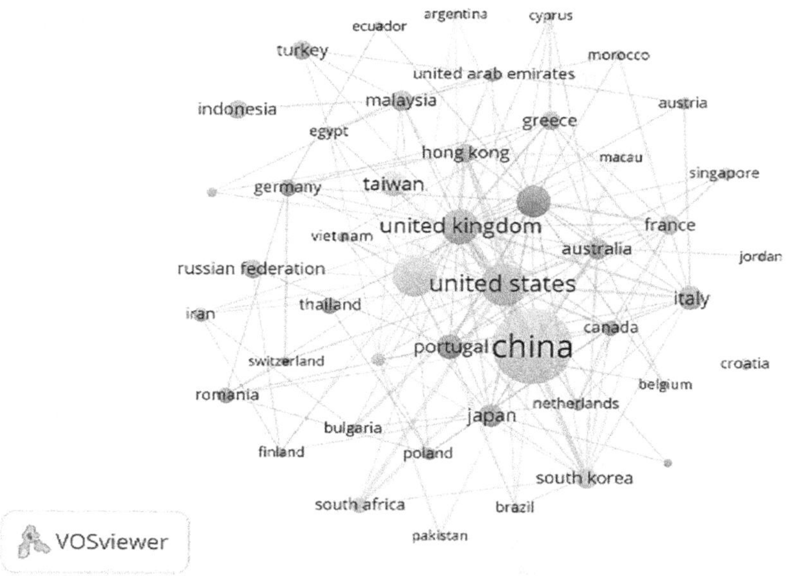

FIGURE 15.14   Bibliographic Coupling of Analysis of Countries.

Source: Scopus database; software: VOSviewer

FIGURE 15.15   Bibliographic Coupling of Analysis of Organizations.

Source: Scopus database; software: VOSviewer

15.5   RESULTS AND DISCUSSION

The present study is a step toward investigating the trends and demand for AI in T&HI, especially when everything is becoming touchless after COVID-19 (Rana et al., 2021).

Human touch cannot be replaced with technology because of various factors, such as personalization, customization, and empathy, but technology is becoming helpful in generating efficient and effective businesses.

Artificial intelligence is becoming very effective for T&HI. A bibliometric analysis using Biblioshiny (using R) and VOSviewer was performed to investigate its growing importance.

Metadata were extracted from the Scopus database (July 5, 2022) from its inception 1987 to 2022. After filtering the data, 1152 clean datasets were used for the final study. Various research gaps were identified, and the research objectives were framed based on these gaps.

15.5.1 Objective 1: To Investigate the Research Trend, Cluster Research, and the Evolution of Recent Research Domains in Tourism and the Hospitality Industry

Research trends were investigated with the help of Biblioshiny, and 1152 documents published in the last 35 years were found in 629 sources (journals, conference proceedings, books, etc.), with an annual growth rate of 12.44%.

A total of 3030 authors were identified, of which a single author wrote 179 documents. Single-author document collaboration was found at 195, and co-authors per document were seen at 3.11. In contrast, the international co-authorship percentage or collaboration index (CI) was found to be 3.472, which was found stable and exciting compared to other research themes.

"Social media" and "sustainable tourism" have a CI of 3.4, which is equivalent to that of the present study (Martí-Parreño & Gómez-Calvet, 2020). In contrast, "Airbnb" has a CI of 2.26 (Andreu et al., 2020), "revenue management in airlines" has a CI of 1.85 (Raza et al., 2020), and "information technologies" has the lowest CI of 0.71 (Khaparde Professor & Pawar, 2013).

Owing to the multidisciplinary and interdisciplinary nature of the authors, as it covers engineering, management, hospitality, and tourism, the CI is probably high.

Using MCA, a conceptual analysis of the author keywords generated five clusters, as shown in Figure 15.7.

The most significant and impactful cluster is in red, showing the collaboration of AI, smart tourism, the COVID-19 pandemic, tourism, marketing, the hospitality industry, big data, etc., in an emerging area. The cluster in blue is the second most impactful, followed by green, violet, and orange (least collaborative).

The main keywords associated with the research domain from 1987 to 2013 are artificial intelligence, collaboration with the IoT, AHP, e-tourism, tourism, and artificial intelligence (Bhambri et al., 2022).

From 2014 to 2018, modeling (biggest) collaborated with forecasting, service robot, the COVID-19 pandemic, and AI. Forecasting was the most significant contributor between 2019 and 2020, collaborating with bibliometrics, text mining, smart city, AI, and chatbots (Khang et al., 2022b).

From 2020 to 2021, bibliometrics was the most significant keyword contributor, collaborating with AI and COVID-19, as shown in Figure 15.16.

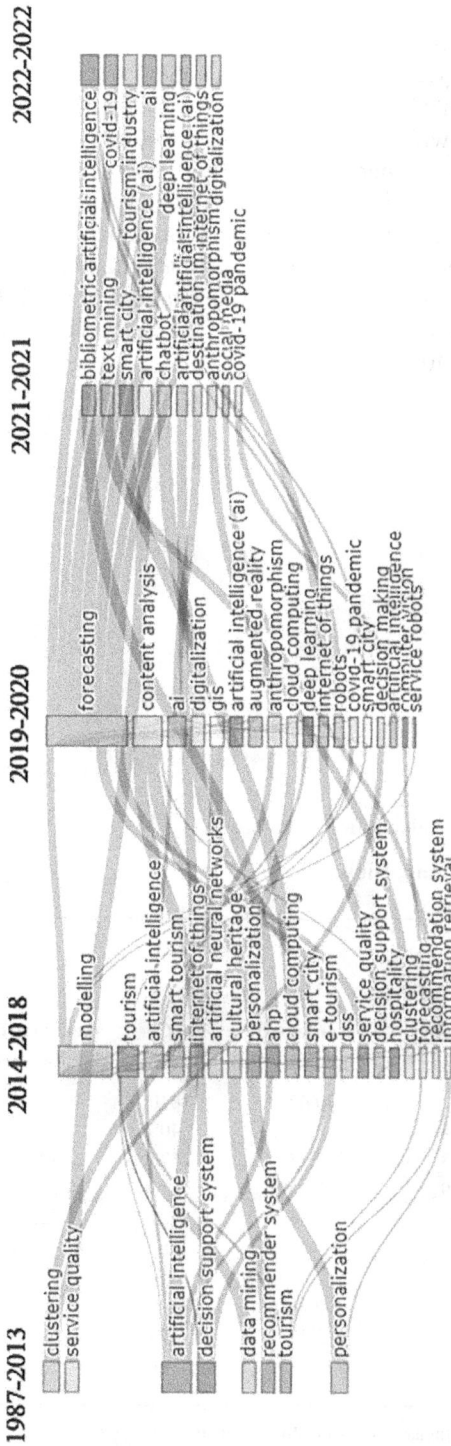

FIGURE 15.16   Thematic Evolution.

Source: Scopus database; software: Biblioshiny

15.5.2 OBJECTIVE 2: TO INVESTIGATE THE SCIENTIFIC PRODUCTION BY COUNTRIES

Since 1987, various countries have published research articles on AI in T&HI. China, USA, India, Spain, and the UK are the most significant contributors in this area.

China is the biggest contributor with 2575 publications, followed by USA with 630 publications and India with 595 publications, as shown in Figure 15.17.

USA has the maximum number of citations (i.e., 2678), and the average article citations were 49.59. China ranked number two in the maximum number of citations (1840), and the average article citations were 7.86. Spain has 873 citations, and the average article citations were 20.79 (more than China's citations).

Germany has 609 citations and average article citations of 67.67. Hong Kong has 654 citations and average citations of 59.45, and the Netherlands has 297 citations and average citations of 59.40 (Table 15.6).

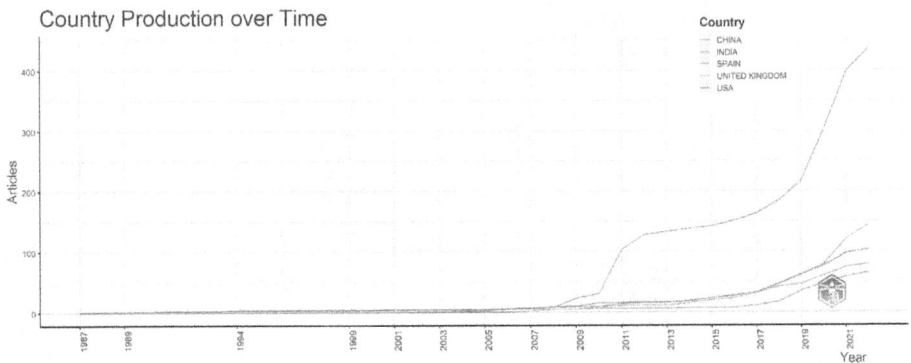

FIGURE 15.17   Countries' Production Overtime.

Source: Scopus database; software: Biblioshiny

TABLE 15.6
Most Cited Countries.

Country	Frequency	Percentage of frequency (%)	TC	Average article citations
China	437	30.67	1840	7.86
India	144	10.11	408	10.46
USA	103	7.23	2678	49.59
Spain	79	5.54	873	20.79
UK	65	4.56	707	22.81
Portugal	41	2.88	80	3.64
Turkey	38	2.67	50	3.33
Italy	35	2.46	399	23.47
Indonesia	30	2.11	73	3.64
Japan	30	2.11	66	6.00

(Source: Scopus database; software: Microsoft Excel)

15.5.3 OBJECTIVE 3: TO INVESTIGATE THE SCIENTIFIC PRODUCTION BY AUTHORS

The present study identified the involvement of 3030 authors in writing 1152 documents (articles) from 1987 to 2022 (July 5, 2022). Zhan et al., who published 14 articles (maximum), have a 6.23 article fractionalized value, followed by Liu Y, Wang S, Wang Y, and Li X, who published 12, 12, 10, and 10 articles and have fractionalized values of 5.80, 4.69, 4.37, and 2.79, respectively.

When assessing an author's relevance to a specific topic, productivity and influence are two critical factors. Both metrics are considered to provide a general summary of the top 10 most productive authors over the past 36 years.

The author's publication over a specified period was used to measure the productivity. This impact was assessed by considering the number of annual citations. Zhang Y, Liu Y, and Wang S are the authors who produced the most work, and Law R and Wang S are the authors who received the most citations each year.

Furthermore, Law R has a continuous stream of papers on the subject and is a crucial figure in the field, as shown in Figure 15.18. According to Lotka's law (Lokta, 1926) of author's contribution to the publication, only a few authors have published more than 10 documents, whereas 2725 authors published only one document. It justifies Lotka's law (the more publications, the fewer authors can be found).

15.5.4 OBJECTIVE 4: TO INVESTIGATE THE SCIENTIFIC PRODUCTION BY INSTITUTIONS

Various institutions are involved in the study of the field of AI. Still, the most significant institutions are the "Hong Kong Polytechnic University (China)," the "University of Surrey (UK)," and the "School of Hotel and Tourism Management (China)," which indicate the dominance of Chinese institutions in the field of AI in the T&H sector.

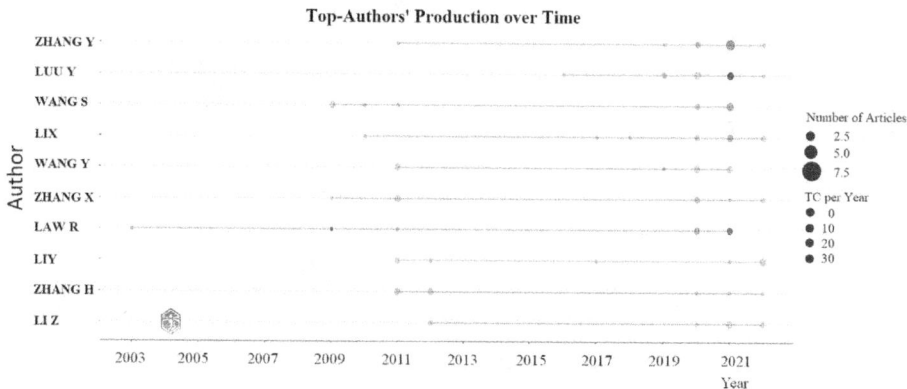

FIGURE 15.18 Authors' Production Overtime.

Source: Scopus database; software: Biblioshiny

15.5.5 Objective 5: To Investigate the Scientific Collaboration of the Countries

The USA was the most cited country with 2678 citations, and the average article citations were found at 49.59 with a betweenness of 193 and appeared in the seventh cluster.

The UK ranked second with a total citation of 707. The average article citations were 20.81 and 157 betweenness and appeared in the eighth cluster.

China ranked third with 1840 citations, and 7.86 average article citations have a betweenness of 121 and appeared in the first cluster, which is relatively less than the USA.

Germany ranked sixth in the total citations (i.e., 609) but has maximum average article citations of 67.67 in the fifth cluster and has 44 betweenness. The results indicate that American countries are leading from the front, but Europe and Asia are still catching up with their approach. A European country's (Germany) average article citations are at the top.

15.5.6 Objective 6: To Investigate Scientific Production by Sources and Dissemination by Sources

While investigating scientific production by source (journals), 433 journals were identified with published articles related to AI in tourism/hospitality/scientific journals.

These sources were identified based on their impact ("H-index," "G-index," "M-index," and "total citations") since their inception.

International Journal of Contemporary Hospitality Management started in 2002 and published 17 publications; it has an H-index of 10 and total citations of 531. *Tourism Management* began in 2003, has an entire citation (i.e., 975), and the same H-index (i.e., 10) but published less than the former.

Annals of Tourism Research ranked third in the top 20 sources list, which started publishing in 2019 but has an H-index and G-index of 7 and an M-index of 1.757, higher than the top two journals with a total citation of 437. These three sources cover the hospitality and tourism area of articles.

In contrast, *Expert Systems with Applications*, which is not related to hospitality and tourism, published the current area of articles and started publishing in 2009 with the same H-index and G-index of 7 and has less M-index (i.e., 0.5) but with maximum total citations (i.e., 996).

However, it has fewer publications; only seven are exhibited in Table 15.2. Recently, an upward trend has been seen in *Sustainability (Switzerland)* publications in the current area of research.

15.5.7 Objective 7: To Investigate the Content Based on the Author's Keywords, Keywords Plus, Titles, and Abstracts

The authors' keywords in the co-occurrence network identified eight clusters with 46 keywords. Three clusters dominated the co-occurrence network, with AI, tourism, and ML as the dominating areas.

Under AI, cluster robots, service robots, technology, automation, and sustainable development have a strong relationship. Tourism clusters have shown the relationship between big data, IoT (Rani et al., 2021), decision support systems, forecasting, innovation, and marketing. This suggests that the future of tourism is hidden in these areas; therefore, more attention should be given to them.

Machine learning is another cluster that holds keywords, such as deep learning, smart tourism, neural language processing, sentiment analysis, social media, and recommendation system.

The KeyWords Plus co-occurrence network found six clusters, but only three of them held the maximum number of keywords: AI, tourism, and sales. This is equivalent to the keywords used by the authors. Central dominance in the AI cluster has 29 keywords, including decision-making, decision support systems, learning systems, big data, forecasting, and commerce.

The second major cluster is tourism, with 12 keywords: information systems, tourism development, sustainable development, information use, and leisure industry. The third major cluster revolves around sales keyword with only six keywords: robotics, hotels, hospitality, intelligent robots, and service industry.

The title keyword co-occurrence network identified 50 keywords in a total of four clusters. The "tourism" keyword dominates the most significant cluster and has 28 keywords, including based, study, research, analysis, and data.

Another cluster is dominated by the "intelligence" keyword, which has 15 keywords, such as artificial, hospitality, service, and industry, and the AI "system" keyword dominated another cluster five keywords. Another cluster is dominated by "learning," which has only two keywords.

The abstract keyword co-occurrence network identified 50 keywords in only two clusters. The most significant cluster has 38 keywords dominated by "tourism," strongly associated with keywords, such as paper, industry, development, information, technology, and data. The second cluster has only 12 keywords dominated by "artificial," which strongly bonds with intelligence, purpose, AI, study, and hospitality.

15.5.8   OBJECTIVE 8: TO INVESTIGATE THE CONTENT BASED ON CITATIONS (MOST CITED REFERENCES)

The most globally cited document, "Modelling the world from Internet photo collections," had a total citation of 1567 (maximum), a total citation per year of 104.47, and a normalized citation score of 20.15.

"Sentiment classification of online reviews to travel destinations by supervised machine learning approaches" was the second most cited reference, with a total citation of 440, a total citation per year of 31.43, and an entire normalized citation of 11.90.

"Learning concept hierarchies from text corpora using formal concept analysis" was the third most cited document, with a total citation of 412, a total citation per year of 22.89, and a normalized total citation score of 2.96.

15.5.9   OBJECTIVE 9: TO INVESTIGATE THE LESS RESEARCHED KEYWORDS BASED ON CENTRALITY AND DENSITY

The investigation of over- and under-researched keywords helps to explore different study areas. The present study was conducted using a network approach—a thematic map—KeyWords Plus.

The primary keywords were categorized into four quadrants, namely, niche theme, motor theme, basic theme, and emerging or declining theme.

A two-dimensional thematic map was created that mapped the field's most important subjects based on the authors' keywords, as shown in Figure 15.19. The map shows how strongly the exterior (density) and internal relationships are connected (centrality).

According to Solomon et al. (2016), the map is divided into four quadrants: high density and centrality in quadrant 1, low density and centrality in quadrant 2, high density and low centrality in quadrant 3, and motifs with common values on both axes in quadrant 4.

The literature considers that the research topics in quadrant 1 have common themes because they have strong internal and external links. This quadrant contains forecasting, a decision support system, and decision-making. These issues are pertinent to this body of research because AI helps the hospitality and tourism industry forecast their business and can be used as a decision support system.

The study subjects in quadrant four must be stronger, emerging or declining themes that need further exploration because they have low ratings for density and centrality. The major subjects of this cluster are education and remote sensing.

It can be concluded that this research stream is distinct from other issues in the area based on our earlier findings, which are provided in this section. The organizations and journals that publish these findings might be more related to other fields, including the hospitality and tourism industry.

Keywords moving toward the intersection of density and centrality explain their importance. In Figure 15.19, remote sensing and education move toward the

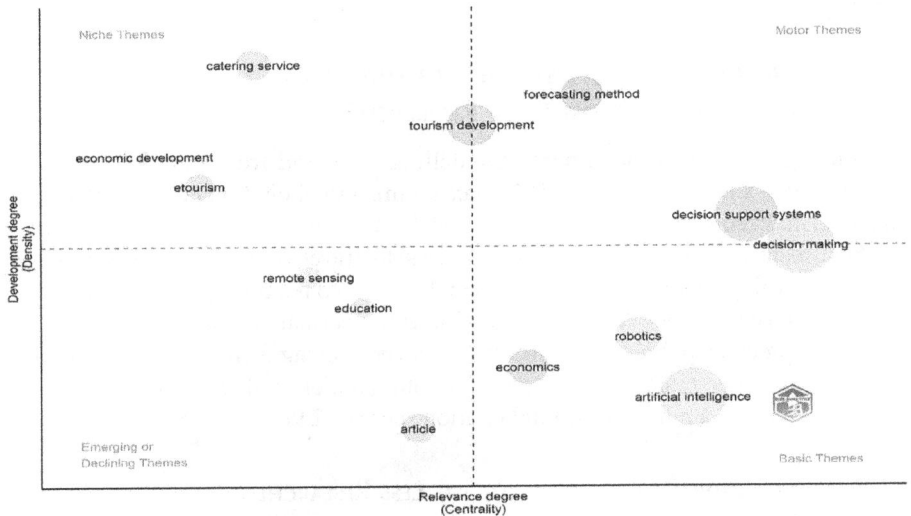

FIGURE 15.19   Thematic Map Using KeyWords Plus.

Source: Scopus database; software: Biblioshiny

intersection of density and centrality. This indicates that these areas are emerging research areas, and more attention is required.

COVID-19 has forced people to rely more on technology in forecasting or providing a safe and comfortable experience for guests in the hospitality and tourism sectors.

15.6 CONCLUSION

COVID-19 taught us a lesson to maintain social distancing; improved the importance of hygiene, cleanliness, and sanitation in our daily routine; and increased an individual's privacy, especially hotel guests. It brought one of the worst conditions for the hospitality and tourism industry because of its real, not virtual, experience.

People have become trapped in their homes during the COVID-19 pandemic. Still, they slowly started availing themselves of the virtual facility to enjoy their dream destinations, and the demand for online food deliveries also increased due to safety concerns.

Slowly, when people started moving from their homes to outstations for leisure, pleasure, sports, medical, or business purposes, the government also imposed guidelines for the safety of guests and their employees.

Here comes touchless check-in, barcode menu scanning, mobile room key, and so on. All of these guest activities were being tracked/noticed by the mobile or software cookies. It started by suggesting an area of concern and interest to the guest with the help of AI/ML. This study proposes a systematic review of AI/ML based on statistical methodologies to identify dominant publications, authors, and research groups.

To examine the social, intellectual, and conceptual structures of knowledge, it was decided to carry out network visualization using VOSviewer and data exploration (Biblioshiny using R). After applying the qualifying requirements and eliminating duplicates, inadequate, and missing data, 1152 publications, 3030 authors, 629 sources, and 109 countries, published between 1987 and July 5, 2022, were examined.

Over time, the increasing number of keywords demonstrates the importance of AI and power phrases, such as robots, service robots, technology, automation, and sustainable development.

Because of the importance of AI in improving the guest experience while staying in a hotel room, scanning menu cards, paying an online bill, maintaining social distancing, and enjoying touchless expertise are considered the most conventional and significant research paths (Khang et al., 2023b).

According to our study, there has been a significant increase in publications on AI in hospitality and tourism. The most critical and pertinent factor for the hospitality and tourism industry's success lies in AI/ML to understand guests' happiness, satisfaction, trust, and revisit intention.

The relationship between big data, IoT, decision support systems, forecasting, innovation, and marketing was significant. Machine learning is a cluster with keywords, such as deep learning, smart tourism, neural language processing, sentiment analysis, social media, and recommendation system, which can attract people's attention in T&HI (Khang et al., 2024).

The journal that has expanded the most throughout the years is the *International Journal of Contemporary Hospitality Management*. With a total of 12,906 citations and an average of 11.2 citations per document, the article with the most citations is "Modelling the world from Internet photo collections" (Snavely et al., 2008).

The three nations with the most significant numbers of articles are China, India, and Spain, but the spread reveals that the UK and USA are just slightly behind.

15.7 LIMITATIONS

This study had several limitations. Even though Scopus is one of the most reputed databases, some journals still need to be indexed, which means that publications in these journals may have to be recovered. Some pre-processing options, such as removing duplicate documents, are unavailable in the Bibliometrix package of R software.

Moral-Muñoz et al. (2020) claimed that pre-processing elements in bibliometric tools are underdeveloped. To overcome this restriction and eliminate duplicate documents, we used a two-stage data extraction strategy in this study.

To obtain helpful, complete, and comparable metadata, we narrowed the search by source. Only 10 journals were used in our study. The objective was to choose the top journals for the research. The outcomes would have been different if we had utilized various journals and an alternative index, like the Web of Science.

Furthermore, some authors may use numerous names, initials, or names in various publications. This restriction may lead to inaccuracies in the output of these institutions or authors and may result in discrepancies in the bibliographic analysis. The most currently available technologies exhibit issues when working with large datasets, whereas constrained datasets can provide better output.

Metadata were extracted from the Scopus database; however, these international databases may need to include relevant studies from emerging and underdeveloped nations. We also noted that the authors cannot read full articles and abstracts in languages other than English, which were not considered.

15.8 FURTHER RESEARCH

This study can be expanded in future research to incorporate the literature from emerging and developing nations. We have included additional pertinent publications, books, and executive insights to make the investigation more thorough. As a result, our study contributes to the application of AI in the hospitality and tourism industry in emerging and developing countries (Khang et al., 2022a).

The present study used AI in T&HI, and further research may involve the same in the field of other service industries by following the same methodology and can be compared with the output of the present study. A conclusion can be drawn based on the same results.

- **Acknowledgment:** The authors acknowledge the support of software, previous authors, and metadata provider for improving the quality of this chapter.
- **Declaration of conflicting interests:** There are no conflicts of interest in authorship concerning the research or anything else.

REFERENCES

Abdi H., Valentin D., (2007). "Multiple correspondence analysis," *Encyclopedia of Measurement and Statistics*, 2(4), 651–657. www.utdallas.edu/~Herve/Abdi-MCA2007-pretty. pdf

Akehurst G., (2009). "User-generated content: The use of blogs for tourism organizations and tourism consumers," *Service Business*, 3(1), 51–61. https://doi.org/10.1007/ s11628-008-0054-2

Andreu L., Bigne E., Amaro S., Palomo J., (2020). "Airbnb research: An analysis in tourism and hospitality journals," *International Journal of Culture, Tourism, and Hospitality Research*, 14(1), 2–20. https://doi.org/10.1108/IJCTHR-06-2019-0113

Aria M., Cuccurullo C., (2017). "Bibliometrix: An R-tool for comprehensive science mapping analysis," *Journal of Informetrics*, 11(4), 959–975. https://doi.org/10.1016/j. joi.2017.08.007

Aria M., Cuccurullo C., (2020). "Science mapping analysis with Bibliometrix R-package: An example," *Bibliometrix. Org*, 15 September. https://jscires.org/sites/default/files/JScientometRes-8-3-156_0.pdf

Bhambri P., Rani S., Gupta G., Khang A., (2022). *Cloud and Fog Computing Platforms for Internet of Things*. CRC Press. https://doi.org/ 10.1201/9781003213888

Borràs J., Moreno A., Valls A., (2014). "Intelligent tourism recommender systems: A survey," *Expert Systems with Applications*, 41(16), 7370–7389. https://doi.org/10.1016/j. eswa.2014.06.007

Bradford S. C., (1934). "Sources of information on specific subjects," *Engineering*, 137, 85–86. https://journals.sagepub.com/doi/pdf/10.1177/016555158501000406

Buhalis D., Law R., (2008). "Progress in information technology and tourism management: 20 years on and 10 years after the Internet," *The state of eTourism research. Tourism Management*, 29(4), 609–623. www.sciencedirect.com/science/article/pii/ S0261517718301134

Buhalis D., Leung R., (2018). "Smart hospitality—interconnectivity and interoperability towards an ecosystem," *International Journal of Hospitality Management*, 71, 41–50. https://doi.org/10.1016/j.ijhm.2017.11.011

Cambria E., Speer R., Havasi C., Hussain A., (2010). "SenticNet: A publicly available semantic resource for opinion mining," *AAAI Fall Symposium—Technical Report*, FS-10–02, 14–18. https://cdn.aaai.org/ocs/2216/2216-9491-2-PB.pdf

Cho V., (2003). "A comparison of three different approaches to tourist arrival forecasting," *Tourism Management*, 24(3), 323–330. www.sciencedirect.com/science/article/pii/ S0261517702000687

Cuccurullo C., Aria M., Sarto F., (2016). "Foundations and trends in performance management. A twenty-five years bibliometric analysis in business and public administration domains," *Scientometrics*, 108(2), 595–611. https://doi.org/10.1007/s11192-016-1948-8

Fagan J., Eddens K. S., Dolly J., Vanderford N. L., Weiss H., Levens J. S., (2018). "Assessing research collaboration through co-authorship network analysis," *The Journal of Research Administration*, 49(1), 76. www.ncbi.nlm.nih.gov/pmc/articles/PMC6703830/

Florencio B. P., Roldán L. S., Pineda J. M. B., (2020). "Communication, trust, and loyalty in the hotel sector: The mediator role of consumer's complaints," *Tourism Analysis*, 25(1), 183–187. www.ingentaconnect.com/content/cog/ta/2020/00000025/00000001/art00013

Geisel A., (2018). "The current and future impact of artificial intelligence on business," *International Journal of Scientific & Technology Research*, 7(5), 116–122. https://doi. org/10.1201/ 9781003145011

Gössling S., Scott D., Hall C. M., (2020). "Pandemics, tourism and global change: A rapid assessment of COVID-19," *Journal of Sustainable Tourism*, 0(0), 1–20. https://doi.org/ 10.1080/09669582.2020.1758708

Hotho A., Staab S., Cimiano P., (2005). "Learning concept hierarchies from text corpora using formal concept analysis," *Journal of Artificial Intelligence Research*, 24, 305–339. www.aaai.org/Papers/JAIR/Vol24/JAIR-2409.pdf

Hoyer W. D., Kroschke M., Schmitt B., Kraume K., Shankar V., (2020). "Transforming the customer experience through new technologies," *Journal of Interactive Marketing*, 51, 57–71. https://doi.org/10.1016/j.intmar.2020.04.001

Hu H. H., Kandampully J., Juwaheer D. D., (2009). "Relationships and impacts of service quality, perceived value, customer satisfaction, and image: An empirical study," *Service Industries Journal*, 29(2), 111–125. https://doi.org/10.1080/02642060802292932

Huang M. H., Rust R. T., (2018). "Artificial intelligence in service," *Journal of Service Research*, 21(2), 155–172. https://doi.org/10.1177/1094670517752459

Ii B. J. P., Disney W., (2002). "Dtirentiatig hospitality operations via experiences up experiences," *Cornell University*, 1(3), 87–96. https://doi.org/10.1201/9781003145011

Jain A., Pandey A. K., (2017). "Multiple quality optimizations in electrical discharge drilling of mild steel sheet," *Materials Today: Proceedings*, 4(8), 7252–7261. www.sciencedirect.com/science/article/pii/S2214785317313421

Jiang Y., Wen J., (2020). "Effects of COVID-19 on hotel marketing and management: A perspective article," *International Journal of Contemporary Hospitality Management*, 32(8), 2563–2573. https://doi.org/10.1108/IJCHM-03-2020-0237

Juaneda-Ayensa E., Mosquera A., Murillo Y. S., (2016). "Omnichannel customer behavior: Key drivers of technology acceptance and use and their effects on purchase intention," *Frontiers in Psychology*, 7(JUL), 1–11. https://doi.org/10.3389/fpsyg.2016.01117

Kau A. K., Lim P. S., (2005). "Clustering of Chinese tourists to Singapore: An analysis of their motivations, values and satisfaction," *International Journal of Tourism Research*, 7(4–5), 231–248. https://doi.org/10.1002/jtr.537

Khang A., Gupta S. K., Hajimahmud V. A., Babasaheb J., Morris G., (2023a). *AI-Centric Modelling and Analytics: Concepts, Designs, Technologies, and Applications* (1st ed.). CRC Press. https://doi.org/10.1201/9781003400110

Khang A., Gupta S. K., Shah V., Misra A., (Eds.). (2023b). *AI-Aided IoT Technologies and Applications in the Smart Business and Production*. CRC Press. https://doi.org/10.1201/9781003392224

Khang A., Rani S., Sivaraman A. K., (2022a). *AI-Centric Smart City Ecosystems: Technologies, Design and Implementation* (1st ed.). CRC Press. https://doi.org/10.1201/9781003252542

Khang A., Ragimova N. A., Hajimahmud V. A., Alyar V. A., (2022b). "Advanced technologies and data management in the smart healthcare system," In *AI-Centric Smart City Ecosystems: Technologies, Design and Implementation*, Vol. 16, No. 10 (1st ed.). CRC Press. https://doi.org/10.1201/9781003252542-16

Khang A., Vugar A., Vladimir H., Vrushank S., (2024). *Advanced IoT Technologies and Applications in the Industry 4.0 Digital Economy* (1st ed.). CRC Press. https://doi.org/10.1201/9781003434269

Khaparde Professor V., Pawar S., (2013). "Authorship pattern and degree of collaboration in information technology," *Journal of Computer Science & Information Technology*, 1(1), 46–54. www.researchgate.net/profile/Vaishali-Khaparde/publication/ 311351094_ Authorship_Pattern_and_Degree_of_Collaboration_in_Information_Technology/ links/58426f4b08ae2d2175622936/Authorship-Pattern-and-Degree-of-Collaboration-in-Information-Technology.pdf

Krce Miočić B., Zekanović Korona L., Matešić M., (2012). "Adoption of smart technology in Croatian hotels," *MIPRO 2012–35th International Convention on Information and Communication Technology, Electronics and Microelectronics—Proceedings*, 1440–1445. https://ieeexplore.ieee.org/abstract/document/6240879/

Kumar A., Sharma R., Chuah C., (2021). "Turning the blackbox into a glassbox: An explainable machine learning approach for understanding hospitality customer," *International Journal of Information Management Data Insights*, 1(2), 100050. https://doi.org/10.1016/j.jjimei.2021.100050

Liu Y., Teichert T., Rossi M., Li H., Hu F., (2017). "Big data for big insights: Investigating language-specific drivers of hotel satisfaction with 412,784 user-generated reviews," *Tourism Management*, 59, 554–563. https://doi.org/10.1016/j.tourman.2016.08.012

Lokta A., (1926). "The frequency distribution of scientific distribution," *Journal of the Washington Academy of Sciences*, 16, 317–323. https://doi.org/10.1201/9781003145011

Lu L., Cai R., Gursoy D., (2019). "Developing and validating a service robot integration willingness scale," *International Journal of Hospitality Management*, 80, 36–51. www.sciencedirect.com/science/article/pii/S0278431918306455

Lutz C., Newlands G., (2018). "Consumer segmentation within the sharing economy: The case of Airbnb," *Journal of Business Research*, 88, 187–196. https://doi.org/10.1016/j.jbusres.2018.03.019

Lyu J., Khan A., Bibi S., Chan J. H., Qi X., (2022). "Big data in action: An overview of big data studies in tourism and hospitality literature," *Journal of Hospitality and Tourism Management*, 51, 346–360. https://doi.org/10.1016/j.jhtm.2022.03.014

Martí-Parreño J., Gómez-Calvet R., (2020). "Social media and sustainable tourism: A literature review," *Proceedings of the International Conference on Tourism Research*, 148– 153. https://doi.org/10.34190/IRT.20.067

Moed H., Glänzel W., Schmoch U., Schubert A., (2004). "Analyzing scientific networks through co-authorship," 1963, 257–276. https://link.springer.com/chapter/10.1007/1-4020-2755-9_12

Moral-Muñoz J. A., Herrera-Viedma E., Santisteban-Espejo A., Cobo M. J., (2020). "Software tools for conducting bibliometric analysis in science: An up-to-date review," *Profesional de La Información*, 29(1) (Software tools for conducting bibliometric analysis in science)

Murphy J., Hofacker C., Gretzel U., (2017). "Dawning of the age of robots in hospitality and tourism: Challenges for teaching and research," *European Journal of Tourism Research*, 15(2017), 104–111. www.google.com/books?hl=en&lr=&id=zkonDwAAQBAJ&oi=fnd&pg=PA104&dq=Dawning+of+the+age+of+robots+in+hospitality+and+tourism:+Challenges+for+teaching+and+research&ots=avcA4fIlnK&sig=CIdcB2J0ZIjCQdPftoedd7cwPY0

Navigli R., Velardi P., (2004). "Structural semantic interconnection: A knowledge-based approach to word sense disambiguation," In *Proceedings of the SENSEVAL@ACL 2004: 3rd International Workshop on the Evaluation of Systems for the Semantic Analysis of Text—Held in Cooperation with ACL 2004*, July, 179–182. https://ieeexplore.ieee.org/abstract/document/1432741/

Nkengasong J., (2020). "China's response to a novel coronavirus stands in stark contrast to the 2002 SARS outbreak response," *Nature Medicine*, 26(3), 310–311. https://doi.org/10.1038/s41591-020-0771-1

Perini V., (n.d.). *How Artificial Intelligence Makes Hotel Upselling Smart – and Effective.* www.hospitalitynet.org/opinion/4104243.html

Rana G., Khang A., Sharma R., Goel A. K., Dubey A. K., (Eds.). (2021). *Reinventing Manufacturing and Business Processes through Artificial Intelligence.* CRC Press. https://doi.org/10.1201/9781003145011

Rani S., Bhambri P., Kataria A., Khang A., Sivaraman A. K., (2023). *Big Data, Cloud Computing and IoT: Tools and Applications* (1st ed.). Chapman and Hall/CRC. https://doi.org/10.1201/9781003298335

Rani S., Chauhan M., Kataria A., Khang A., (Eds.). (2021). "IoT equipped intelligent distributed framework for smart healthcare systems," In *Networking and Internet Architecture*, Vol. 2, p. 30. CRC Press. https://doi.org/10.48550/arXiv.2110.04997

Raza S. A., Ashrafi R., Akgunduz A., (2020). "A bibliometric analysis of revenue management in airline industry," *Journal of Revenue and Pricing Management*, 19(6), 436–465. https://doi.org/10.1057/s41272-020-00247-1

Roy P., Ramaprasad B. S., Chakraborty M., Prabhu N., Rao S., (2020). "Customer acceptance of use of artificial intelligence in hospitality services: An Indian hospitality sector perspective," *Global Business Review*. https://doi.org/10.1177/0972150920939753

Rushdi Saleh M., Martín-Valdivia M. T., Montejo-Ráez A., Ureña-López L. A., (2011). "Experiments with SVM to classify opinions in different domains," *Expert Systems with Applications*, 38(12), 14799–14804. https://doi.org/10.1016/j.eswa.2011.05.070

Sánchez-Franco M. J., Navarro-García A., Rondán-Cataluña F. J., (2019). "A naive Bayes strategy for classifying customer satisfaction: A study based on online reviews of hospitality services," *Journal of Business Research*, 101(June), 499–506. https://doi.org/10.1016/j.jbusres.2018.12.051

Small H., (1999). "Visualizing science by citation mapping," *Journal of the American Society for Information Science*, 50(9), 799–813. https://doi.org/10.1002/(SICI)1097-4571(1999)50:9<799::AID-ASI9>3.0.CO;2-G

Snavely N., Seitz S. M., Szeliski R., (2008). "Modeling the world from Internet photo collections," *International Journal of Computer Vision*, 80(2), 189–210. https://doi.org/10.1007/s11263-007-0107-3

Solomon G. T., Fayolle A., et al., (2016). "The role of development entrepreneurial orientation and market orientation in improving the performance of creative industry SMEs in denpasar," *Journal of Small Business Management*. https://pdfs.semanticscholar.org/fca6/6fcfc27f4cd70107f9657546007f8171fcd1.pdf

Tung V. W. S., Au N., (2018). "Exploring customer experiences with robotics in hospitality," *International Journal of Contemporary Hospitality Management*. www.emerald.com/insight/content/doi/10.1108/IJCHM-06–2017–0322/full/html

Tung V. W. S., Law R., (2017). "The potential for tourism and hospitality experience research in human-robot interactions," *International Journal of Contemporary Hospitality Management*, 29(10), 2498–2513. www.emerald.com/insight/content/doi/10.1108/IJCHM-09-2016-0520/full/html

Van der Veer Martens B., (2007/2008). "Mapping research specialties," *Annual Review of Information Science and Technology*, 42, 213. www.google.com/books?hl=en&lr=&id=aPg6CW0V0YEC&oi=fnd&pg=PA213&dq=Mapping+research+specialties.+Annual+Review+of+Information+Science+and+Technology&ots=GFmJEvlIHh&sig=TsDDow44rlX_WjQjYC4Jc39dwI8

Velázquez B. M., Blasco M. F., Saura I. G., (2015). "ICT adoption in hotels and electronic word-of-mouth," *Academia Revista Latinoamericana de Administración*. www.emerald.com/insight/content/doi/10.1108/ARLA-10-2013-0164/full/www.internetworldstats.com&

Vishnu Sai Kumar D., Ritik C., Anuradha M., Praveen Kumar M., Khang A., (Aug 24, 2023). "Heart disease and liver disease prediction using machine learning," In *Data-Centric AI Solutions and Emerging Technologies in the Healthcare Ecosystem*, p. 4 (1st ed.). CRC Press. https://doi.org/10.1201/9781003356189-13

Waltman L., van Eck N. J., Noyons E. C. M., (2010). "A unified approach to mapping and clustering of bibliometric networks," *Journal of Informetrics*, 4(4), 629–635. https://doi.org/10.1016/j.joi.2010.07.002

Ye Q., Zhang Z., Law R., (2009). "Sentiment classification of online reviews to travel destinations by supervised machine learning approaches," *Expert Systems with Applications*, 36(3), 6527–6535. www.sciencedirect.com/science/article/pii/S0957417408005022

Zeng Z., Chen P. J., Lew A. A., (2020). "From high-touch to high-tech: COVID-19 drives robotics adoption," *Tourism Geographies*, 22(3), 724–734. https://doi.org/10.1080/146 16688.2020.1762118

Zhang J., Yu Q., Zheng F., Long C., Lu Z., Duan Z., (2016). "Comparing keywords plus of WOS and author keywords: A case study of patient adherence research," *Journal of the Association for Information Science and Technology*, 67(4), 967–972. https://asistdl. onlinelibrary.wiley.com/doi/abs/10.1002/asi.23437

Zollo L., Filieri R., Rialti R., Yoon S., (2020). "Unpacking the relationship between social media marketing and brand equity: The mediating role of consumers' benefits and experience," *Journal of Business Research*, 117(April), 256–267. https://doi.org/10.1016/j. jbusres.2020.05.001

16 Data-Centric Predictive Analytics for Solving Environmental Problems

Elmina Gadirova Musrat and Vugar Abdullayev

16.1 INTRODUCTION

One of the ecotoxicants is phenol and its derivatives, which enter water with effluents from organic synthesis enterprises. Chlorophenols, which are formed during the chlorination of drinking water and serve as starting materials for the formation of dioxins or a group of tricyclic aromatic compounds classified as the most highly toxic persistent organic pollutants with mutagenic and carcinogenic properties (Alekseev, 2007), pose an increased risk.

Therefore, water purification from phenol and its derivatives is a serious environmental challenge, and this topic is applied to solve environmental problems in the building stage of smart cities (Khang et al., 2023a).

Existing methods for cleaning organic pollutants include biological, physical, chemical, and adsorption methods (Marsh, 2006). The latter are more often used in the post-treatment stages and are most effective when using materials with high adsorption activity for smart city environments (Khang et al., 2022).

In addition, adsorption is a fairly simple technological process in contrast to chemical and biological methods that can be implemented at rather high flow rates of the media to be purified. The most widespread process of phenol adsorption is activated carbon. There are also scientific studies using other sorbents (Isaeva, 2009).

The adsorption purification of phenol is an area of research aimed at solving environmental problems (SEPs). The adsorption purification of water from phenol is the most important chemical direction because sorption treatment is widespread. According to the literature, adsorption processes in the presence of TiO_2 are known contain information on the adsorption of titanium oxide.

The use of TiO_2 as an adsorbent with photocatalytic properties provides many advantages. TiO_2 is chemically and thermally more stable than TiO_2, which has a photocatalytic effect and is widely used for the purification of various materials from toxic substances (Rani et al., 2022).

In Japan and China, TiO_2 is added to concrete reinforcement and wall masonry in closed tunnels to purify the air from nitrogen gases (Ghafari, 2019).

The use of photocatalysis for the purification of wastewater from organic pollutants is an interesting alternative that has attracted great interest from many

 DOI: 10.1201/9781003400110-16

researchers in recent years. An experimental study of the adsorption of phenol selected as a model pollutant on a photocatalyst, titanium oxide anatase (Degussa P25/Evonik Aeroxide TiO_2 P25), a titania photocatalyst that is widely used because of its relatively high levels of activity in many photocatalytic reaction systems, was presented.

The amount of phenol adsorbed was measured using UV spectroscopy. The equilibrium of adsorption was reached after 1 h; the adsorption kinetics was slow and obeyed the Lager grain model. Adsorption is monolayer chemisorption that follows the Langmuir model.

This study also demonstrated the advantages of operating at high stirring speeds and natural pH. Stirring with ultrasound leads to a small increase in the amount of adsorbed phenol (5%) because this mode of stirring reduces the agglomeration of titanium oxide particles and therefore increases the interfacial area of the catalyst (Gundogdu et al., 2012).

There are various methods of phenol adsorption (Goto, 1986). New composite materials have been developed by impregnating cotton materials made from cellulose with Al_2O_3/TiO_2; these composites are widely used for the treatment of toxic solutions.

Cotton is very good in absorbing the solution; therefore, toxic substances in the solution are absorbed together with the solution (Salari et al., 2019). It was found that this method can remove any toxic substance from the system.

The adsorption of phenol by graphite oxide nanoparticles and activated carbon has also been studied (Shabanova, 2007). For this purpose, graphite oxide (GO) has recently been used (Gadirova et al., 2019). Graphite oxide is a graphite oxidizing product with carbonyl and carboxyl groups, as well as epoxy and hydroxyl groups, along the edges of its layers (Fabing et al., 2005).

Graphite oxide is a highly valued membrane material because of its cheap and simple production process, good chemical stability, mechanical strength, and high pollutant purification capability (Shang R., 2017). Studies have shown that GO membranes possess very good ionic and molecular selectivity and water permeability (Gadirova et al., 2024).

The use of GO/Al_2O_3 composite membranes for water purification is a subject of increasing research interest because of its simple and energy-efficient approach. It is possible to purify phenol from wastewater by 99.9% on the obtained composite (Hu et al., 2018). In addition, there are articles on the role of TiO_2 in wastewater treatment (Huang et al., 2016).

16.2 MATERIALS AND METHODS

Processes were carried out in the presence of various nanocomposites and TiO_2 (anatase phase). However, no cases involving pure rutile-phase nanoparticles have been reported. In this study, only TiO_2 nanoparticles in the rutile phase were used.

The process was performed at room temperature, and TiO_2 with rutile modification was used. The adsorption process was continued for 2 hours with periodic stirring. In the process, 0.05 g of TiO_2 and 20 ml of 1 mg L^{-1} phenol solution were taken, and the process was carried out at 250°C.

In contrast to the literature, the adsorbent properties of TiO$_2$ nanoparticles with a complete rutile phase have been studied. TiO$_2$ nanoparticles were prepared using the sol–gel method. TiO$_2$ nanoparticles were analyzed according to the structural equation modeling (SEM) methods.

The TiO$_2$ nanoparticles have a spherical shape with sizes varying from 10 to 30 nm. The SEM data correlated well with the results obtained from the X-ray powder diffraction (XRD) analysis.

The purity and crystalline properties of the TiO$_2$ nanoparticles were investigated by the powder XRD method. X-ray structure analysis graphs of the studied nanocomposite materials were recorded on a Rigaku Mini Flex 600 powder diffractometer.

X-ray tube with a copper anode (Cu Kα radiation, 30 kV and mA) was used to draw the diffraction spectra at room temperature. At $2\theta = 20°–80°$ in the discrete growth mode, these spectra were obtained at $2\theta = 0.05°$, and the exposure time was $\tau = 5$ seconds (Elmina et al., 2020).

In this work, TiO$_2$ nanoparticles were used to remove phenols from wastewater. For this, a solution (5 ml) of 0.05 g of TiO$_2$ in 10 ml of distilled water was prepared. The nanoparticles were completely mixed in the presence of X-ray radiation for a uniform discharge in distilled water. The resulting TiO$_2$ was used to absorb a phenol solution with a concentration of 1 mg L^{-1}.

A mixture of 5 ml of the rutile-phase TiO$_2$ composite with 20 ml of a 1 mg L^{-1} phenol solution was absorbed for 2 hours at 25°C. Based on the curves obtained, adsorption was found to be slow. The process was investigated using a "Varian Cary 50" device (Khang et al., 2024).

16.3 DISCUSSION AND RESULTS

The following graph shows the dependence of the absorption coefficient of 1 mg L^{-1} phenol solution on the wavelength, as shown in Graph 16.1. There was no adsorption in the process; therefore, curves compatible with phenol were obtained in the 270 nm area.

As shown in Figure 16.1, phenol-similar curves were obtained at 200–300 nm. It is also known from the literature that the curve obtained at 270 nm wavelengths corresponds to that of phenol, as shown in Figure 16.2.

After 2 hours of adsorption process, the phenol solution was filtered off. This solution was then transferred to the "Varian Cary 50" spectrophotometer. In this case, curves corresponding to phenol (270 nm) were again observed; however, according to the measurements on the degree graph, a decrease in phenol (30%) was observed in the solution, indicating weak adsorption.

After 60 minutes of the process, the adsorption curves practically did not change, which is associated with the capture of the surface of the TiO$_2$ nanoparticles attributable to chemisorption. Consequently, adsorption occurred over a period of time and then did not occur. Figure 16.3 shows a comparison of the adsorption curves for 30, 60, 90, and 120 minutes.

Figure 16.4 shows a comparison of the adsorption curves for 30 and 120 minutes. As shown in Figures 16.3 and 16.4, phenol-specific peaks were observed in the obtained curves.

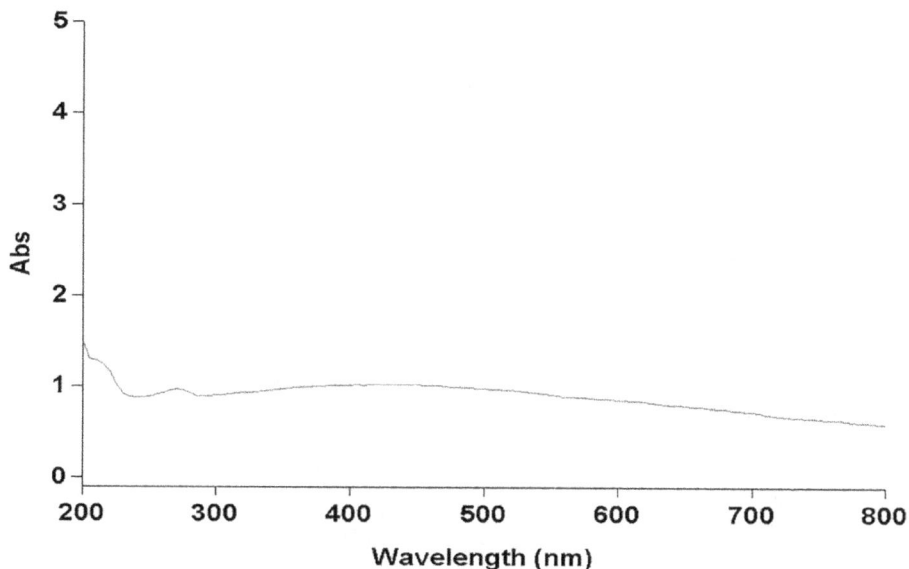

FIGURE 16.1   Adsorption Process of 1 mg L^{-1} Phenol Solution.

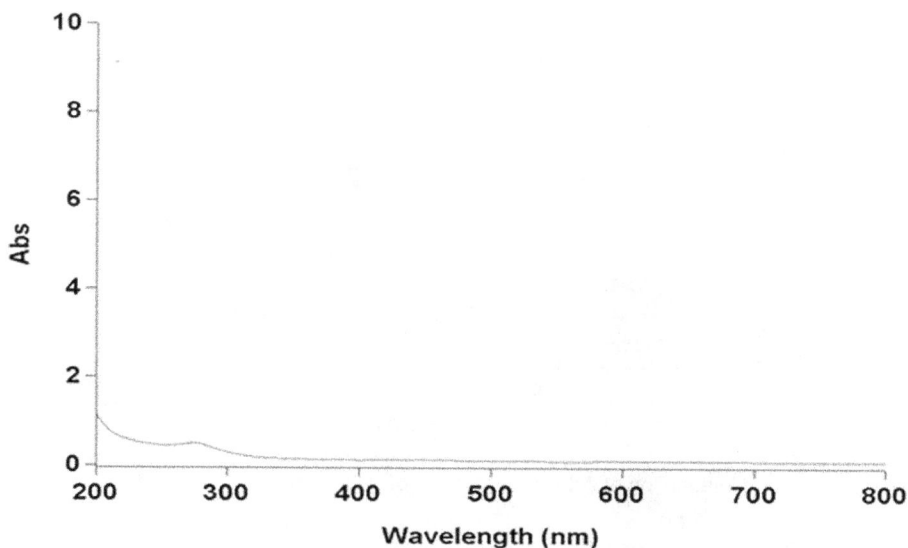

FIGURE 16.2   After Absorption of 1 mg L^{-1} Phenol + TiO$_2$ Nanoparticles.

The following graph shows the decrease in phenol concentration during the first 60 minutes. After a while, this difference did not diminish. Gradually, the phenol concentration should decrease. However, this was not observed, indicating incomplete adsorption, as shown in Figure 16.5.

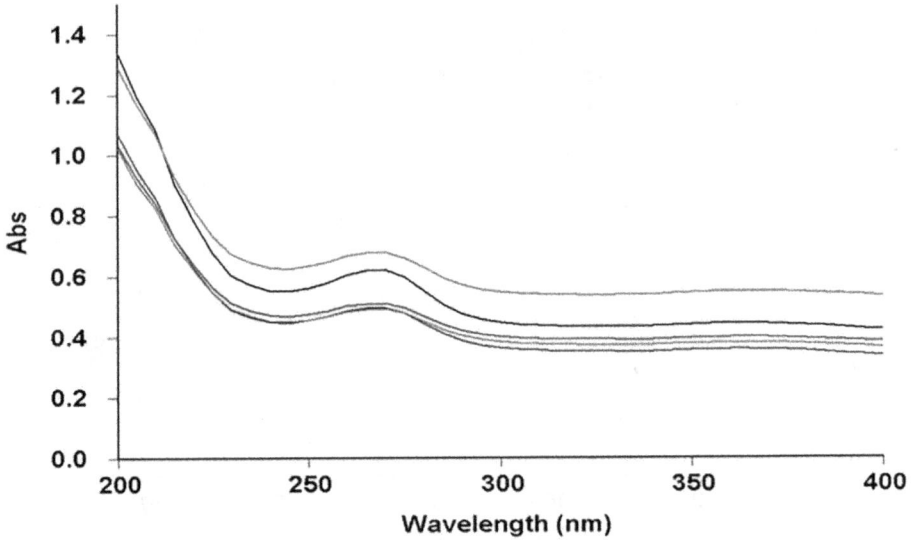

FIGURE 16.3   Comparison of Adsorption Curves of Phenol in the Presence of TiO_2.

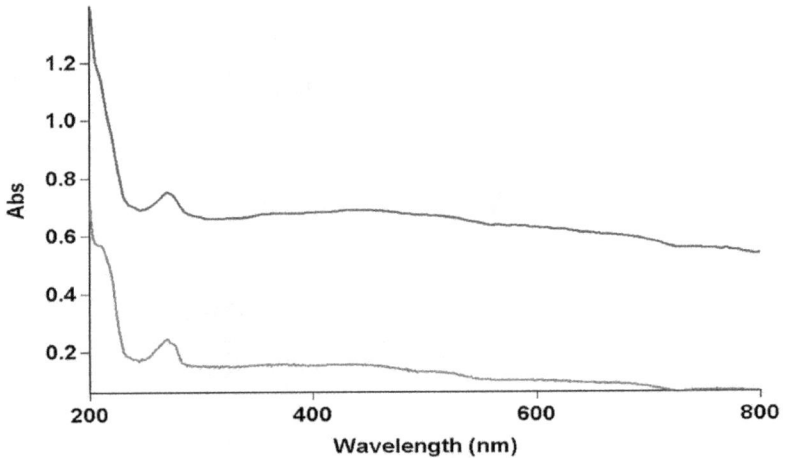

FIGURE 16.4   Comparison of Adsorption Curves of Phenol in the Presence of TiO_2.

As shown in Figures 16.3 and 16.4, the phenol curve declined from top to bottom but was still observed. This indicates that adsorption started over a period of time, after which adsorption stopped. It is believed that chemisorption and aggregation of nanoparticles occurred during the process.

The presence of phenol in the mixture was analyzed using the gas chromatography/mass selective detector (6890N/Agilent 5975) method, which is an accurate method of analysis. The presence of 72% phenol in the solution after the adsorption was confirmed.

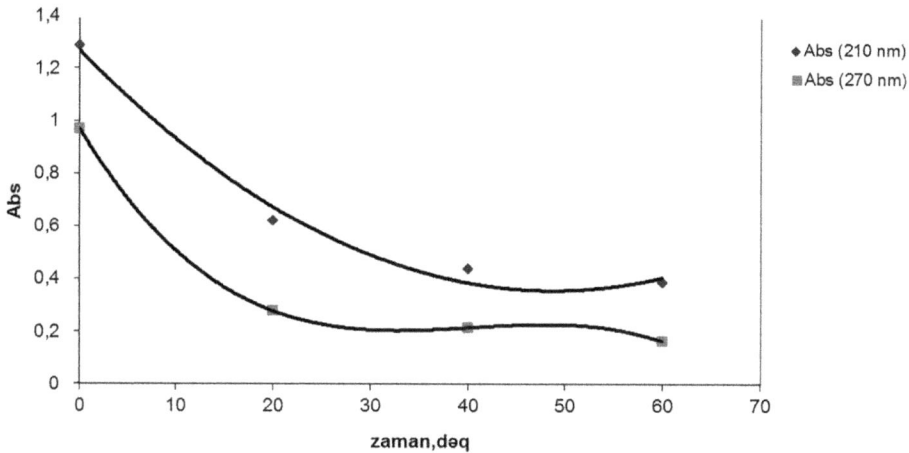

FIGURE 16.5 Decrease in Phenol Concentration during the First 60 Minutes.

According to the degree graph, the calculations coincided with the analytical method. The mass spectrometry (MS) spectrum of phenol is shown in Figure 16.6.

16.3.1 EXPONENTIAL DECAY AND LOGISTIC MODELS FOR REMOVAL OF PHENOL

First, the adsorption of phenol in the presence of rutile-phase TiO_2 nanoparticles was considered. To evaluate phenol adsorption in the presence of TiO_2 nanoparticles, we performed an experiment.

The amount of phenol remaining in the solution was measured after 2 hours of phenol adsorption. We used an exponential decay and a logistic model to estimate this process.

Further information on these models can be found in any mathematical modeling or calculus book (Hughes-Hallett et al., 2009). For both simulations, we estimated the amount of phenol in the solution after adsorption for reference values of TiO_2.

The results for the exponential model are as follows:

The exponential growth or decline model assumes constant growth rate (Equation 16.1):

$$\frac{dP}{dt} = cP \tag{16.1}$$

where P is the population at time t, and c is the constant. The solution of Equation 16.2 is

$$P(t) = P_0 e^{rt}, t \geq 0 \tag{16.2}$$

where P_0 is the initial population, $P(t_0) = P_0$.

The time required for the complete removal of phenol from a 1 mg L^{-1} phenol solution with the addition of TiO_2 is shown in Figure 16.6.

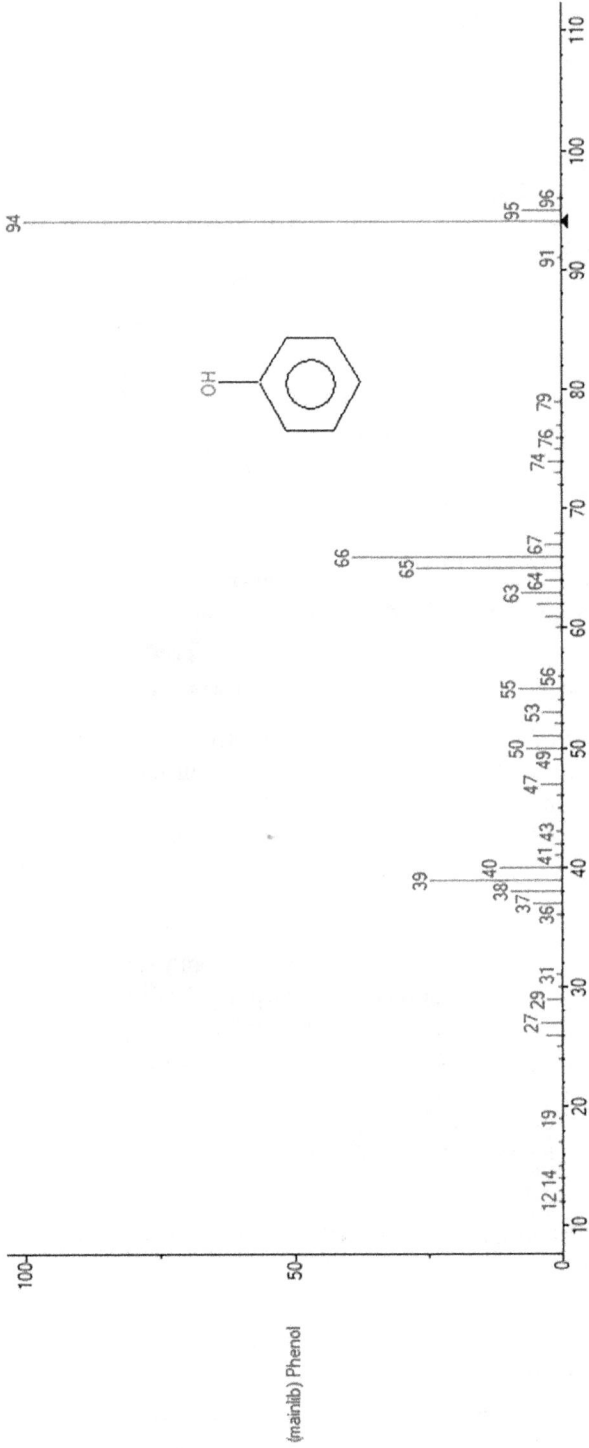

FIGURE 16.6 MS Spectrum of Phenol in a Mixture After Adsorption According to the GC–MSD Spectroscopy.

By applying the curve fitting tool in MATLAB for the values in Table 16.1, we obtained Equation 16.3:

$$P(t) = e^{-1.539t}, t \ge 0 \tag{16.3}$$

We plotted the exponential function obtained by the model with the data in Figure 16.7 and provide some estimation results via the exponential decay solution in Table 16.1.

As shown in Table 16.2, we can say that the phenol concentration reached 0.5, 0.1, and 0.01 approximately 0.46, 0.85, and 1 hour, respectively, after the process started.

TABLE 16.1

Experimental Results for the Time Required for the Complete Removal of Phenol from a 1 mg L^{-1} Phenol Solution with the Addition of TiO$_2$.

Time (hours)	TiO$_2$ (g)
0	1
2.58	0.02
2.17	0.03
2.00	0.05

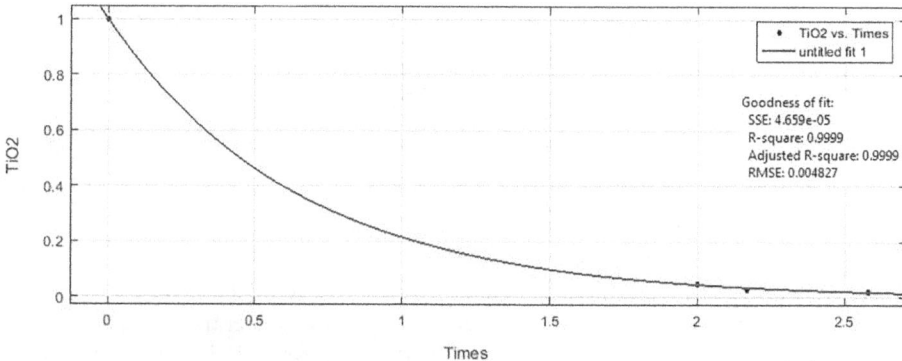

FIGURE 16.7   Decrease in Phenol Concentration with the Exponential Decay Solution.

TABLE 16.2

Estimation Times Required for the Complete Removal of Phenol from a 1 mg L^{-1} Phenol Solution with the Addition of TiO$_2$.

Time (hours)	TiO$_2$ (g)
0.4632	0.50
0.8573	0.10
0.9847	0.01

16.3.2 RESULTS FOR THE LOGISTIC MODEL

The logistic model assumes the growth rate proportional to the factor of $P\left(1-\dfrac{P}{L}\right)$, which means that the growth rate is close to zero when P is too small or P is close to L. In other words, for the logistic model, the relative growth rate is proportional to $(1 - P/L)$, whereas the growth rate of the exponential growth model is a constant.

Here, L is the carrying capacity (i.e., the maximum population that lives in the area for the population growth problem), and the resulting model is shown in Equation 16.4:

$$\frac{dP}{dt} = rP\left(1-\frac{P}{L}\right) \tag{16.4}$$

The solution of Equation 16.4 is Equation 16.5:

$$P(t) = \frac{L}{1+Ae^{rt}}, t \ge 0 \tag{16.5}$$

where $A = \dfrac{L-P_0}{P_0}$ depends on the initial population $P(t_0) = P_0$. We used the data in Table 16.1 to obtain logistic solution for the problem. By using regression analysis in GeoGebra, which is a dynamic mathematics software, we obtained the following solution for the solution in Equation 16.5 as Equation 16.6:

$$P(t) = \frac{2.4202}{1+1.4202e^{1.7826t}}, t \ge 0 \tag{16.6}$$

We plotted the logistic solution with data in Figure 16.8 and provide some estimation results via the logistic solution in Table 16.3.

As shown in Table 16.3, the results obtained with the logistic model are similar to the results of the exponential decay model. The phenol concentration reached 0.01 and 0.5 within approximately an hour and 0.54 hour, respectively, after the process started (Gadirova et al., 2024).

TABLE 16.3
Estimation Times by Logistic Solution Required for the Complete Removal of Phenol from a 1 mg L⁻¹ Phenol Solution with the Addition of TiO₂.

Time (hours)	TiO₂ (g)
0.5422	0.50
0.8972	0.10
0.9895	0.01

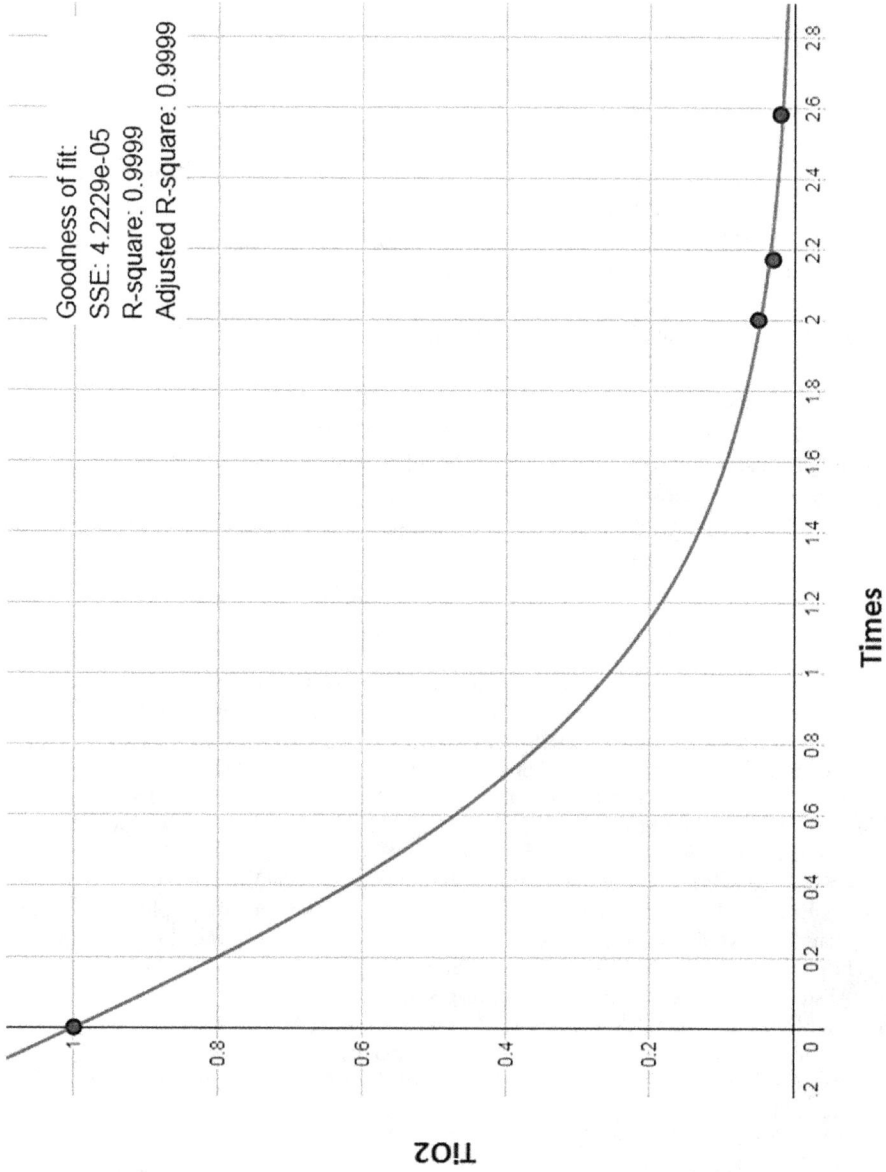

FIGURE 16.8 Decrease in Phenol Concentration with the Logistic Solution.

16.4 CONCLUSION

For the first time, the adsorption properties of rutile-phase TiO_2 nanoparticles (10– 30 nm) were studied. It was determined that TiO_2 had the same adsorption properties as in the anatase form.

- The adsorption process was performed for 2 hours at 250°C. It was found that the adsorption of phenol in the presence of rutile-phase TiO_2 was incomplete.
- The adsorption curves of the process were drawn using a "Varian Cary 50." The rutile-phase TiO_2 nanoparticles were characterized using the XRD method. X-ray structure graphs of the studied nanocomposite materials were recorded on a Rigaku Mini Flex 600 powder diffraction.
- TiO_2 nanoparticles used in the adsorption process were 10–30 nm in size. The nanoparticles were identified by SEM analysis. Mathematical modeling of the process was developed using both logistic and exponential methods.

The results for the exponential model were as follows: the phenol concentration reached 0.5, 0.1, and 0.01 within approximately 0.46, 0.85, and 1 hour, respectively, after the process started, as shown in Table 16.2.

The results for the logistic model are as follows: the phenol concentration reached 0.01 and 0.5 within approximately an hour and 0.54 hour, respectively, after the process started (Khang, 2023b).

REFERENCES

Alekseev E. V., (2007). "Physical and chemical wastewater treatment," *M. Association of Building Universities*, 248. www.sciencedirect.com/science/article/pii/S0959652617326045

Elmina G., Sevinj H., Afsun S., (2020). "Investigation of photocatalytic properties of TiO_2 nanoparticles belonging to rutile phase," *Journal of Molecular Structure*, 1227, 129534. https://sciencedirect.com/science/article/pii/S0022286020318494/10.1016.j/molstruc

Fabing S., Lu L., Tee Meng H., Zhao X. S., (2005). "Phenol adsorption on zeolite-templated carbons with different structural and surface properties," *Carbon*, 43(6), 1156–1164. www.sciencedirect.com/science/article/pii/S0008622305000084

Gadirova Elmina M., (2019). "Adsorption of phenol in the presence of graphine oxide nanoparticles. Az.TU.Scientific works," *Fundamental Sciences*, 1, 166–171 (1815–1779). https://staff-beta.najah.edu/media/conference/2021/11/27/1688_EnviroChem_2021_Abstract_and_proceedings_Book.pdf#page=45

Gadirova Elmina M., Abdullayev Vugar H., Abuzarova Vusala A., (2024). "Treatment solution of the environmental water pollution for smart city," In *Smart Cities: IoT Technologies, Big Data Solutions, Cloud Platforms, and Cybersecurity Techniques* (1st ed.). CRC Press. https://doi.org/10.1201/9781003376064-22

Ghafari M., Cui Y., Alali A., Atkinson J. D., (2019). "Phenol adsorption and desorption with physically and chemically tailored porous polymers: Mechanistic variability associated with hyper-cross-linking and amination," *Journal of Hazardous Materials*, 361, 162– 168. www.sciencedirect.com/science/article/pii/S0304389418307568

Goto M., Hayashi N., Goto S., (1986). "Adsorption and desorption of phenol on anion-exchange resin and activated carbon," *Environmental Science & Technology*, 20, 463–467. https://pubs.acs.org/doi/pdf/10.1021/es00147a004

Gundogdu A., et al., (2012). "Adsorption of phenol from aqueous solution on a low-cost activated carbon produced from tea industry waste: Equilibrium, kinetic, and thermodynamic study," *Journal of Chemical & Engineering Data*, 57, 2733–2743.

Hu X., Yu Y., Ren S., Lin N., Wang Y., Zhou J., (2018). "Highly efficient removal of phenol from aqueous solutions using graphene oxide/Al_2O_3 composite membrane," *Journal of Porous Materials*, 25(3), 719–726. https://link.springer.com/article/10.1007/s10934-017-0485-z

Huang L., Chen J., Gao T., Zhang M., Li Y., Dai L., Qu L., Shi G., (2016). "Reduced graphene oxide membranes for ultrafast organic solvent nanofiltration," *Advanced Materials*, 28(39), 8669–8674. https://onlinelibrary.wiley.com/doi/abs/10.1002/adma.201601606

Hughes-Hallett D., Gleason A. M., Mcallum W. G., et al., (2009). *Calculus: Single variable*. John Wiley & Sons. www.google.com/books?hl=en&lr=& id=dd9FEAAAQBAJ&oi=fnd&pg=PP1&dq=Calculus:+Single+variable,+By+John+Wiley+%26+Sons

Isaeva L. N., Simonova V. V., Tamarkina Y. V., Bovan D. V., Kucherenko V. A., Shendrik T. G., (2009). "Phenol adsorption by chemically activated brown coal // Pratsi of Donetsk National Technical University," *Series: Chemistry and Chemistry Technology*, 12(144), 122–127. https://pdfs.semanticscholar.org/27a8/cafd0d6e58acfe63f52a5013391d5247f4af.pdf

Khang A., Gupta S. K., Rani S., Karras D. A., (2022). *Smart Cities: IoT Technologies, Big Data Solutions, Cloud Platforms, and Cybersecurity Techniques* (1st ed.). CRC Press. https://doi.org/10.1201/9781003376064

Khang A., Gupta S. K, Rani S., Karras D. A., (2023a). *Smart Cities: IoT Technologies, Big Data Solutions, Cloud Platforms, and Cybersecurity Techniques* (1st ed.). CRC Press. https://doi.org/10.1201/9781003376064

Khang A., (2023b). *Advanced Technologies and AI-Equipped IoT Applications in High-Tech Agriculture* (1st ed.). IGI Global Press. https://doi.org/10.4018/9781668492314

Khang A., Gadirova Elmina M., Abdullayev Vugar H., Abuzarova Vusala A., (2024). "Treatment solution of the environmental water pollution for smart city," In *Smart Cities: IoT Technologies, Big Data Solutions, Cloud Platforms, and Cybersecurity Techniques* (1st ed.). CRC Press. https://doi.org/10.1201/9781003376064-22

Marsh H., Rodrigues-Reinoso F., (2006). *Activated Carbon*, p. 536. Elsevier Science & Technology Books. www.sciencedirect.com/science/article/pii/S0008622398003248

Rani S., Bhambri P., Kataria A., Khang A., (2022). "Smart city ecosystem: Concept, sustainability, design principles and technologies," In *AI-Centric Smart City Ecosystems: Technologies, Design and Implementation* (1st ed.). CRC Press. https://doi.org/10.1201/9781003252542-1

Salari M., et al., (2019). "High performance removal of phenol from aqueous solution by magnetic chitosan based on response surface methodology and genetic algorithm," *Journal of Molecular Liquids*, 285, 146–157. www.sciencedirect.com/science/article/pii/S0167732219313832

Shabanova N. A., (2007). *Chemistry and Technology of Nanodispersed Oxides* (N. A. Shabanova, V. V. Popov, P. D. Sarkisov. M.: ICC "Acad. book", 2007, 309). https://search.proquest.com/openview/652e432ef1b0d4615f4d15d43def72ff/1?pq-origsite=gscholar&cbl=2044430

Shang R., Goulas A., Tang C. Y., Serra X. F., Rietveld L. C., Heijman S. G. J., (2017). "Atmospheric pressure atomic layer deposition for tight ceramic nanofiltration membranes: Synthesis and application in water purification," *Journal of Membrane Science*, 528, 163–170. www.sciencedirect.com/science/article/pii/S0376738816322414

17 Phishing Attack and Defense

An Exploratory Data Analytics of Uniform Resource Locators for Cybersecurity

*Taiwo O. Olaleye, Oluwasefunmi T. Arogundade,
Agbaegbu Johnbosco, Olayemi O. Sadare,
Adekunle M. Azeez, Azeez A. Opatunji, Ayobami
A. Tewogbade, and Saminu Akintunde*

17.1 INTRODUCTION

The Internet has developed into a crucial infrastructure that greatly facilitates human and societal coexistence (Sekiya, 2021). However, the Internet also has inevitable security issues, such as phishing, dangerous malware, and privacy revelation, which have already posed severe dangers to users (Sekiya, 2021).

The venture of mimicking a legitimate website by fraudsters, referred to as phishing, is a cybercrime that continues to evolve with respect to the dynamic nature of the World Wide Web. It is a criminal tactic that uses both social engineering and technological deception to steal users' personal identifying information and log-in credentials to defraud them of tangible and intangible assets (Qurashi et al., 2021a).

In this scenario, unsuspecting Internet users take a fake website as authentic, and hence become vulnerable during the course of interaction with the fictitious website. Moreover, phishing tactics have expanded in variety over the years and may now be more harmful than ever.

17.1.1 BACKGROUND

With the incorporation of social media and log-in options, an attacker may be able to steal pieces of personal information about a person with a single phished password, leaving them open to ransomware assaults (Khang et al., 2023b).

In another approach, a perpetrator uses a fake domain name or a uniform resource locator (URL) to pose as a reliable organization or individual (Salah and Zuhair,

DOI: 10.1201/9781003400110-17

2021). Attackers also employ phishing emails to disseminate malicious URLs or attachments that are capable of performing a number of tasks.

Hence, phishing is now more widely propagated than through conventional channels, such as email, short message service (SMS), and pop-up windows. Although the popularity of mobile Internet and social networks has made life easier for users, they have also been used to spread phishing, including spear phishing, QR code phishing, and the spoofing of mobile applications (Morais, 2022).

In addition, because many people assume that hypertext transfer protocol (HTTPs) websites are probably authentic, many crafty phishing attacks are located on websites that have HTTPs and secure socket layer certificates (Kalabarige et al., 2022). Therefore, phishing is an evolving trend that brings new obstacles for its detection.

Hence, security professionals and researchers have invested considerable time and energy in the detection of phishing websites despite the poisonous and covert nature of phishers. Blacklists and whitelists are frequently employed in the identification of phishing websites to shield users from phishing attacks, as Google makes a constantly updated blacklist of harmful websites available (Gillis, 2020).

However, studies show that 63% of phishing websites barely last for 2 hours, and that 47% to 83% of phishing websites are added to blacklists after 12 hours (Kalabarige et al., 2022). As a result, updating the blacklist lags far behind the creation of phishing websites.

17.1.2 METHODOLOGY

The machine learning use case of artificial intelligence (AI) is employed for phishing website detection with a higher accuracy level by depending on extracted features from both legitimate and fraudulent website URLs.

Several studies, including Sekiya (2021), Yang et al. (2019), Odeh et al. (2021), and Tang and Mahmoud (2021), have employed the predictive analytics approach relying on several extracted features of their employed URL datasets.

However, the recently released phishing website dataset by Mendeley (Vrbančič, 2020) has been widely employed in detection studies, including Sekiya (2021), Qurashi et al. (2021b), Salah and Zuhair (2021), and Morais (2022), without a prior exploratory data analysis (EDA) to unravel actionable insights from the various attributes that make up the dataset.

This important research gap has been noted as a threat to the internal validity of the proposed machine learning-based models in the literature (Olaleye et al., 2023) and forms the motivation of this chapter.

The result of the EDA would help dictate the pace of the ensuing prediction while revealing indispensable patterns that determine the status of a feature website as either phishing or legitimate.

Therefore, this study employed statistical data analysis to identify trends and patterns in the Mendeley Phishing Websites Dataset prior to predictive analytics using naive Bayes (NB) and neural network learner algorithms (Khanh & Khang, 2021).

17.1.3 Chapter Organization

The rest of the chapter is organized as follows. Section 2.2 discusses the related existing studies on cybersecurity analytics using data science methodologies. Section 2.3 describes the conceptual framework designed for the chapter, while section 2.4 discusses the experimental result. Section 2.5 concludes with recommendations.

17.2 RELATED WORK

Feature selection techniques, including information gain (IG), gain ratio, chi-square, and correlation-based approaches, were employed by Adi et al. (2019) on the UCI phishing dataset.

When applied to the chosen feature subset, the classification algorithms NB, *k*-nearest neighbor (KNN), support vector machine, decision tree (DT), and iterative dichotomiser 3 algorithm show a decline in accuracy when compared to the full set of 30 features.

On the Mendeley dataset, Almseidin et al. (2022) used feature selection techniques, such as IG and ReliefF, along with classification algorithms, including J48, RF, and multilayer perceptron (MLP).

Le et al. (2010) suggested a technique for identifying phishing websites by collecting lexical information from URL strings and applying adaptive regularization of weights. This approach ensures higher detection accuracy while overcoming noise in the training data.

The URL size, number of dots, number of hyphens, number of numeric characters, a discrete variable that corresponds to the presence of an IP address in the URL, and similarity index were the only six URL features used by Zouina and Outtaj (2017) in their proposed lightweight phishing website detection method. Owing to their few features and quick identification, the features collected are entirely reliant on the URLs.

However, inadequate experimental data are available. Xiang et al. (2011) proposed a CANTINA+ phishing website detection system based on CANTINA.

Before extracting 15 highly distinct features from the URL vocabulary, HTML DOM, WHOIS data, and search engine information, the method first filtered out highly similar phishing websites and webpages without log-in forms.

Finally, phishing website prediction was implemented using a machine learning algorithm. A strategy for detecting phishing websites that is scalable and independent of language was proposed by Marchal et al. (2016).

In their study, 212 characteristics were chosen for the URL and HTML, and gradient boosting was applied to identify fraudulent websites with high accuracy. Based on the combined attributes, phishing detection performs better because it accurately depicts a website.

Qurashi et al. (2021b) proposed a method that creates adversarial phishing assaults by identifying the best subsets of attributes that increase the rate of evasion. Recursive feature elimination, least absolute shrinkage and selection operator, and cancel out were utilized as feature engineering techniques to create and evaluate attack vectors that have a higher potential to elude phishing detection.

The techniques were evaluated before classifying various evasion tests as passed or failed, based on their evasion rates. In contrast to the original generative adversarial deep neural network, which randomly perturbs features, their threat model has a greater capacity for evasion detection.

In the work of Abdulrahman et al. (2023), cloud infrastructure and deep learning frameworks are employed for web phishing detection, while bidirectional transformers and machine learning models were employed for spam detection in another study (Bhambri et al., 2022).

Advances in data science have disrupted the cybersecurity industry, with far-reaching mitigation potential. The capacity to collect and analyze high-resolution data from every computer, virtual machine, and network tap in even the largest global computer networks in close to real time has been made possible by improvements in the collection, storage, and big data infrastructure (Khang et al., 2022).

We now have tremendous visibility into the operating network, resolved down to individual assets reporting nearly continually, as opposed to the data of the past, which were measured at the perimeter of a cyber system and merely aggregate summaries. These data can be produced and stored in petabyte storage facilities and can be condensed in a variety of logs, metadata, low-level operating system, and network telemetry (Khang et al., 2023a).

The operating systems of machines connected to the network are an important source of data. Endpoint detection and response technology, recently introduced by the cybersecurity sector, can collect these data (Hero et al., 2023).

Endpoint detection and response entails the addition of monitoring software that runs on operating systems, collects various telemetry relevant to security, and uploads it to either on-premises or cloud servers for analysis (Vrushank et al., 2023).

Among the many behaviors are changes to the registry and file system, command-line usage, process use, and network telemetry. These data can be used to detect malicious execution by file-based malware; in fact, this is how antivirus software functions.

As mentioned previously, data science approaches employed in the literature to mine useful datasets exhibit various threats to the internal validity of the models used (Rani et al., 2021).

For those that used machine learning tools, EDA was not implemented prior to their predictive analytics.

Exploratory analytics is expected to help infer actionable insights into the nature and status of the predictive attributes contained in training sets (Olaleye et al., 2023).

This will make informed decisions on the choice of learner algorithm and feature engineering techniques that could enhance the capacity of the predictive model, especially the feature selection technique, which has been proven to enhance the performance metrics of machine learning algorithms (Olaleye et al., 2022).

17.3 PROPOSED METHODOLOGY

The conceptual framework employed in this study is discussed in this section, which addresses some of the threats to the internal validity of the proposed methodologies in the literature for phishing detection.

This approach is an experimental operation, which is a three-phase strategy. It includes descriptions of the phases, including historical data acquisition, statistical data analysis, and predictive analytics, as shown in Figure 17.1.

17.3.1 DATA ACQUISITION

This study employed the phishing website dataset from Mendeley (Vrbančič, 2020) with 112 extracted attributes and 88,647 instances of 58,000 legitimate (labeled 1) and 30,647 phishing (labeled 0) websites.

A description of the selected 11-no website attributes from the Mendeley set used in this study is presented in Table 17.1. Categorized into three subgroups of URL-based, domain-based, and parameter-based, attributes in each category are used for the statistical evaluation.

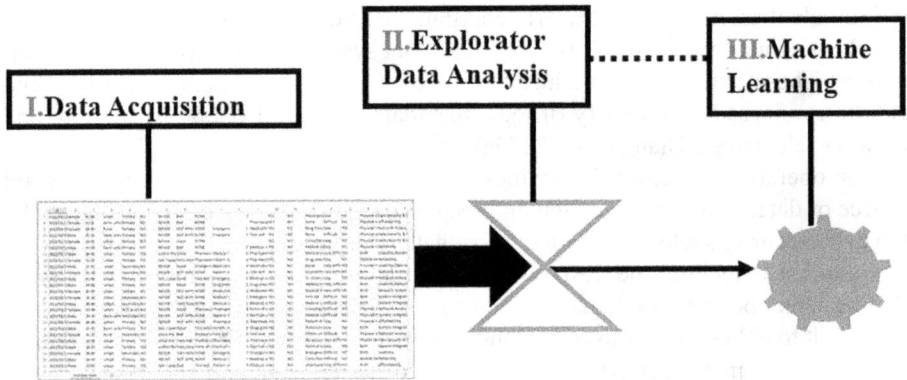

FIGURE 17.1   Conceptual Framework of the Study.

TABLE 17.1
Features Employed for the Statistical Data Analysis.

Description	Feature	Remarks
URL length	length_url	
Number of resolved IPs	qty_ip_resolved	Used for the statistical
Number of redirects	qty_redirects	evaluation of URL-based
Number of resolved name servers (NameServers—NS)	qty_nameservers	features
Parameter length	params_length	Used for the statistical
Top-level domain (TLD) presence in arguments	tld_present_params	evaluation of parameter-based
Quantity of parameters	qty_params	features
Domain length	domain_length	
Time (in days) of domain activation	time_domain_activaton	Used for the statistical
Time (in days) of domain expiration	time_domain_expiration	evaluation of domain-based
Count vowels in domain	qty_vowels_domain	features

For the purpose of this study, the URL length, parameter length, and domain length are further classified as basic, high, higher, and highest, depending on the data distribution.

However, after feature selection to identify the most significant attributes out of 112, the identified features were employed for predictive analytics in the third phase of this study.

The selected attributes do not require pre-processing, as the set has been annotated and labeled for a supervised machine learning task.

17.3.2 EXPLORATORY DATA ANALYSIS

The initially selected 11-no attributes were evaluated using statistical analysis of the interquartile range (IQR) through the box plot and correlation analysis using the heat map plot.

However, IG was employed to identify the most significant attributes, out of the entire 112, for the detection of phishing or legitimate websites during the machine learning phase.

The IQR serves as a gauge for where a dataset's "middle 50" lies, just as it measures where the majority of the values lay. Data points in the dataset are first organized in ascending order from the smallest to the largest, and then the first and third quartile locations are used, as computed in Equations 17.1 and 17.2, to calculate the entire range:

$$Q1 = \left\{ \frac{n+1}{4} \right\} \text{th} \tag{17.1}$$

as the most centered value in the first half of the rank-organized dataset, as computed in Equation 17.2:

$$Q3 = \left\{ 3\frac{n+1}{4} \right\} \text{th} \tag{17.2}$$

as the most centered value in the second half of the rank-organized dataset.

Q2 is the median and computed as Equation 17.3:

$$Q2 = Q3 - Q1 \tag{17.3}$$

By altering a dataset, IG is the decrease in entropy or surprise and is determined by contrasting the dataset's entropy before and after a change. For the website data with C ground truths as Equation 17.4:

$$E = -\sum_{i}^{C} pi \log 2pi \tag{17.4}$$

where pi is the likelihood that a class i element will be chosen at random (i.e., the proportion of the dataset made up of class i).

The correlation coefficient () is a metric that assesses how closely the movements of two different variables are related, with range of potential values from –1.0 to 1.0. Therefore, the numbers cannot be more than 1.0 or lower than –1.0.

A perfect negative correlation is represented by a correlation of –1.0, and a perfect positive correlation is represented by a correlation of 1.0. A positive link exists between the status of a website and any of the 112 attributes if the correlation coefficient between them is greater than zero.

However, a negative association exists if the value is less than zero. If the value is zero, then there is no correlation between the two variables. Before determining the correlation, the covariance of the two variables must first be calculated, and hence, the standard deviation of each variable.

By dividing the covariance by the sum of the standard deviations of the two variables, the correlation coefficient was computed as Equation 17.5:

$$\text{Correlation } = P = \frac{\text{COV}(x, y)}{\sigma X \sigma Y} \tag{17.5}$$

where X and Y are the attribute and ground truth, respectively.

The IQR and correlation evaluation analysis, through box and heat map plots, were implemented using the object-oriented Python programming libraries of Seaborn, Matplotlib, and DataFrame in a Jupyter notebook.

The IG computation for feature selection was implemented using the Orange data mining toolkit (Borondics et al., 2020) with Python libraries. The result of feature selection determines the dataset features that will be used to train the algorithm for the next phase of the conceptual framework.

17.3.3   PREDICTIVE ANALYTICS OF UNIFORM RESOURCE LOCATORS

This subsection describes the supervised machine learning approach of the conceptual framework for detecting phishing in website URLs.

The machine learning algorithms to be trained with the prominent parameters emerging from the IG analysis are presented and discussed.

The training set, which consists of the prominent parameters returned by IG, is used to train the learner algorithms of NB and a neural network.

Bayes theorem and strong (naive) independence assumptions between the features serve as the foundation for the collection of simple probabilistic classifiers known as NB (Olaleye et al., 2023). The mathematical model of the supervised NB learning is presented as follows.

17.3.3.1   Naive Bayes Mathematical Model

The NB algorithm assumes that all independent variables are conditionally independent given the response variable.

Y: the target variable (1 if the URL is phishing; 0 if otherwise)

$X1, X2 \ldots X11$: the URL predictive attributes to approximate the target variable Y

Bayes theorem calculates the probability of the target variable given the independent variables, as in Equation 17.6:

$$P(Y|X1, X2 \ldots X11) = P(X1, X2 \ldots X11|Y) * P(Y)/P(X1, X2 \ldots X11) \quad (17.6)$$

To further simplify the model, we can assume that the prior probability of Y ($P(Y)$) is equal to the proportion of phishing URLs in the training data.

We can also assume that the conditional probabilities of each independent variable, given the response variable ($P(Xi|Y)$), can be estimated using the training data.

With these assumptions, we calculate the probability of a URL being phishing as Equation 17.7 or not with Equation 17.8:

$$P(Y = 1|X1, X2 \ldots X11) =$$

$$P(X1|Y = 1) * P(X2|Y = 1) * \ldots * P(X11|Y = 1) * P(Y = 1)/P(X1, X2 \ldots X11) \quad (17.7)$$

$$P(Y = 0|X1, X2 \ldots X11) =$$

$$P(X1|Y = 0) * P(X2|Y = 0) * \ldots * P(X11|Y = 0) * P(Y = 0)/P(X1, X2 \ldots X11) \quad (17.8)$$

To classify a new URL, we compared its probability of being phishing ($P(Y = 1|X1, X2 \ldots X11)$) to the probability of it not being phishing ($P(Y = 0|X1, X2 \ldots X11)$).

If the probability of it being phishing is high, it is classified as phishing; otherwise, it is classified as not phishing.

Note that the denominator $P(X1, X2 \ldots X11)$ is the same for both Equations 17.9 and 17.10, so it can be omitted when comparing the probabilities.

17.3.3.1.1 Logarithm Approach

$$\log (P(Y = 1|X1, X2, \ldots, X11)) = \log (P(X1|Y = 1)) + \log (P(X2|Y = 1))$$
$$+ \ldots + \log (P(X11|Y = 1)) + \log (P(Y = 1)) \quad (17.9)$$

$$\log (P(Y = 0|X1, X2, \ldots, X11)) = \log (P(X1|Y=0)) + \log (P(X2|Y = 0))$$
$$+ \ldots + \log (P(X11|Y = 0)) + \log (P(Y = 0)) \quad (17.10)$$

17.3.3.1.2 Frequency Count Approach

To estimate the conditional probabilities of each independent variable given the response variable as Equation 17.11:

$$P(Xi|Y = j) = \text{count } (Xi = j \text{ and } Y = j)/\text{count } (Y = j) \quad (17.11)$$

where count ($Xi = j$ and $Y = j$) is the number of times $Xi = j$ and $Y = j$ occur together in the training set; count ($Y = j$) is the number of times $Y = j$ occurs in the training data.

If count ($Xi = j$ and $Y = j$) is zero, Laplace smoothing is used to avoid zero probabilities as Equation 17.12:

$$P(Xi = j|Y = j) = (\text{count } (Xi = j \text{ and } Y = j) + \quad (17.12)$$

On the other hand, a synthetic neural network called an artificial neural network (ANN) was used to overcome problems with AI (Olaleye et al., 2022).

Biological neuron connections are modeled by ANNs as weights between nodes, where an excitatory link represents a positive link, and an inhibitory link represents a negative link (Kassaymeh, 2022).

Each input of the prominent attributes is assigned a weight before being added together. The mathematical model of a free-forward ANN is presented as follows.

17.3.3.2 Free-Forward Artificial Neural Network Mathematical Model

Let:

$x1, x2, . . ., x11$: predictive attributes to approximate the target variable y
y: the target variable (1 if the URL is phishing; 0 if otherwise)

The feed-forward neural network consists of three layers. The input layer has n neurons corresponding to the n input features, where $n = 11$ in this case.

17.3.3.2.1 Hidden Layer

The hidden layer has m neurons, where m is a hyper-parameter that must be chosen. We use a hyperbolic tangent (tanh) activation function for the hidden layer neurons.

17.3.3.2.2 Output Layer

The output layer has one neuron corresponding to the binary response variable y, and we used a sigmoid activation function for this neuron.

17.3.3.3 Mathematical Representation of the Artificial Neural Network Model

17.3.3.3.1 Input layer

For a given input feature vector $x = [x1, x2 . . . x11]$, the input layer neurons are just the input features themselves. Therefore, the input layer can be represented as Equation 17.13:

$$a\,(0) = x \tag{17.13}$$

17.3.3.3.2 Hidden Layer

The output of each neuron in the hidden layer can be calculated by applying a weighted sum of the input features and adding a bias term, followed by the activation function.

We used a tanh activation function for the hidden layer neurons. Therefore, the hidden layer can be represented as Equation 17.14:

$$z\,(1) = W\,(1)\,a\,(0) + b\,(1)\,a\,(1) = \tanh\,(z\,(1)) \tag{17.14}$$

where $W\,(1)$ is an $m \times n$ weight matrix that contains the weights of the connections between the input layer and the hidden layer, and $b\,(1)$ is an m-dimensional bias vector.

17.3.3.3.3 Output Layer

The output of the output layer neuron can be calculated by applying a weighted sum of the hidden layer outputs and adding a bias term, followed by a sigmoid activation function.

The sigmoid function maps the output to a probability value between 0 and 1, which can be interpreted as the probability of the URL being phishing. The output layer can be represented in Equation 17.15:

$$z(2) = W(2)\, a(1) + b(2)\, a(2) = \text{sigmoid}(z(2)) \tag{17.15}$$

where $W(2)$ is a $1 \times m$ weight matrix that contains the weights of the connections between the hidden layer and the output layer, and $b(2)$ is a scalar bias term.

The output of the neural network can be interpreted as the probability of the URL being phishing or not. We define a threshold value of 0.5 and classify URLs with outputs above the threshold as phishing and those below the threshold as not phishing.

To train the neural network, we define a loss function that measures the difference between the predicted and true outputs for a given input feature vector.

We used the binary cross-entropy loss function for this task. The objective of the training process was to determine the weights and biases that minimized the loss function.

To find the optimal values of the weights and biases, we used a backpropagation algorithm with stochastic gradient descent (SGD) optimization (Altschuler & Talwar, 2022).

The backpropagation algorithm computes the gradient of the loss function with respect to the weights and biases, and updates them accordingly.

The SGD optimization updates the weights and biases after each batch of input feature vectors, where a batch is a subset of the URL training set.

The ANN model can be summarized as follows:

Input layer as Equation 17.16

$$a(0) = x \tag{17.16}$$

Hidden layer as Equation 17.17

$$z(1) = W(1)\, a(0) + b(1), a(1) = \tanh(z(1)) \tag{17.17}$$

Output layer as Equation 17.18

$$z(2) = W(2)\, a(1) + b(2), a(2) = \text{sigmoid}(z(2)) \tag{17.18}$$

where $x = [x1, x2 \ldots x11]$ is the input feature vector;

W (1) is an $m \times n$ weight matrix that contains the weights of the connections between the input layer and the hidden layer.
b (1) is an m-dimensional bias vector;
z (1) is the output of the hidden layer before activation;

a (1) is the output of the hidden layer after activation, using the tanh activation function;

W (2) is a $1 \times m$ weight matrix that contains the weights of the connections between the hidden layer and the output layer;

b (2) is a scalar bias term;

z (2) is the output of the output layer before activation;

a (2) is the output of the output layer after activation, using the sigmoid activation function.

Therefore, to classify a new URL as phishing or not phishing, we fed the input feature vector x into the neural network and obtained the output a (2), which represents the probability of the URL being phishing.

We can then compare the output a (2) to a threshold value of 0.5 and classify the URL as phishing if a (2) $\geq$ threshold and not phishing if otherwise.

The machine learning phase of this study was also implemented on the Jupyter notebook using the object-oriented Python programming language.

17.4   RESULTS

The experimental results of this study are presented in this section. The IG analysis conducted in the second phase of the framework returned five most significant feature attributes from the 112 encapsulated in the Mendeley training dataset, as shown in Table 17.2.

The statistical evaluations of the three categories of features presented in Table 17.1 for the URL, domain, and parameter-based evaluations are presented in the box plots of Figures 17.2–17.7.

The statistical summary produced by Python EDA (including the mean, 50th percentile, standard deviation, count, and quartile values) of the five most prominent attributes returned by the IG evaluation is presented in Table 17.3 for the legitimate websites, while that of the malicious websites used for phishing is presented in Table 17.4.

The correlation distribution analysis of the URL, domain, and parameter attributes is likewise presented in the heat map plots of Figures 17.8, 17.9, and 17.10, respectively.

According to the IG, the five most prominent attributes in the dataset include the quantity of slash (/) in a URL, the direction of the slash, the length of the directory, the quantity of dot (.) sign included in the directory, and the quantity of the dot sign

TABLE 17.2
Results of IG Feature Selection.

Attribute	IG
qty_slash_url	0.535
qty_slash_direction	0.534
directory_length	0.528
qty_dot_directory	0.505
qty_dot_file	0.499

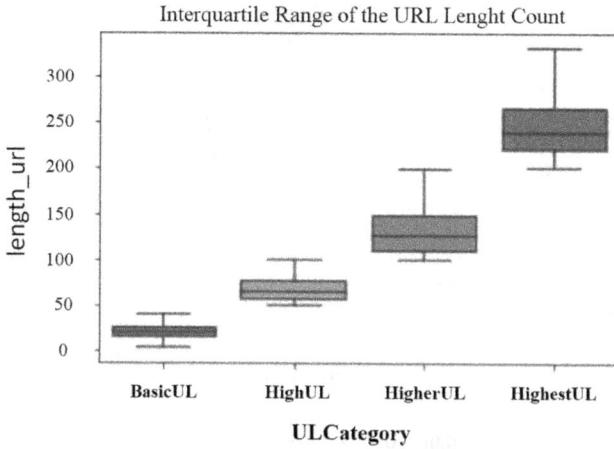

FIGURE 17.2 IQR of URL-Based Attributes.

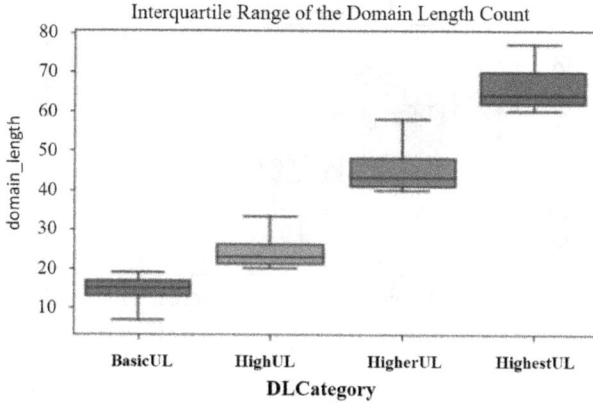

FIGURE 17.3 IQR of Domain-Based Attributes.

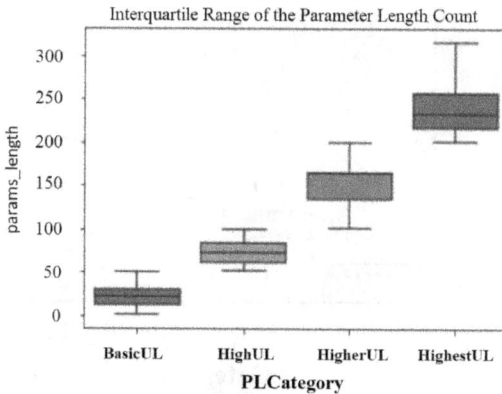

FIGURE 17.4 IQR of Parameter-Based Attributes.

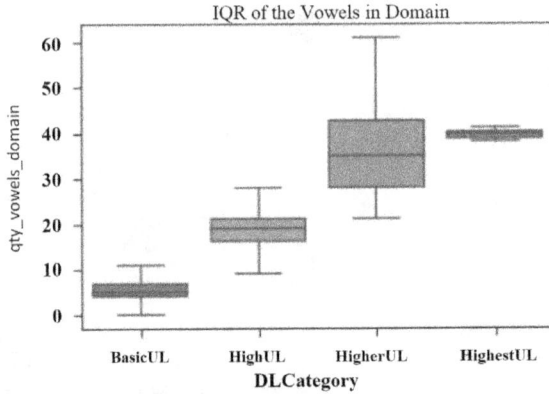

FIGURE 17.5 IQR of the Domain Vowels.

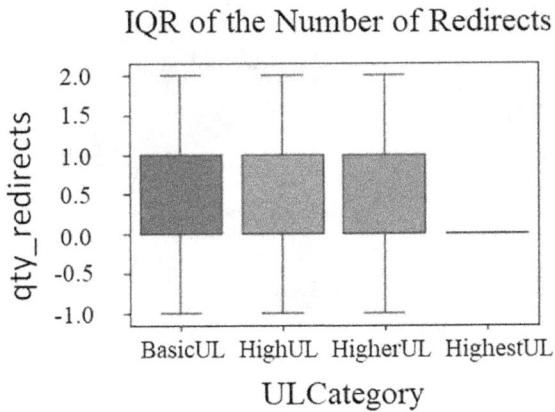

FIGURE 17.6 IQR of the Number of Redirects.

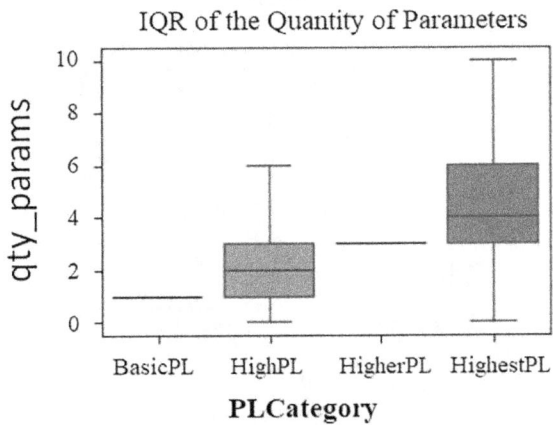

FIGURE 17.7 IQR of Parameter Quantity.

TABLE 17.3

Descriptive Statistical Summary of the Legitimate Websites.

	qty_vowels_domain	domain_length	length_url	qty_slash_url	qty_slash_irectory	directory_length	qty_dot_directory	qty_dot file
Mean	5.234931	18.51412	21.24569	0.319379	−0.48879	1.551828	−0.775	−0.1533
Standard deviation	2.236183	5.311816	12.17683	0.804454	1.144055	10.30515	0.4933	0.804150
Min	0	4	4	0	−1	−1	−1	−1
Q1	4	15	15	0	−1	−1	−1	−1
Q2	5	18	19	0	−1	−1	−1	−1
Q3	7	22	23	0	−1	−1	−1	−1
Max	23	55	546	10	10	530	10	12

TABLE 17.4

Descriptive Statistical Summary of Malicious Phishing Websites.

	qty_vowels_domain	domain_length	length_url	qty_slash_url	qty_slash_irectory	directory_length	qty_dot_directory	qty_dot file
Mean	5.829804	18.6492	64.92828	3.103142	2.989395	28.46921	0.530492	0.404575
Standard deviation	2.997466	8.517043	68.13703	2.020008	1.952827	32.25514	0.874476	0.589524
Min	0	4	5	0	−1	−1	−1	−1
Q1	4	14	31	2	2	10	0	0
Q2	6	17	47	3	3	23	0	0
Q3	7	22	72	4	4	39	1	1
Max	61	231	4165	44	22	1286	19	12

included in the file, in that order. This returns slash (/) as the most prominent discriminative factor used in both the domain of the malicious email and its directory.

The box plots show the location, skewness, and dispersions of the data points across the IQR. Figure 17.4 indicates that the median URL lengths of websites are determined by the size of the URL involved, as the URLs in the highest categories tower above the other subcategories.

The same observations are returned on the plot of length of domains and parameters in Figures 17.4 and 17.5, respectively. It is also observed that the range of data distribution across the length categories for URL, domain, and parameter is directly proportional to the sizes of their lengths.

The skewness observed in the highest, higher, and high categories of the plots indicates that the means are close to the median, which is a symmetric relationship, unlike the basic category, which is often of normal distribution.

The descriptive statistical summary of the prominent attributes is quite revealing. Table 17.3 indicates that an average of five vowels are contained in the domain

Correlation distribution plot for URL parameter

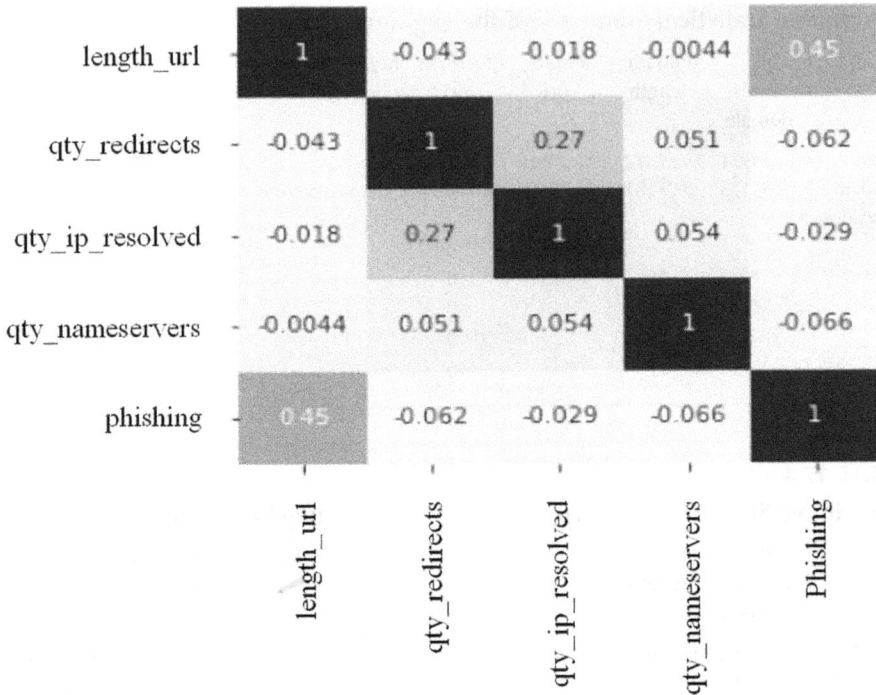

	length_url	qty_redirects	qty_ip_resolved	qty_nameservers	Phishing
length_url	1	-0.043	-0.018	-0.0044	0.45
qty_redirects	-0.043	1	0.27	0.051	-0.062
qty_ip_resolved	-0.018	0.27	1	0.054	-0.029
qty_nameservers	-0.0044	0.051	0.054	1	-0.066
phishing	0.45	-0.062	-0.029	-0.066	1

FIGURE 17.8   Correlation of URL Attributes.

names of the legitimate websites, while the length of the domain is, on average, 16 across the 58,000 legitimate websites.

Likewise, the average length of 21 characters makes up their URLs, when an average of 0.3 slash (/) characters are included in their URLs. A negative average of −0.5 quantities of the slash sign is noticed in their directories, as the average directory length is 1.5.

Likewise, the legitimate websites have −0.8 average number of dots (.), while the average number of dots in their files is −0.1533.

For the malicious phishing websites, however, Table 17.4 indicates that an average of six vowels are contained in their domains, which is often made up of 19 characters.

Unlike the 21 length of URL characters observed in the legitimate websites, the malicious websites contain an average of five, while the number of slash (/) is higher in malicious websites.

The quantity of the slash sign is likewise higher in their directories, with a far higher directory length of 28 compared with the average length of 1.5 in the legitimate websites.

The quantity of dots is higher in the malicious websites with an average number of 0.5 (compared with the −0.8 recorded for the legitimate websites).

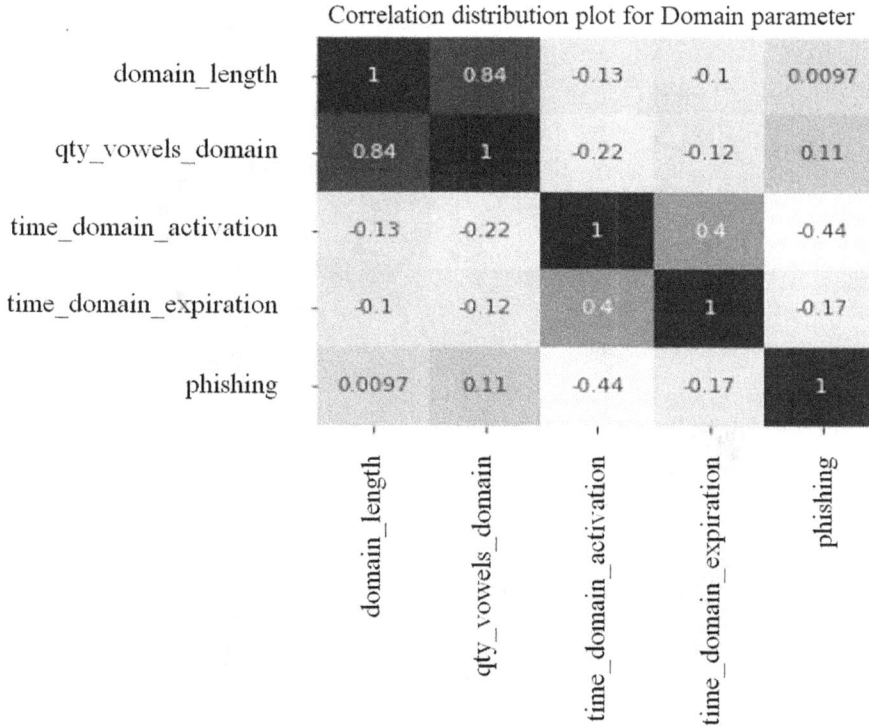

Correlation distribution plot for Domain parameter

FIGURE 17.9   Correlation of Domain Attributes.

TABLE 17.5
Performance Metrics of the Learner Algorithms on 10-Fold Cross Validation.

Algorithm	Accuracy	F1	Precision	Recall
NB	92.4	91.13	90.37	91.91
Neural network	95.65	94.95	95.45	95.65

The quantity of dots in the file of malicious websites is higher with 0.404 compared with −0.1533 of the legitimate websites.

From the statistical analysis, it was observed that there is a strong positive correlation between the length of a URL and the eventual status of the website, as shown in Figure 17.8, just as it is for the quantity of vowels present in a domain name, as shown in Figure 17.9.

However, the length, quantity, and presence of TLD in arguments have negative correlations with whether a website is malicious or not, as shown in Figure 17.10.

The KNN performed best in the predictive analytics of the five prominent attributes, returning a 95.65 accuracy compared with 92.4 weighted average returned by the NB in Table 17.5.

Correlation distribution plot for Parameter

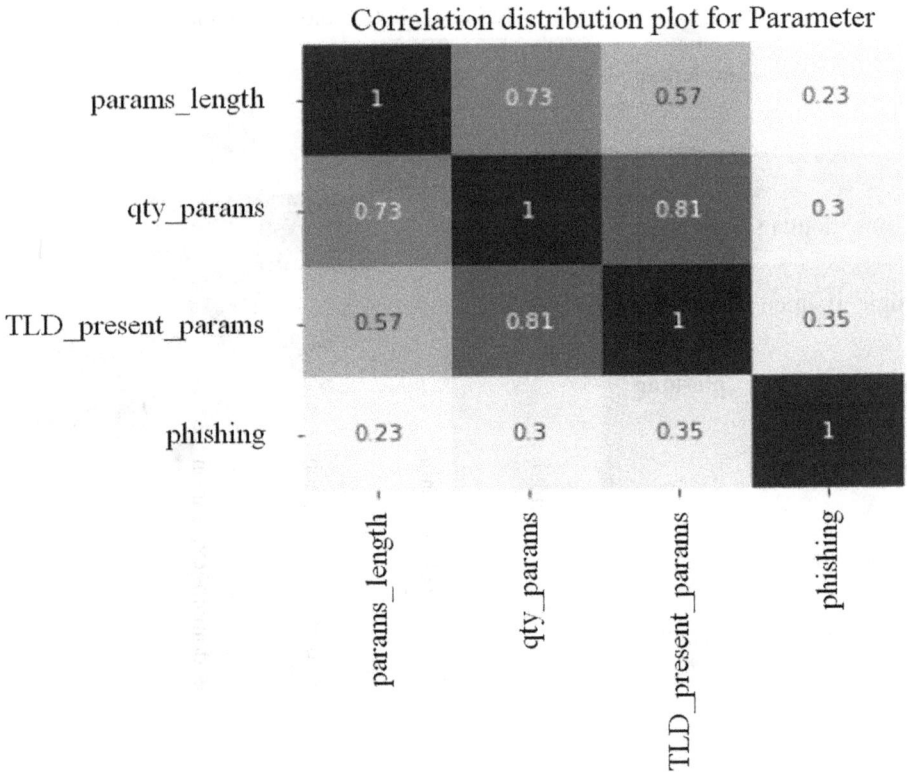

FIGURE 17.10   Correlation of Parameter Attributes.

17.5   CONCLUSION

This study employed predictive analytics and statistical data analysis to evaluate the Mendeley dataset for detecting phishing websites.

The IG feature selection technique was employed to identify five prominent feature attributes from the entire 88,647 websites in the dataset.

Prominent attributes are used to train a neural network and the naive learner algorithms to predict the status of a particular website as either positive (phishing) or negative (legitimate).

Other data attributes that relate to the URL, domain name, and parameters were further analyzed to identify their location, skewness, and dispersions on the interquartile range and their correlation coefficient with the ground truths of being malicious or legitimate.

The experimental results revealed that the slash (/) character in the file directory or domain has the highest IG, and hence determines the status of a website. The results show that the slash character is more prominent on phishing websites than on legitimate websites.

The length of website URLs and domains as well as the number of vowels in the domain of a website have a positive correlation with their phishing status. The neural network likewise outperformed NB in detecting phishing websites based on the five prominent attributes identified by the feature selection technique employed in this study (Rana et al., 2021).

17.6 RECOMMENDATION

Future studies could employ ensemble machine learning on the entire training set to juxtapose the results with the study based on feature selection. In addition, the out-of-sample data sampling approach can be further used to evaluate the performance of the predictive models in comparison with the in-sample approach employed in this chapter (Khang et al., 2023c).

REFERENCES

Abdulrahman L. M., Ahmed S. H., Rashid Z. N., Jghef Y. S., Ghazi T. M., Jader U. H., (2023). "Web phishing detection using web crawling, cloud infrastructure and deep learning framework. *Journal of Applied Science and Technology Trends*, 4(1), 54–71. www.jastt. org/index.php/jasttpath/article/view/144

Adi S., Pristyanto Y., Sunyoto A., (2019). "The best features selection method and relevance variable for web phishing classification," *Proceedings of the International Conference on Application of Information and Communication Technologies (ICOIACT)*, pp. 578–583. https://ieeexplore.ieee.org/abstract/document/8938566/

Almseidin M., Alkasassbeh M., Alzubi M., Al-Sawwa J., (2022). "Cyber-phishing website detection using fuzzy rule interpolation," *Cryptography*. www.mdpi. com/2410-387X/6/2/24

Altschuler J., Talwar K., (2022). "Privacy of noisy stochastic gradient descent: More iterations without more privacy loss," *Advances in Neural Information Processing Systems*, 35, 3788–3800. https://proceedings.neurips.cc/paper_files/paper/2022/hash/18561617ca0b-4ffa293166b3186e04b0-Abstract-Conference.html

Bhambri P., Rani S., Gupta G., Khang A., (2022). *Cloud and Fog Computing Platforms for Internet of Things*. CRC Press. https://doi.org/ 10.1201/9781003213888

Borondics F., Vitali F., Shaulsky G., (2020). *Data Mining*. Retrieved September 17, 2022. https://orangedatamining.com/

Gillis A. S., (2020). *TechTarget*. Retrieved September 18, 2022. www.techtarget.com/ searchsecurity/definition/phishing#:~:text=A%20phishing%20website%20is%20 a,potentially%20more%20dangerous%20than%20before

Hero A., Kar S., Moura J., Neil J., Poor H. V., Turcotte M., Xi B., (2023). "Statistics and data science for cybersecurity," *Statistics and Data Science for Cybersecurity*. https://hdsr. mitpress.mit.edu/pub/koyzu1te

Kalabarige L. R., Rao R. S., Abraham A., Gabralla L. A., (2022). "Multilayer stacked ensemble learning model to detect phishing websites," *IEEE Access*, 10, 79543–79552. https:// ieeexplore.ieee.org/abstract/document/9843994/

Kassaymeh, S. A.-L.-B., (2022). "Backpropagation neural network optimization and software defect estimation modelling using a hybrid Salp swarm optimizer-based simulated annealing algorithm," *Knowledge-Based Systems*, 244, 108511. www.sciencedirect. com/science/article/pii/S0950705122002209

Khang A., Hahanov V., Abbas G. L., Hajimahmud V. A., (2022). "Cyber-physical-social system and incident management," In *AI-Centric Smart City Ecosystems: Technologies, Design and Implementation*, Vol. 2, No. 15 (1st ed.). CRC Press. https://doi. org/10.1201/9781003252542-2

Khang A., Vrushank S., Rani S., (2023a). *AI-Based Technologies and Applications in the Era of the Metaverse* (1st ed.). IGI Global Press. https://doi.org/10.4018/9781668488515

Khang A., Rani S., Gujrati R., Uygun H., Gupta S. K., (Eds.). (2023b). *Designing Workforce Management Systems for Industry 4.0: Data-Centric and AI-Enabled Approaches.* CRC Press. https://doi.org/10.1201/9781003357070

Khang A., Vrushank S., Rani S., (2023c). *AI-Based Technologies and Applications in the Era of the Metaverse* (1st ed.). IGI Global Press. https://doi.org/10.4018/9781668488515

Khanh H. H., Khang A., (2021). "The role of artificial intelligence in blockchain applications," *Reinventing Manufacturing and Business Processes through Artificial Intelligence,* 2(20–40) (CRC Press). https://doi.org/10.1201/9781003145011-2

Le A., Markopoulou A., Faloutsos M., (2010). "Phish Def: URL names say it all," *Proceedings of the IEEE International Professional Communication Conference (INFOCOM),* pp. 191–195. https://ieeexplore.ieee.org/abstract/document/6027869/

Marchal S., Saari K., Singh N., Asokan N., (2016). "Know your phish: Novel techniques for detecting phishing sites and their targets," In *2016 IEEE 36th International Conference on Distributed Computing Systems (ICDCS),* pp. 323–333. IEEE. https://ieeexplore.ieee.org/abstract/document/7536531/

Morais V. B., (2022). "Analysis of selection bias in online adversarial aware machine learning systems," *Louisiana Tech University ProQuest Dissertations Publishing.* https://search.proquest.com/openview/5d5b89e0ef78db1286a7d8fcef64a471/1?pq-origsite=gscholar&cbl=18750&diss=y

Odeh A., Keshta I., Abdelfattah E., (2021). "Machine learning techniques for detection of website phishing: A review for promises and challenges," In *2021 IEEE 11th Annual Computing and Communication Workshop and Conference (CCWC).* IEEE. https://ieeexplore.ieee.org/abstract/document/9375997/

Olaleye T., Abayomi-Alli A., Adesemowo K., Arogundade O. T., Misra S., Kose U., (2022). "SCLAVOEM: Hyper parameter optimization approach to predictive modelling of COVID-19 infodemic tweets using smote and classifier vote ensemble," *Soft Computing,* 27(6), 3531–3550. https://doi.org/10.1007/s00500-022-06940-0

Olaleye T. O., Arogundade O. T., Misra S., Abayomi-Alli A., Kose U., (2023). "Predictive analytics and software defect severity: A systematic review and future directions," *Scientific Programming,* 2023, 18. https://doi.org/10.1155/2023/6221388

Qurashi A. R., AlEroud A., Saifan A. A., Alsmadi M., Alsmadi I., (2021a). "Generating optimal attack paths in generative adversarial phishing," *2021 IEEE International Conference on Intelligence and Security Informatics (ISI).* IEEE. https://ieeexplore.ieee.org/abstract/document/9624751/

Qurashi A. R., AlEroud A., Saifan A. A., Alsmadi M., Alsmadi I., (2021b). "Generating optimal attack paths in generative adversarial phishing," *Antonio: IEEE.* https://ieeexplore.ieee.org/abstract/document/9624751/

Rana G., Khang A., Sharma R., Goel A. K., Dubey A. K., (Eds.). (2021). *Reinventing Manufacturing and Business Processes through Artificial Intelligence.* CRC Press. https://doi.org/10.1201/9781003145011

Rani S., Chauhan M., Kataria A., Khang A., (Eds.). (2021). "IoT equipped intelligent distributed framework for smart healthcare systems," In *Networking and Internet Architecture,* Vol. 2, p. 30. CRC Press. https://doi.org/10.48550/arXiv.2110.04997 (the Metaverse. (1st ed.). IGI Global Press. https://doi.org/10.4018/9781668488515)

Salah H., Zuhair H., (2021). "Catching a Phish: Frontiers of deep learning-based anticipating detection engines," *International Conference of Reliable Information and Communication Technology,* pp. 483–497. Springer. https://link.springer.com/chapter/10.1007/978-3-030-98741-1_40

Sekiya Y., (2021). *Feature Selection Approach for Phishing Detection Based on Machine Learning.* University of Tokyo. https://link.springer.com/chapter/10.1007/978-3-030-95918-0_7

Tang L., Mahmoud Q. H., (2021). "A survey of machine learning-based solutions for phishing website detection," *MDPI*, 3, 672–694. www.mdpi.com/2504-4990/3/3/34

Vrbančič G. I. J., (2020). "Datasets for phishing websites detection," *Mendeley*, 33. https://doi.org/10.1016/j.dib.2020.106438

Vrushank S., Vidhi T., Khang A., (2023). "Electronic health records security and privacy enhancement using blockchain technology," In *Data-Centric AI Solutions and Emerging Technologies in the Healthcare Ecosystem*, p. 1 (1st ed.). CRC Press. https://doi.org/10.1201/9781003356189-1

Xiang G., Hong J., Rose C. P., Cranor L., (2011). "Cantina+ a feature-rich machine learning framework for detecting phishing web sites," *CM Transactions on Information and System Security (TISSEC)*, 14, 1–28 (ACM). https://dl.acm.org/doi/abs/10.1145/2019599.2019606

Yang P., Zhao G., Zeng P., (2019). "Phishing website detection based on multidimensional features driven by deep learning," *IEEE Access*, 7, 15196–15209. https://ieeexplore.ieee.org/abstract/document/8610190/

Zouina M., Outtaj B., (2017). "A novel lightweight URL phishing detection system using SVM and similarity index," *Human-Centric Computing and Information Sciences*, 7(1), 1–13. https://hcis-journal.springeropen.com/articles/10.1186/s13673-017-0098-1

18 Analysis of Deep Learning-Based Approaches for Spam Bots and Cyberbullying Detection in Online Social Networks

Santhosh Kumar A. V., Suresh Kumar N., Kanniga Devi R., and Muthukannan M.

18.1 INTRODUCTION

Online social networks (OSNs) have become a current interest for individuals to meet new people and collect views on different problems. Some applications are frequently used worldwide because of the creation of free accounts and usage (Shetty et al., 2022). However, some OSN users fraudulently handle the platform to profit by posting false content.

Without proper prediction tools for posting false content and trustworthiness of other social network users, common people with a good mindset are addicted to many problems (Alkhamees et al., 2021).

Bots are computer programs designed to execute many tasks that run over the Internet. A botnet is a group of Internet devices connected over the Internet, which controls many bots. The term social botnet refers to botnets that use OSNs as a platform. Botnets are used to send spam messages or any unsolicited communication sent in bulk.

Spam is usually sent via email, short message service (SMS), social media, or phone calls. Spam messages often come in the form of harmless (although annoying) promotional emails. However, spam can sometimes be fraudulent or malicious, as it takes data and helps hijackers access user information (Latah, 2020).

A spam bot helps send or display false content over social networking sites where people interact. Spam refers to an unrequested false report on various products and advertisements in the form of links, images, or videos that attract social network users (Latah, 2019).

DOI: 10.1201/9781003400110-18

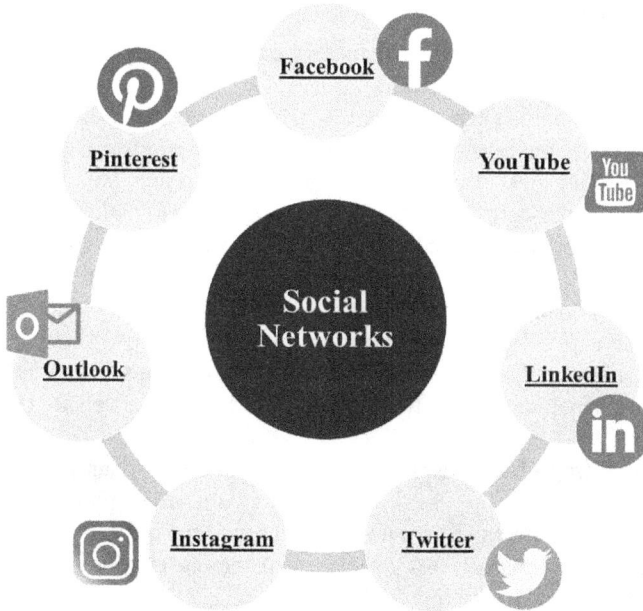

FIGURE 18.1   Online Social Networks

Source: Khang (2021)

Spam bots operate by creating fake accounts on social media, emails, and other platforms. Some spam bots send fake mail IDs and contact numbers to collect user information. There are various types of spam bots, such as email, comment, and social media spam bots (Chang et al., 2021).

Many bots actively work over social media platforms, such as Instagram, X/Twitter, and Facebook.

Many bots like, share, or post unrelated comments for posts on social media (Orabi et al., 2020). There are both good and bad bots. Good bots include search engines, copyright bots, and chatbots. Bad bots include content scraping and click fraud bots, as shown in Figure 18.1.

18.1.1 SPAM

Some researchers of scientific papers have proposed that spam is an acronym that stands for "stupid pointless annoying malware." Spam refers to fake messages, news, images, links, or video formats that are usually sent to OSN users (Roy et al., 2020).

Spam may be categorized as an Internet spam, in which unwanted fake news or messages are sent over the Internet to users. To send these unwanted messages, spam bots are used (Ahmed et al., 2018). A spam bot sends spam messages to the users.

Spam bots can also send messages over networks both directly and indirectly (Masood et al., 2019). There are many types of direct and indirect network community spams, such as email, comments, and tags, as shown in Figure 18.2.

18.1.1.1 Types of Spam

There are many types of spam. Spammers can use any platform to send spam messages.

Some types of spam, such as email, search engine optimization (SEO), social networking, messaging, and mobile spams, are shown in Figure 18.2. Currently, in the world of the Internet, email spam is the greatest issue. This type of spam can affect the financial background of an organization and also makes email users angry (Sharma & Uma, 2018). This can be detected using various techniques and methods.

The spam type, SEO spam, is also known as web spam. This allows the spammer's webpage to reach a higher position in search engines (Oskouei et al., 2018).

The two types of SEO spam are content and link spams. Content spam refers to the unauthorized and unwanted use of content from one's own website on third-party websites, which can affect the reputation of the entire website (Risch & Ralf, 2020).

Link spam refers to links displayed on a website that appear as legitimate links. This may be included on the user comment page and other places on the website. Such links can attract users to give a website a high ranking (Shahzad et al., 2021).

Social networking spam refers to unwanted messages, links, or video formats that appear on OSNs (Almadhoor, 2021).

18.1.1.2 Spam Message

Spam is any unsolicited communication sent in bulk, usually through email, but also distributed via text messages (SMS), social media, or phone calls.

Spam messages often come in the form of harmless (although annoying) promotional emails. However, spams are sometimes fraudulent or malicious. Messaging spam refers to spam messages sent in SMS from unknown numbers (Wei & Trang, 2020).

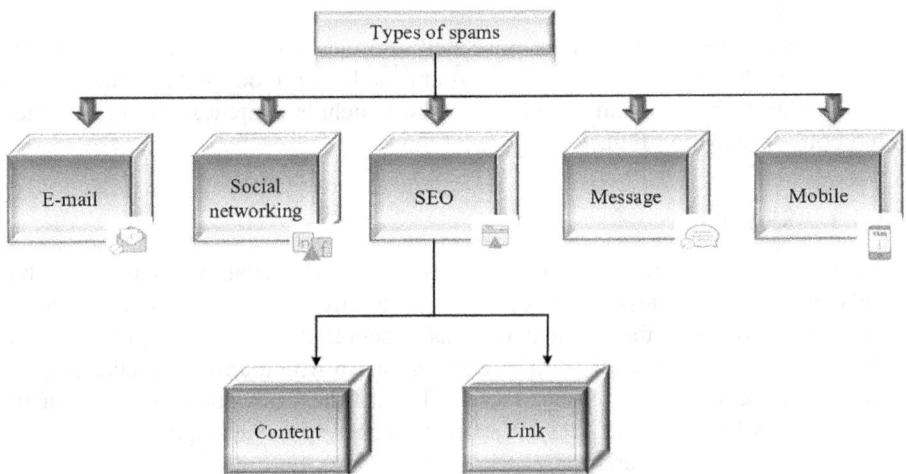

FIGURE 18.2 Types of Spam.

Mobile spam refers to misleading information or messages sent on a mobile message platform or other online platforms (Yang et al., 2021).

18.1.2   SPAM BOT DETECTION METHODS

Researchers have introduced many methods to reduce the spread of spam bots on online social media, as shown in Figure 18.3.

18.1.2.1   Crowdsourcing-Based Method

The crowdsourcing-based method is a spam bot detection (SBD) method proposed by Wang et al. (2019), which is defined using two layers. The first is the filtration layer, the function of which is to segregate a doubtful profile from genuine spam bots using certain techniques.

The other layer is the crowdsourcing layer, which evaluates doubtful profiles and identifies spam bots. Tuskers are people involved in the crowdsourcing layer. Compared with other methods, this method yields very low false-negative and false-positive rates.

The accuracy of this method can be maximized by eliminating tuskers using a voting system. The main risk of using this method is that it can affect the privacy of the user (Aroyo et al., 2019).

18.1.2.2   Graph-Based Method

The graph-based method is used to construct a social network structure. It focuses on social network graphs to distinguish social spam bots from genuine users. Three other methods are integrated to detect spam bots in OSN.

The first method is trust propagation, in which the honest relationship between the objects of the two graphs is analyzed whether strong or weak. The next method

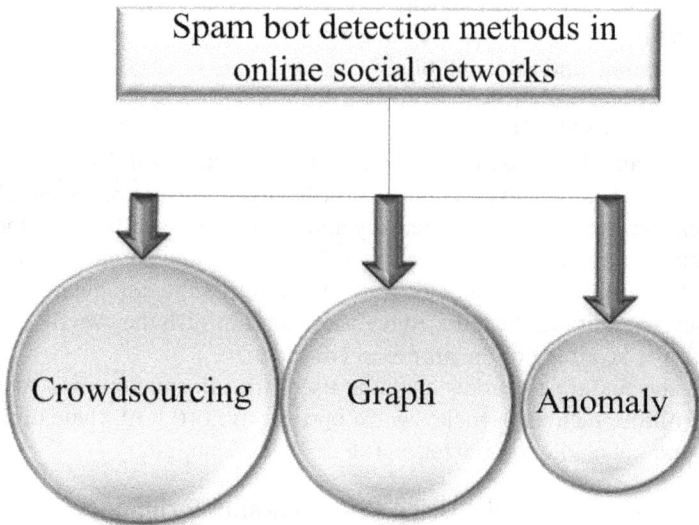

FIGURE 18.3   Types of Spam Bot Detection Methods.

is the clustering of a graph, in which the nodes of the graph are grouped based on the user's similar characteristics. The final method involves studying the performance metrics of the graph (Noekhah et al., 2020).

18.1.2.3 Anomaly-Based Method

The anomaly-based method finds anomalies in every parameter or common pattern within a group, which is not similar to the real behavior of the network (Vengatesan, 2018). It is divided into two types: action- and interaction-based models.

18.1.2.4 Stages for Spam Bot Detection in Online Social Network Using Deep Learning

The basic methodology shown in Figure 18.4 for detecting spam bots in OSNs using deep learning (DL) includes data collection, pre-processing, feature extraction, training the model using DL techniques, testing the model, and evaluating the performance measure.

18.1.2.5 Data Collection

Online social network datasets can be either non-graph-based or graph-based. If graphs are constructed from a social network, those datasets are known as graph-based datasets. However, if datasets are taken from the extracted features of OSNs using APIs, such datasets are known as non-graph-based datasets (Wang et al., 2019). Every OSN community has its own APIs, such as Facebook APIs and Twitter4J APIs, to create datasets.

18.1.2.6 Data Pre-Processing

After gathering all the data from the social networks, the next step is to pre-process the data. The data gathered must be cleaned or filtered using pre-processing mechanisms before applying it to DL models (Cheng et al., 2019). The pre-processing stage involves several stages, such as cleaning or filtering, stemming, removal of stop words, tokenization, and segmentation.

18.1.2.7 Feature Extraction

The model is trained only based on the best feature extracted data. Therefore, the best feature extraction techniques must be utilized in further processes. This is the main concept for improving the efficiency and effectiveness of the DL model.

The feature extraction techniques can effectively remove unnecessary features from a set of features (Jogin et al., 2018). This reduces the complexity of the proposed model. In addition, the reduced features can diminish the overfitting problem and increase the accuracy of the proposed DL mode.

Some of the feature selection methods used are information gain and optimization algorithms, such as particle swarm optimization (PSO), whale optimization algorithm, and ant colony optimization (Sharma & Manu, 2021).

18.1.2.8 Training the Model Using Deep Learning Techniques

Using the extracted features, the proposed model is trained using DL techniques. Some of the major DL techniques are convolutional neural network (CNN), recurrent

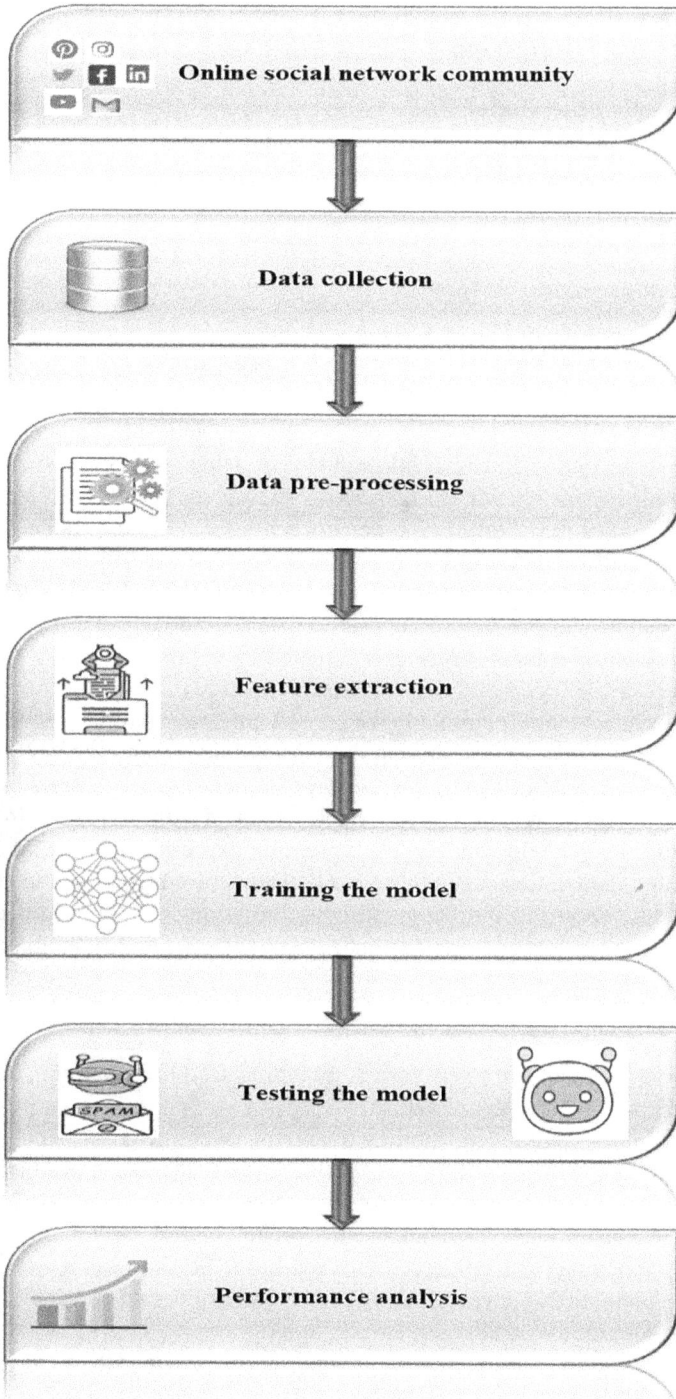

FIGURE 18.4 Methodology of SBD Using DL.

neural network (RNN), long short-term memory (LSTM), and bidirectional long short-term memory (Bi-LSTM).

18.1.2.9 Testing the Model

The model performance is tested using the following performance parameters: precision, accuracy, recall, F-measure, G-measure, equal error rate, and normalized mutual information.

18.1.3 CYBERBULLYING

Online social networks are considered one of the most popular social tools utilized by adolescents and account for their daily Internet activity. Currently, these platforms offer a chance for young people to be subjected to cyberbullying.

Cyberbullying involves one person repeatedly harassing and threatening another person using electronic devices. It takes place in various platforms, such as instant messages, online games, text messages, and social media (Bozyiğit et al., 2021).

The major issues in cyberbullying detection are lack of identifiable parameters, clearly quantifiable standards, and definitions that are used to classify tweets as bullying or non-bullying (Kumari & Jyoti Prakash, 2021). With social networks consuming people's entire time, cyberbullying has emerged as a social issue that demands the best solution. Therefore, automated detection is essential for identifying the occurrence of cyberbullying (Kumar et al., 2021).

Cyberbullying is not commonly observed among Spanish school students (Castellanos et al., 2021). Conventional methods for detecting cyberbullying activities comprise the guidelines and development of standards that all users must adhere to, the use of profane word lists, the employment of human editors to manually check for bullying behavior, and the use of regular expressions (Abarna, 2022).

As a result, the maintenance of these methods is time consuming and labor intensive. Therefore, because of widespread cyberbullying, it is important to develop detection methods to identify new instances of cyberbullying (Mahbub et al., 2021).

18.1.4 TYPES OF CYBERBULLYING

Cyberbullying can occur in many forms, including flooding, flaming, harassment, cyberstalking, cyber threats, and racism.

There is a high likelihood of flooding occurring in online groups and chat rooms, as reported by Xu and Paula (2021). In online games, sending unwanted text messages to individuals in a chat room is referred to as flaming (Zhang, 2021).

Cyberbullying through harassment can spread quickly by sending abusive or insulting text messages. Compared with flaming, persistence is identifiable in harassment (Song, 2021). Cyberstalking is defined as posting threatening, hurtful, or destructive text messages.

Cyber stalkers often seek to destroy their target's friendships and reputation (Kaur et al., 2021). Cyberthreats can be categorized into two types: direct threats and distressing materials. Direct threats are threats that cause a person to commit suicide, as shown in Figure 18.5.

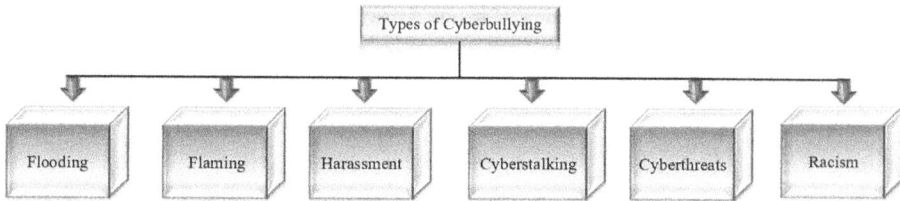

FIGURE 18.5   Types of Cyberbullying.

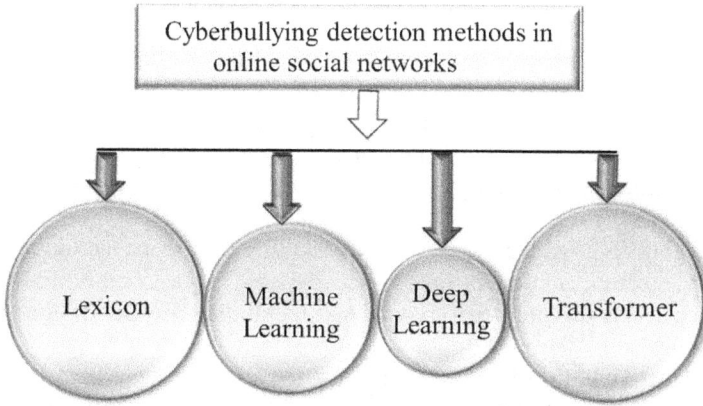

FIGURE 18.6   Types of Cyberbullying Detection Methods.

Information related to actual planned events is usually involved in the threat. People with emotional distress are provided with some clues by online content, known as distressing materials (Herath et al., 2022).

Racism can be found on the Internet in various forms, such as racist images, videos, blogs, websites, and online comments. This term applies to discrimination due to different races or religions, nationalities, and ethnicities (Ortiz-Marcos et al., 2021).

18.1.5   CYBERBULLYING DETECTION METHODS

The most commonly used cyberbullying detection methods are machine learning (ML), lexicon-based, DL-based, and transformer-based methods, as shown in Figure 18.6.

18.1.5.1   Lexicon-Based Methods

These text classification strategies for cyberbullying detection are based on the traditional bag-of-words (BoW) model. It generates a group of abusive, hateful, and sensitive words. Then, it uses algorithms to identify the words in online content that must be analyzed (Hang, 2019). The lexicon-based methods work using N-gram techniques and BoW.

- Bag of words: This is a classical model in which a document is regarded as a vector, and its weights indicate the occurrence of words in the document. One major limitation of BoW is that it fails to capture semantic information (Shannag, 2022).
- N-gram technique: This is a traditional method that considers the occurrence of N words in a tweet and determines formal expressions (Atoum, 2021).

18.1.5.2 Machine Learning-Based Methods

Because of the challenge of identifying between bullying and non-bullying behaviors, ML-based methods have evolved for accurate classification and detection of cyberbullying. These methods were trained using effective ML algorithms on various datasets (Khang et al., 2023a).

In addition to ML algorithms, extraction and selection are performed for accurate detection and classification of cyberbullying (Chia et al., 2021). The most commonly used feature extraction technique with ML is term frequency–inverse document frequency (TF-IDF).

Birunda and Kanniga Devi (2021) states that this feature extraction technique is effective in detecting genuine or fake news articles. This is a weighting-based technique that converts text into features, as ML techniques are not designed to work with raw text.

The ML technique has been used not just for cyberbullying detection but also recently in predicting shipment times for therapeutics (Mariappan et al., 2022a), vaccines, and diagnostics in e-pharmacy supply chains during both pre- and post-COVID-19 pandemic (Mariappan et al., 2022b).

18.1.5.3 Deep Learning-Based Methods

Deep learning techniques are used to enhance the accuracy of cyberbullying detection by analyzing patterns through multiple layers in neural networks, based on sample data.

A neural network is composed of input layers, which are connected to multiple hidden layers, and a set of output layers (Zaib et al., 2021). These networks also use additional pre-processing, extraction, and selection techniques for accurate detection and classification.

18.1.5.4 Transformer-Based Methods

Earlier DL models, such as LSTM and gated recurrent unit (GRU), were improved by using a transformer-based neural network model (Ahmed et al., 2022). While these methods are designed to handle large sequential data, they do not require sequential data, compared with RNN.

18.1.5.5 Stages for Cyberbullying Detection in Online Social Network Using Deep Learning

The basic workflow for detecting cyberbullying in OSN is shown in Figure 18.7. The model includes the following steps: data collection and pre-processing, feature

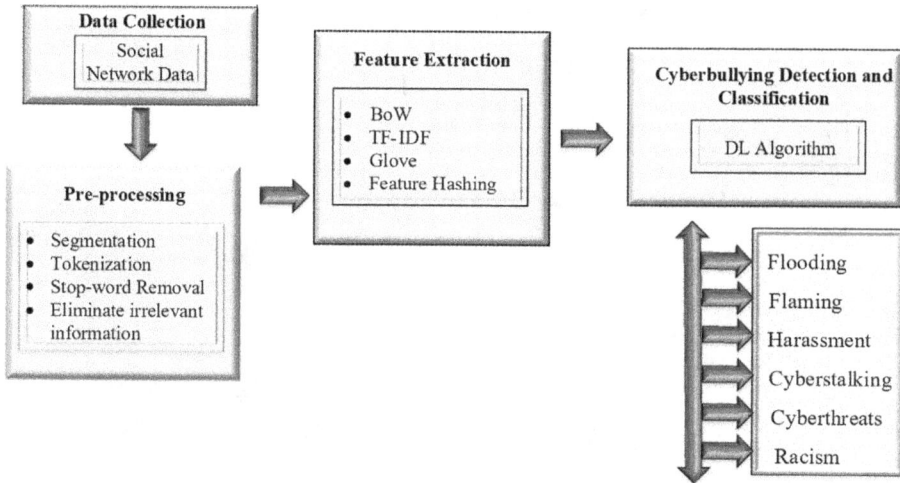

FIGURE 18.7   Methodology of Cyberbullying Detection Using DL.

extraction, training the model using a DL algorithm, testing the model, and evaluating its performance.

- **Data pre-processing**: Pre-processing is the first stage in cyberbullying detection. It includes the following processes: tokenization, removal of stop words, removal of punctuation, stemming, conversion of uppercase to lowercase characters, replacement of URLs, and removal of extra unwanted symbols (Yuvaraj et al., 2021).
- **Feature extraction**: This process involves deriving a set of words from text and converting them into a set of features. This step is useful for analyzing of text data, which can be applied to various purposes, such as data mining, generating reports, identifying research subjects, managing automated terminology, and maintaining clinical records. Feature extraction may function as a dimensionality reduction procedure, extracting only the most essential features. The pre-processed data are represented as a vector consisting of features. Some of the commonly used feature extractors are BoW, TF-IDF (Rahman et al., 2021), and feature hashing (San et al., 2021).
- **Classification model for cyberbullying detection**: DL models are commonly used for the classification of cyberbullying detection because of their advanced capabilities in real-time applications, such as image/text classification, object detection, and various other tasks in image analysis. To identify the type of cyberbullying the DL algorithm is trained on, the features extracted through the process are used.

18.1.5.6   Contribution of This Chapter

This systematic review of SBD and cyberbullying using DL allows future researchers to innovate new systems using optimized DL algorithms. In previous studies,

Spam bot and Cyberbullying
Detection in OSN using DL

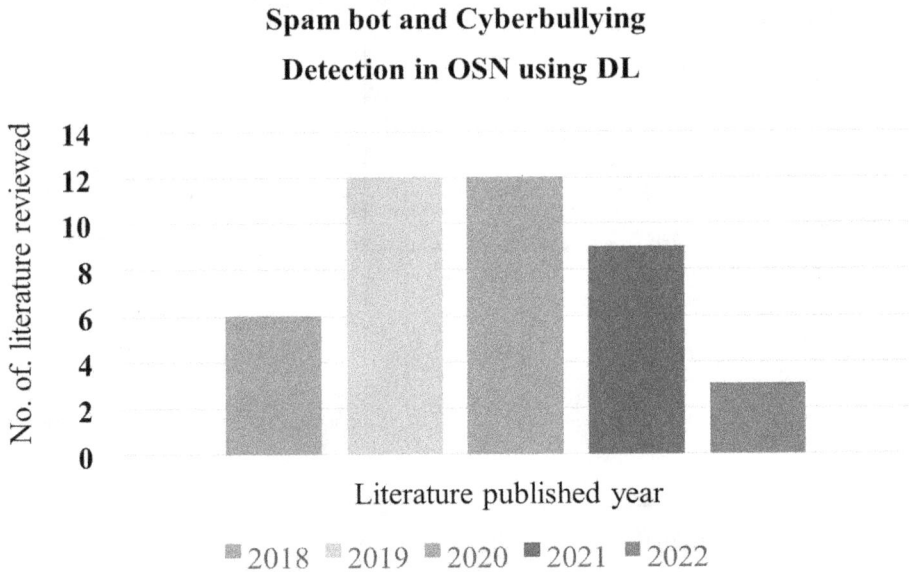

FIGURE 18.8   Literature Reviewed.

researchers have used outdated detection methods using various techniques. Our systematic review was purely based on recent papers from 2018 to 2022.

Several review papers have not provided clarifications on how spam bots and cyberbullying are detected using various DL algorithm techniques. This systematic review briefly explains all recently used DL algorithms for detecting spam bots and cyberbullying in OSN.

Since 2022, few studies have been conducted on the detection of spam bots and cyberbullying in OSNs using DL. Many studies have been directed toward detecting spam bots and cyberbullying through the use of ML.

In 2019, some research papers related to DL techniques were published. This systematic review aimed to gather all research papers related to DL techniques to provide ideas for future inventions, as depicted in Figure 18.8.

In the systematic review, a detailed description was provided for OSNs, botnets, spam and its types, cyberbullying and its types, social media bots, various DL techniques, spam bot and cyberbullying detection methods, and the basic stages for detecting spam bots and cyberbullying in OSNs using DL. This systematic review is then organized into the following sections:

- Section 18.2 consists of the review methodology.
- Section 18.3 includes the basic performance metrics.
- Section 18.4 contains the literature survey of spam bot and cyberbullying detection in OSNs using DL, while section 18.5 includes the search gap analysis and outcome of the literature survey.
- Section 18.6 presents the conclusion of this review.

18.2 PROPOSED METHODOLOGY

The methodology to explain a systematic review, as shown in Figure 18.9, is divided into three stages: planning, identifying, and reporting stages.

During the first stage, or the planning stage, a researcher should emphasize the need to conduct a systematic review and formulate research queries that will be answered by their work by showing the accuracy of the model.

During the second stage, or the identification stage, the researcher should specify the search strategy employed, the journals or research papers consulted, and the analysis of various models used. In the final stage, or the reporting stage, the researcher should present the full concept of the research domain, including the overall mechanism and the results obtained.

The remainder of this chapter includes the previously mentioned methodology. Initially, questions related to spam bots and cyberbullying are identified. The exploration strategies and the papers reviewed are then presented.

Figure 18.9 depicts the discussion of the overall spam bot and cyberbullying detection methods, including the objectives of detecting spam bots and cyberbullying, comparisons with other methods, the models and datasets used, and the parameters evaluated in previous studies.

18.2.1 RESEARCH QUERIES

The primary objective of this system review is to answer the following queries:

- Q1. What are the models proposed to detect spam bots and cyberbullying in OSNs?
- Q2. What are the metrics calculated to identify the effectiveness of the proposed models?

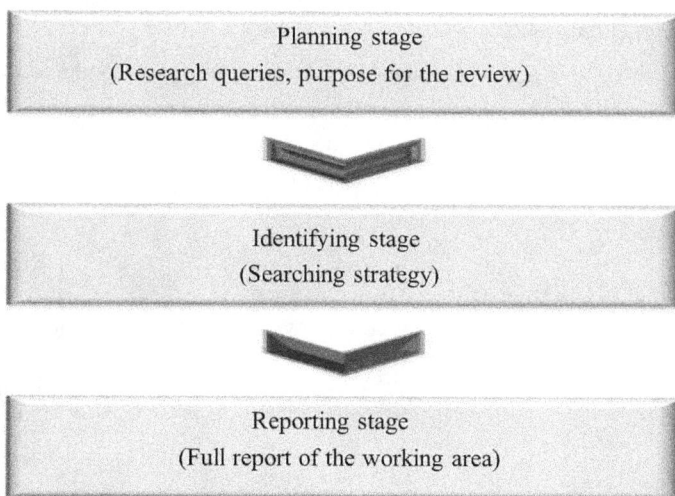

FIGURE 18.9 Review Methodology.

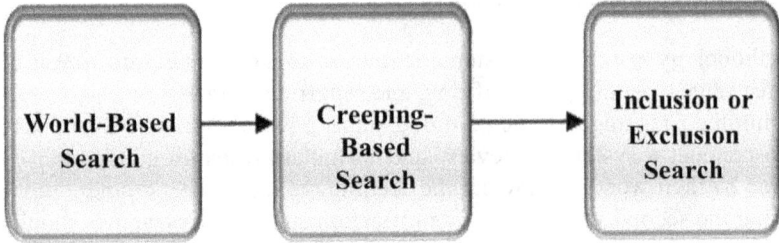

FIGURE 18.10   Searching Strategies.

- Q3. What are the datasets used to train the model?
- Q4. What are the performance metrics analyzed?
- Q5. What are the stages used in DL for spam bot and cyberbullying detection?

18.2.2   SEARCH STRATEGY

Based on the research queries and other targets, the search process is divided into three stages: word-based, creeping-based, and inclusion-/exclusion-based searching (Figure 18.10).

Word-based searching is a strategy for collecting data using words related to the domain. This involves exploring an inbuilt source, such as Google Scholar, to retrieve information. Creeping-based searching is a strategy for exploring previous literature in the same domain.

Inclusion-/exclusion-based searching is conducted to determine whether the papers reviewed are relevant to a particular domain.

18.2.2.1   Word Search

Based on the technique and model used, the word search was conducted.

- SBD and cyberbullying
- Deep learning
- CNN model
- RNN model

18.2.2.2   Resources Referred

The following information-gathering centers were found using Google Scholar and other search engines:

- IEEE
- Springer
- Elsevier
- Research Gate
- Science Direct

18.2.3 Study Choice

A total of 1500 published studies were taken using the mentioned search strategy. Of these, 730 papers were confirmed and selected for further validation, as depicted in Figure 18.11. Additional filtration of these published studies was necessary because several of these publications were outside the scope of the current study. For this purpose, the following inclusion and exclusion rules were applied:

1. Inclusion rules
   - Include surveys conducted from 2018 to 2022.
   - Include surveys in the domain of spam bot and cyberbullying detection in OSNs.
   - Include surveys written only in the English language.
2. Exclusion rules
   - Exclude surveys without proposed methodologies.
   - Exclude surveys and review papers.
   - Exclude surveys with unclear information about the publication.

18.2.4 Quality Valuation Tools

The quality of the survey papers was evaluated using the following quality valuation tools:

- Was the flow of writing in the literature presented well?
- Did the publication identify the contribution of the research domain?

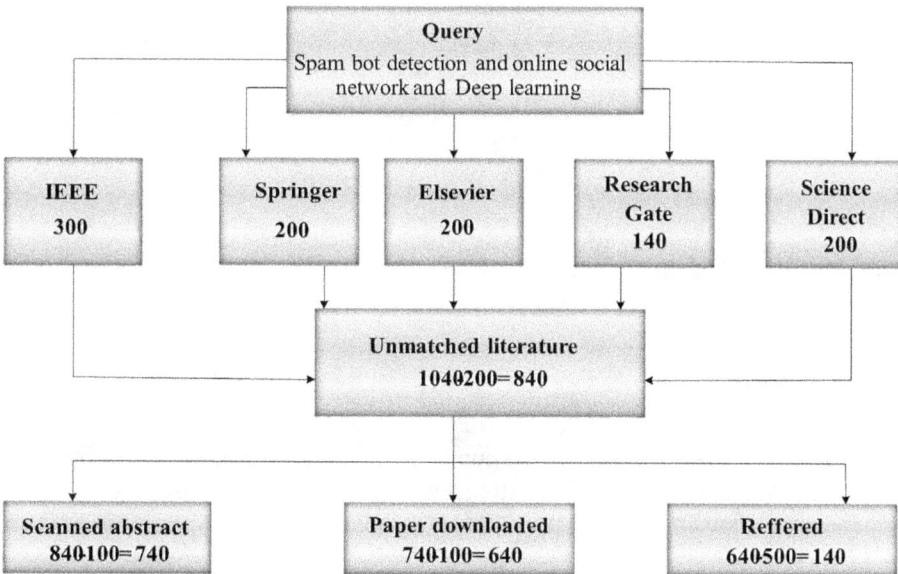

FIGURE 18.11 Study Choice.

- Was there sufficient information about spam bot and cyberbullying detection provided in the literature?
- Did the publication briefly describe terminologies, such as abbreviations used?
- Were all OSNs defined?
- Were the results of the proposed models revealed?
- Did the proposed models attain optimal solutions for various performance evaluations?

18.3 PERFORMANCE METRICS FOR SPAM BOT AND CYBERBULLYING DETECTION

The most commonly used performance metrics for the methods proposed by different researchers to detect spam bots and cyberbullying in OSNs using a DL approach are accuracy and average accuracy, precision and average precision, recall and average recall, F-measure and average F-measure, G-measure, specificity, and average specificity.

18.3.1 ACCURACY AND AVERAGE ACCURACY

The degree to which the number of predicted genuine and fake tweets in a dataset is correct is referred to as accuracy. If the accuracy level is 100%, then the proposed method will be of value. Equations 18.1 and 18.2 were used to calculate accuracy and average accuracy, respectively:

$$A = \frac{S_p + S_n}{S} \tag{18.1}$$

$$AA = \frac{1}{n} \cdot \sum_{k=1}^{n} A_k \tag{18.2}$$

where S is the total population. The S denoted here represents

$$S_p + S_n + R_p + R_n$$

18.3.2 PRECISION AND AVERAGE PRECISION

Precision is the measurement of the number of accurately classified genuine bots among all the information in the testing set. Equations 18.3 and 18.4 were used to calculate precision and average precision, respectively:

$$A = \frac{S_p}{S_p + R_p} \tag{18.3}$$

$$AP = \frac{1}{n} \cdot \sum_{k=1}^{n} P_k \qquad (18.4)$$

18.3.3 Recall and Average Recall

Recall is the measurement of the number of accurately divided genuine bots among all genuine information in the testing set. Equations 18.5 and 18.6 were used to evaluate recall and average recall, respectively:

$$A = \frac{S_p}{S_p + R_n} \qquad (18.5)$$

$$AR = \frac{1}{n} \cdot \sum_{k=1}^{n} r_k \qquad (18.6)$$

18.3.4 F-Measure

F-measure is the balanced mean of both the recall and precision metrics. Equations 18.7 and 18.8 were used to calculate F-measure and average F-measure, respectively:

$$F = 2 \cdot x \cdot \frac{Pxr}{Pxr} \qquad (18.7)$$

$$AF = \frac{1}{n} \cdot \sum_{k=1}^{n} F_k \qquad (18.8)$$

18.3.5 G-Measure

G-measure is the square root of precision and recall. Equation 18.9 was used to calculate G-measure:

$$g = \sqrt{P \times r} \qquad (18.9)$$

18.3.6 Specificity and Average Specificity

Specificity was used to calculate the capacity of the classifier to accurately identify true-negative results. Equations 18.10 and 18.11 were used to evaluate specificity and average specificity, respectively:

$$Sp = \frac{S_{pn}}{S_n + R_p} \qquad (18.10)$$

$$ASp = \frac{1}{n} \cdot \sum_{k=1}^{n} Sp_k \qquad (18.11)$$

18.4   SPAM BOT DETECTION IN ONLINE SOCIAL NETWORK USING DEEP LEARNING

18.4.1 SOCIAL MEDIA BOTS

The world works over the Internet. The Internet allows people to interact and communicate with others across the globe through online social media platforms, such as Instagram, YouTube, and Facebook. However, spam bots are always involved with social media and are known as social media bots. There are many positive and negative social media bots (Fan et al., 2020).

Several review papers have used ML techniques for SBD on social networks (Khang et al., 2023b). However, these ML techniques are unable to extract additional features to detect fake accounts and other malicious activities in OSNs.

The following studies examined the behavior of spam bots on various social media platforms and explored the use of DL techniques for their detection. Figure 18.12 shows the overall domain, datasets used, and performance metrics evaluated for the proposed model in this study.

Daouadi et al. (2019) proposed SBD in OSN using deep forest, a feature-based SBD technique. Information, such as user profile, post, and time of posting, was collected and produced as two different datasets: manual and honeypot datasets.

The dataset contains recent posts that have been labeled. Thirteen parameters from user information were used. The extracted features were used as the binary and numerical parameters. A filter-based feature selection method, known as information gain, was used to identify the most important and required features.

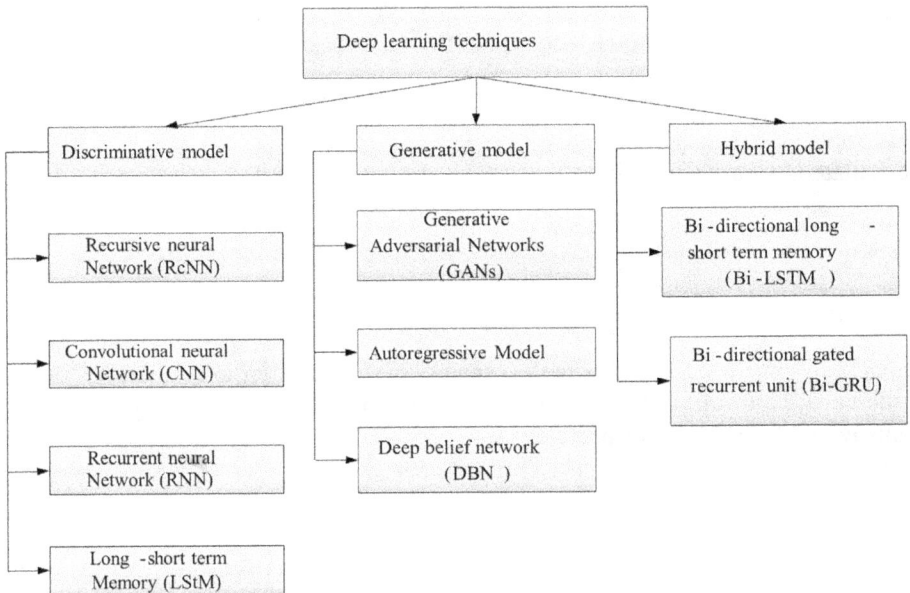

FIGURE 18.12   Deep Learning Techniques.

The implementation was carried out using the Waikato Environment for Knowledge Analysis, and the code was executed on the Python platform to run the deep forest algorithm. The drawback of this method is that it may not select the best features for detecting spam bots.

Fazil et al. (2021) investigated a deep neural network (DNN) model with an attention mechanism for social bot detection. The DeepSBD technique was introduced to examine social bots on social networks. The behavior of social network users can be examined using a hybrid of Bi-LSTM and CNN architectures.

For experimentation, spambot1, spambot2, and genuine account-based datasets were manipulated. This work was compared with other approaches, such as naive Bayes (NB), decision tree, and random forest (RF), and produced comparative results.

In the experimental scenario, the maximum detection rate attained was approximately 99%, precision was 100%, and accuracy was 99%. However, this method suffered from high spam drift. Because this method uses three different models to detect spam bots, it incurs the drawback of increased computational time.

Zhao et al. (2020) proposed the attention-based graph neural network (GNN) for SBD in OSNs. This method is mainly focused on extracting the user's features among the nodes to direct social graphs. For an effective operation, 1K-10KN-based datasets were used.

To understand the efficiency of this work, the developed model was compared with the multilayer perceptron (MLP), BP, RF, graph convolutional network, GraphSAGE, and graph attention network (GAT) techniques. In the experimental scenario, the maximum recall attained was approximately 0.88, precision was 0.93, and F-measure was 0.91. However, this method shows an incorrect identification of social users because of dataset imbalance. This method does not use real-time data for SBD.

Guo et al. (2021a) proposed spammer detection using a deep graph neural network.

The proposed work was composed of three parts: inferring occasional relationships, developing a neural network framework with deep graph integration, and using binary classification for detection, all of which were evaluated through a sigmoid activation function.

The dataset used for the proposed work was the X/Twitter dataset, which consisted of 10,000 users. After validation, 2060 spammers were identified. The support vector machine (SVM), CNN, and non-negative matrix factorization (NMF) methods were used. The proposed Deg-spam method was compared with these methods. Two attributes were gathered: personal and interactive attributes.

The datasets of X/Twitter and Weibo were analyzed, and the precision results, parameter sensitivity, and proportion of training data were evaluated. The method did not perform pre-processing to remove unwanted punctuations and symbols from tweets.

Kumar et al. (2020) proposed an RNN-based spam message detection technique. The RNN model was used with word embeddings, and an LSTM layer was used to execute a sequence learning task. A Kaggle dataset was used, which was extracted using SMS spam analysis.

The performance metrics evaluated were the loss, accuracy, value loss, and value accuracy. The method requires large memory storage for computation.

Lingam et al. (2019) proposed spam bot detection in OSNs using deep Q-learning (DQL) and PSO algorithms.

The proposed DQL–PSO algorithm uses a deep Q-network with various velocities. The global and local optimal positions of the learning action were analyzed with respect to the Q-value function. Using the Q-value function and optimization scheme, the DQL algorithm was framed.

Performance metrics were evaluated by considering recall and precision values. The SBD process is very slow because of the low efficiency of the model.

Ban et al. (2018) proposed a DL technique to study the features of spam and detect spam bots on an X/Twitter platform. The spam bots were detected by performing various processes, such as data pre-processing, feature extraction, selection, and classification.

To learn the features, Keras was used to implement Bi-LSTM, along with TensorFlow as the backend. W2V was used for feature extraction. A continuous BoW was used, which included all existing data in the dataset along with the converted unique vectors.

The dataset used to train the model was labeled the X/Twitter dataset. The performance metrics evaluated were precision, F1-score, and recall. The method uses a small set of data to check the effectiveness of the model.

Alauthman (2020) proposed a deep RNN-based botnet spam detection via email. A GRU–RNN with SVM was proposed for the email SBD.

Spam base dataset taken from the learning repository of the UCI machine was used. The dataset includes 4601 email messages, out of which 1813 have been identified as spam emails, and 2788 are genuine emails. There are three phases involved to detect email spam.

TensorFlow was used as the backend for the KERAS. CART was used as the feature selection algorithm. The results were compared with those of logistic regression, Gaussian NB, SVM, and RF. Only a few results were obtained.

Ali Alhosseini et al. (2019) proposed a model to detect spam bot using inductive representation learning. The proposed model is a graph-based CNN to detect spam bot. This approach does not include feature extraction for detecting spam bots in OSN.

Kumar and Shubham et al. (2021) proposed a bot language model and bidirectional encoder representations from transformers (BERT) embedding for SBD based on content. This study proposed a text-based CNN and LSTM with BERT embedding. To overcome the problem of imbalanced data, synthetic minority oversampling technique was implemented.

The framework of the proposed model is divided into three stages: pre-processing data, data imbalance handling, and training of the ensemble classifier. In the pre-processing stage, the unwanted spaces and words were removed. The tokenization and stemming processes also occur in the first stage.

Word vectors were generated using the BERT embedding technique, fed into the LSTM model, and finally to the CNN model. The performance metrics evaluated were precision, accuracy, recall, and F1-score. The model should be enhanced to achieve the best SBD.

Ping and Sujuan (2018) proposed a DL algorithm for SBD in OSN. The proposed model is called deep bot detection, which comprises three layers. The first layer extracts content and other related features from tweets.

The second layer draws the metadata of the tweet temporal features. The third layer is a combination of processes performed in the first and second layers. A CNN was used for feature extraction from tweets. The performance metrics evaluated were precision, recall, specificity, accuracy F-score, and multiple chronic conditions. The framework of this method is very low.

Wu et al. (2021) proposed DNNs and active learning for SBD in OSNs. Deep neural networks and active learning (DABot) were proposed for predicting social bots in the Sina Weibo social network. The SWLD-300K dataset was preferred for identifying spams.

A web crawler was used to collect data from Weibo. In the feature extraction process, meta-based features, such as nickname length, follower ratio, default nickname and avatar, profile completeness, and comprehensive level, were used. The return goods authorization (RGA) model is designed to detect spam bots. The performance metrics evaluated were accuracy, recall, precision, and F1 score. The selection of the model was insufficient to detect spam bots more accurately.

Guo et al. (2021b) proposed BERT and a graph convolution network-based model to detect spam bots in OSN. A combination of both BERT and graph convolutional network (BGSRD) was proposed.

The datasets used were Cresci-rtbust, botometer-feedback, gilani, Crescistock-2018, and midterm datasets. The proposed model used the robustly optimized BERT approach (RoBERTa) and GAT. A large number of features are available that degrade the network performance.

Najari et al. (2022) proposed a generative adversarial network (GAN)-based framework for SBD. A GAN is used to accurately identify the data behavior pattern.

18.4.2 SPAM BOT DETECTION IN X/TWITTER USING DEEP LEARNING

Among all the OSNs, X/Twitter is considered the most preferred social network community for interacting with all users. X/Twitter users can post messages with more than 230 characters, termed tweets (Harrigan et al., 2021).

As a result of the large number of X/Twitter users, the platform has become a target for spam bots and other cyber criminals. To gather user profile or information, spammers send unwanted messages, links, and videos. X/Twitter is considered the most spam-ridden social media platform (Cresci et al., 2019).

Many researchers have proposed novel techniques to detect all spam bots on X/ Twitter platform. Figure 18.13 shows the proposed model, the dataset used, and the performance metrics evaluated for SBD in OSN, especially X/Twitter.

An NLP approach was introduced to effectively differentiate spam users. Initially, feature extraction was employed to extract features from the experimental datasets. Subsequently, the feature selection approach was evaluated to remove unwanted noise from the extracted features. For the classification of the outcome, DL approaches, such as RNN, LSTM, and Bi-LSTM, were used. In the experimental scenario, the maximum accuracy attained was approximately 98%, the F measure

Congratulations !!
You won a new
mobile as gift

Click on the link to
know about the
reward

Input layer　　　Convolution layer　Pooling layer　Hidden layer　Output layer

Convolution layer for single tweet

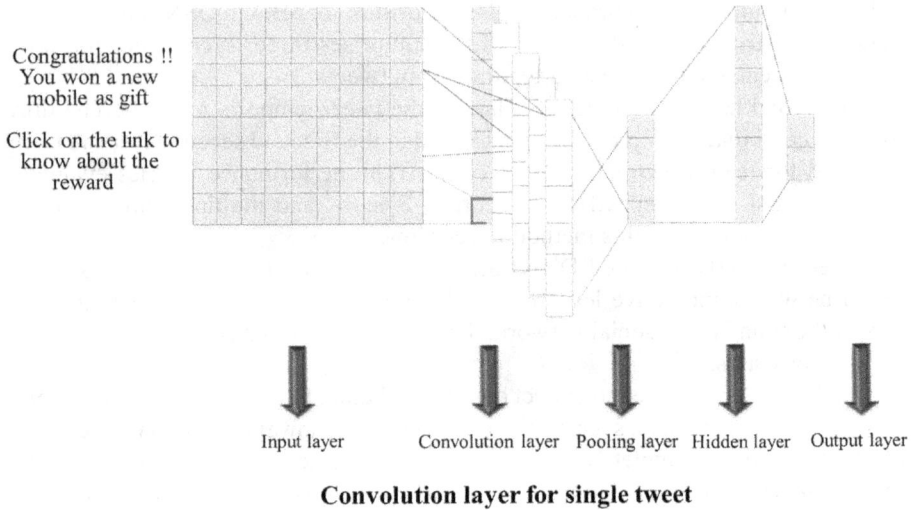

FIGURE 18.13　Convolution Neural Network Model for a Single Tweet.

Source: Rahman et al. (2021)

was 98%, the recall was 99%, the precision was 97%, and the area under the receiver operating characteristic was 99%. However, this method was limited in its ability to address gender and age identification for social media users. The method is computationally more expensive.

Rodríguez-Ruiz et al. (2020) introduced a one-class classification approach for bot detection on X/Twitter, which was based on supervised classification.

The proposed method was compared with multiclass classification and showed that the object classification (OC) achieved better results. For experimentation, bagging-TPMiner and bagging random miner datasets were utilized. In addition, a feature selection approach was introduced, and features were collected based on text, nominal, and numerals.

In the experimental scenario, the maximum area under the ROC curve (AUC) obtained for LR was 0.95. However, this method was highly limited due to poor visibility of the classified output. The method considered only a few features, which led to inaccurate results.

Lingam et al. (2020) proposed a DL model for detecting spam bots in the X/ Twitter network. In this study, an X/Twitter graph network was designed using the behavior and genuine values of users. The behavior was observed among users of the X/Twitter network, based on similarities in URL, tweet content, and interaction on social media.

The genuine value is identified by a model called random walk. Two algorithms were framed: Social Botnet Community Detection (SBCD) and Deep Auto encoder-based SBCD. The first algorithm is based on detecting spam bots with similar behavior, and the second model detects social spam bots more accurately.

Finally, the performance was measured using X/Twitter datasets. The datasets used were fake project and social honeypot datasets. This method attains computational overhead during training.

Alhassun et al. (2022) proposed a text and metadata-based DL framework for detecting spam bots in the X/Twitter OSN community. This study mainly focused on SBD in Arabic on X/Twitter by gathering Arabic datasets.

The data from the X/Twitter platform were collected using the X/Twitter API. Two types of data were used, which include text-based data along with a CNN and meta-based data along with a simple neural network model.

A combination of two model outputs shows whether accounts are spam or not. The text-based model has three phases: embedding, the CNN algorithm, and classification. The data were tweets collected from the X/Twitter Arabic hashtags.

A total of 1.2 million tweets were collected and divided into 250,000 tweets per file. The proposed model was compared with other existing ML and DL models. The performance metric recall is computationally intensive.

Chowdhury et al. (2020) proposed a natural language processing technique for detecting spam bots on X/Twitter social networks. Two datasets were used: social honey pod and custom datasets.

The social honey pod dataset comprises 22,223 and 19,276 polluters and legitimate users. Another dataset was created manually with 1200 and 800 users. The following features were included: account-based, stylistic features, hashtag features, word embedding features, and topic word features.

Two models were designed for word embedding features: global vector (GloVe) and word2vec. For the topic word-based feature, the latent Dirichlet allocation technique was used. This approach uses an ineffective algorithm to train the proposed model.

Madisetty et al. (2018) proposed CNN-based detection of spam bots in X/Twitter social networks. One feature-based model and five CNNs were used as the ensembles.

Different word embedding techniques were used for each CNN model, such as glove, and word2vec to train the model. The methods used for comparison with the existing model are the X/Twitter glove, Edinburgh, Google News, h-spam, and random methods.

The IKS10KN dataset was used to train the model. The performance metrics evaluated were execution time, precision, recall, and F-measure. Sufficient time and data were required to train the model.

Alom et al. (2020) proposed a DL model to detect spam bots on X/Twitter OSN platforms. Convolution neural networks with text-based and combined models have also been proposed.

In the text-based classifier, three steps were employed: embedding, CNN, and classification. The combined classifier contains two major steps: normalization and density. The dataset used was X/Twitter social honeypot, which contains 22,223 spammer accounts and 19,276 non-spammer user accounts.

The other dataset, X/Twitter 1KS-10KN dataset, contains 1000 spammer accounts and 10,000 non-spammer accounts. The Keras DL tool was used, which uses TensorFlow as its backend.

To combine both the text- and metadata-based methods into a single method function, the API was used. The model was trained using an Adam optimization algorithm. The proposed model is more difficult to design and costly.

Wei and Uyen Trang (2019) proposed a Bi-LSTM using neural networks for the detection of spam bots on X/Twitter. BLST and word embedding were performed. Word embedding was performed using the GloVe technique.

The social-bot-1 and social-bot-2 datasets were compared with the Cresci-2017 dataset. The neural network model is designed based on parameters, such as learning rate, network structure, dropout, number of epochs, momentum, and mini-batch.

The performance metrics analyzed were precision, recall, specificity, accuracy, and correlation coefficient. The method requires that the best pre-processing technique must be employed to remove unwanted noise.

Luo et al. (2020) proposed a DNN-based approach for detecting X/Twitter bots. The proposed model was customized using Bi-LSTM to analyze the tweets, data pre-processing, and extraction. It consists of two elements: X/Twitter bot classifier and a web interface, which was developed by a web service for public approach.

An X/Twitter bot classifier was designed using a DNN model to identify whether a tweet was posted by a bot or not. A web interface was created to allow the public to access the X/Twitter bot classifier. The accuracy of the proposed model was calculated. This method requires more implementation results to prove its effectiveness.

Venkateswarlu et al. (2021) proposed an optimized GAN with fractional calculus-based feature fusion for detection of spam bots in X/Twitter platform. Spam detection was performed on the X/Twitter dataset. The Yeo-Johnson transformation method was used to make the dataset suitable for processing in SBD.

The work utilized the Rényi entropy and deep belief network for feature fusion. Finally, SBD was performed using a GAN.

The CAViaR-secret face method based on GAN and the risk-sailfish optimization algorithm were introduced along with the model. The X/Twitter dataset was used to evaluate the performance of the method.

The proposed model does not operate well in a large number of datasets.

18.4.3 CYBERBULLYING DETECTION IN ONLINE SOCIAL NETWORK USING DEEP LEARNING

Cyberbullying is referred to as the utilization of the Internet community to send unwanted tweets and messages by injuring or hurting the Internet users.

The major effect of cyberbullying is that it can make a person commit suicide (Yao et al., 2019). There are two methods involved in the detection of cyberbullying: textual-based and multimodal-based methods.

The textual-based method uses only textual data to detect cyberbullying, whereas a multimodal-based method uses both textual and additional features to detect cyberbullying (Dadvar & Kai, 2020).

Al-Ajlan et al. (2018) proposed a DL-based cyberbullying detection method in the X/Twitter network community. An optimized DL-based cyberbullying detection model is used to detect cyberbullying in social networks.

Without gathering datasets from an Internet source, data collection was performed using Twitter4J API with Java code to fetch 20,000 tweets randomly. The data were then filtered to remove irrelevant, noisy, and duplicate tweets. After filtering, the data were separated into the testing and training sets.

The training data were labeled as either bullying or genuine, using the CrowdFlower platform. Word embedding was also developed using the GloVe technique. After the word embedding process, the tweets were fed into the CNN, which starts learning through iterations and patterns. This method requires an effective algorithm to train the model for cyberbullying detection.

Iwendi et al. (2020) proposed a DL architecture to detect cyberbullying in the OSN community. Deep bidirectional long short-term memory was proposed to detect cyberbullying, and the proposed model was compared with Bi-LSTM, GRUs, RNN, and term LSTM.

A Kaggle dataset was used to detect cyberbullying. Pre-processing of the dataset was performed, which involved text cleaning, tokenization, stemming, lemmatization, and removal of stop words. Thereafter, the results were predicted using the proposed model and the pre-processed data.

The main drawback of this method is that the results obtained are difficult to compare with other existing models.

Cheng et al. (2019) presented a hierarchical attention network (HAN)-based approach for detecting cyberbullying on the Instagram network community. This approach consists of several layers, including a word sequence encoder, a comment sequence encoder, contextual information, a word-level attention layer, and a hidden layer.

To model the proposed design, bidirectional GRURNN was used to encode the sequence of comments and words. The Instagram datasets, consisting of 2218 social media sessions, were collected. Following evaluation, 1540 of the sessions were determined to be genuine, while the remaining 678 are instances of bullying. However, the limitation of the method is that it does not always produce accurate results.

Banerjee et al. (2019) presented a DNN-based approach for detecting cyberbullying in OSNs, using a CNN. The CNN architecture consists of an input layer, some hidden layers, and an output layer.

The Python programming language and TensorFlow were used to implement the proposed model, which was trained using the X/Twitter dataset containing 69,874 tweets. These tweets were converted into vectors using a word embedding technique. The proposed CNN was compared with other technologies, such as ML, data mining, and RNN.

The accuracy of the proposed method was evaluated using existing methods. The drawback of the proposed model is that it requires a large amount of dataset during training.

Alotaibi et al. (2021) proposed a multi-channel DL network to detect cyberbullying in the OSN community. The proposed model is the integration of three advanced models: transformer block, CNN architecture, and bidirectional GRU (BiGRU). These advanced models are based on DL.

A dataset consisting of tweets from three sources has been evaluated and named as offensive and non-offensive datasets. The dataset contains 55,788 tweets in total. The offensive dataset includes bullying tweets, whereas the non-offensive dataset includes genuine tweets. The tweets were then converted into a sequence of tokens in a process known as tokenization.

The proposed method used Python language and various inbuilt libraries for further processing. The evaluation metrics include precision, recall, accuracy, and F-score. The limitation of the proposed method is that the learning speed of the model is slower compared with other methods.

Yadav et al. (2020) proposed a pre-trained BERT model for detection of cyberbullying in OSNs. The datasets were obtained from online sources, such as Formspring and Wikipedia. The Formspring dataset consists of 12,773 question–answer pair comments, whereas Wikipedia consists of 115,864 discussion comments.

A DNN transformer model with 12 layers was used as a base model, with BERT constructed at the top layer. The main task of the BERT model is word embedding. After tokenization, sentences were fed into the other layers of the BERT model. The BERT tokenizer was used to separate each word in a sentence. The main drawback of the proposed method is that it consumes a significant amount of memory to process the model.

Kumar and Nitin (2021a) proposed a capsule network (CapsNet) combined with dynamic routing and a DNN to detect multimodal cyberbullying. Three types of social media data for cyberbullying were detected: visual, textual, and infographic contents.

The proposed model consists of a two-in-one architecture that combines the capabilities of CapsNet and a convolution neural network (ConvNet). CapsNet was used for detecting bullying in textual contents, whereas ConvNet was used for the detecting bullying in visual contents. The infographic containing bullying is detected by separating the text from the image with the help of Google Lens.

The datasets used for this study include Instagram, X/Twitter, and YouTube, each containing 10,000 comments and posts. These datasets were compared with the toxic comment classification challenge dataset. The evaluated performance metrics were precision, recall, and accuracy. The method requires a significant amount of computational resources to detect cyberbullying.

Agrawal and Amit (2018) presented a DNN model for identifying cyberbullying on social media platforms. A hybrid neural network with a binary and multiclass model was proposed for detecting bullying on social media platforms, using a dataset of 44,001 user comments from Facebook, which were categorized into five divisions.

Datasets were collected and then pre-processed. The pre-processed data were used for word embedding, which were then fed into the proposed DL model for training. Finally, the performance of the model was analyzed.

The drawback of this method is that the computational time is very high, as the model requires learning of deeper structures.

Mahat (2021) proposed a DL approach to detect cyberbullying across social media environments.

To train the proposed LSTM model, data were collected from three OSN platforms: Wikipedia, X/Twitter, and Formspring. Each platform provided 3000 samples.

In the pre-processing stage, data were cleaned by removing unwanted punctuations, symbols, and white spaces. The accuracy of the proposed model was the only metric used for evaluation. The drawback of the method is that more features are required to train the proposed model.

Murshed et al. (2022) proposed a hybrid DL approach combining dolphin echolocation algorithm (DEA) and RNN for cyberbullying detection on the X/Twitter social network. The proposed hybrid modal DEA–RNN is a combination of Elman-type RNN and DEA.

Tweets were collected from X/Twitter using the X/Twitter API. To address the issue of class imbalance between bullying and non-bullying classes, synthetic minority oversampling technique was utilized.

Feature extraction was performed using natural language processing tools, such as word2Vec and TF-IDF. The disadvantage of this method is that no optimization technique is used to reduce its complexity.

Kumar et al. (2020) proposed a multi-input integrative learning approach based on a DNN and transfer learning to identify cyberbullying in different languages. Word embedding techniques, such as GloVe and fastText, were used. Glo-Ve was used for English words and fastText for Hindi. The three sub-networks contain capsule networks coupled with dynamic routing.

Two types of datasets were obtained from the Internet: Facebook and X/Twitter. The dataset obtained from Facebook was based on user profiles, whereas that from X/Twitter was based on topics. Data pre-processing was done to remove tags, numbers, and punctuations. Tokenization was carried out using the tree bank word tokenizer, which is a Python language toolkit. The capsule network with dynamic routing was used for processing the English sub-network and Bi-LSTM for processing the Hindi sub-network. Finally, classification was done using the MLP for importing pragmatic features.

The performance metric accuracy was calculated for the proposed model. The limitation of the proposed method is that more iterations are required to get the best optimal results.

Fang et al. (2021) proposed a bidirectional gated recurrent unit (Bi-GRU) with a self-attention mechanism to detect cyberbullying in social networks.

This process involved data collection. Three datasets were used: two X/Twitter datasets and one Wikipedia dataset. The first X/Twitter dataset consists of 16,000 tweets on religious and ethnic minorities. The second X/Twitter database contains terms from the hatebase.org website. The Wikipedia dataset contains 110,000 comments.

The second step involved data pre-processing, which entailed processing of URLs, usernames, and special characters; applying lowercase for the strings, and removing tokens. The proposed model was also compared with several other ML techniques. The performance metrics evaluated were precision, recall, and F1-score. The drawback of this method is that the computational time required for training the model is high.

Chandra et al. (2018) proposed recursive neural network-based cyberbullying detection via an offline repository. This study aimed to detect cyberbullying based on tweets posted in OSN, especially on X/Twitter. The datasets were collected through API from OSN sites and saved in a database known as NoSQL.

The TensorFlow API was used to perform predictive analyses that were saved in the database using recursive networks. The X/Twitter dataset, which is produced in the form of a JSON document, was used. This method does not provide an accurate solution for cyberbullying detection.

Agarwal et al. (2020) proposed an RNN-based detection and classification of cyberbullying tweets via undersampling and class weighting. In addition to the RNN, it uses a max pooling layer with Bi-LSTM and attention layers. Undersampling and class weights were performed to diminish the effect of class imbalance in the dataset during detection and classification.

The Wikipedia dataset with 100,000 posts was used. The performance of the proposed network model was evaluated using the Wikipedia dataset over performance metrics, such as a precision of 0.89, recall of 0.86, and F1-score of 0.88. The drawback of this method is that the learning rate of the model is high because of the large dataset.

Abishak and Kabilash et al. (2020) proposed unsupervised hybrid approaches for cyberbullying detection on the Instagram platform. Because of drawbacks in labeling the data in supervised methods, unsupervised approaches were used.

Two main concepts are involved: a representation learning network that uses a HAN for a multi-model representation graph auto-encoder, and a multi-task learning network that uses GMM-based energy estimation for cyberbullying detection. In addition, abuse was detected using an RNN.

The performance metrics used to evaluate the performance of the method are F-measure, accuracy, and root mean square error. The limitation of the method is that the accuracy rate obtained is lower because of noise interference.

Peiling et al. (2022) proposed a platform-aware adversarial encoding technique for cyberbullying detection in OSNs. An XP-CB framework based on transformers and adversarial learning was proposed for cyberbullying detection.

The transfer model was operationalized with various components, such as embedding alignment, adversarial alignment, input length optimizer, source encoder, target encoder, hidden state selector, discriminator, and encoder measurer.

Three datasets were employed to test the performance of the framework: Formspring, X/Twitter, and Wikipedia. The transformer model was compared with BERT and RoBERTa. The proposed method struggles to identify whether a group of tweets contains bullying or non-bullying content.

Al-Ajlan et al. (2018) proposed a DL algorithm for detecting cyberbullying in OSNs. This algorithm was designed to overcome the issue of researchers continually introducing new features to enhance the detection of cyberbullying. However, the downside is that while new features increase the accuracy of detection, they also make the process more complex.

CNN-CB was proposed to detect cyberbullying without feature extraction and selection. Word embedding is introduced to identify similar words. CNN-CB is based on a CNN and the integration of the word embedding technique.

Data were gathered using X/Twitter streaming API. The performance metrics evaluated were accuracy, precision, and recall. The major drawback of the proposed method is that it does not work well under large computational loads.

Kumar and Nitin (2021b) proposed hybrid models to detect cyberbullying in OSNs: attention-based Bi-GRU and CapsNet. Bi-GRU with self-attention was used to learn the sequential semantic representations and spatial location information. CapsNet was used to detect cyberbullying in the OSN contents.

The datasets used to evaluate the performance of the hybrid model were Formspring.me and Myspace. The performance metrics evaluated were F1-score and accuracy. The disadvantage of the proposed method is that it does not provide accurate solutions for all datasets used.

Paul et al. (2020) proposed the CyberBERT model for cyberbullying detection. Three types of datasets were used to analyze the performance of the model: Formspring, X/Twitter, and Wikipedia. The BERT model was designed to learn the language representations of tweets and other posts.

For cyberbullying classification, along with the BERT model, a fully connected layer was introduced in the hidden layer. The performance metrics evaluated were accuracy, loss, and F1-score. The main drawback of the proposed method is its requirement for more time and a high-end hardware configuration.

Behzadi et al. (2021) proposed a cyberbullying detection method using BERT models. The dataset used to train the model was a hate-speech dataset that contains 85,948 tweets that were labeled using a mechanism named crowdsourcing.

A compact BERT model was used because of its wide growth in language problems. A compact BERT model was used to classify all pre-processed data. Focal loss is considered the cost function of the work. The absence of model robustness is a major disadvantage of the proposed method.

Chen et al. (2020) introduced a heterogeneous neural interaction network (HENIN) for explaining the detection of cyberbullying in OSNs.

HENIN is composed of several components, such as a post-comment co-attention sub-network, a comment encoder, a post-post network, and session interaction extractors to explain cyberbullying. The integration of sentence embedding vectors and post-comment co-attention vectors enables the detection of cyberbullying through a fully connected layer.

Two datasets were used: Instagram, which comprises user comments and image descriptions, and Vine, a mobile application that allows users to edit videos. The performance metrics evaluated were precision@10 and accuracy. The drawback of this method is that the training cost is high.

Gada et al. (2021) proposed LSTM–CNN architecture for cyberbullying detection. A web application was designed to classify tweets as either bullying or non-bullying based on toxicity scores by considering various features. The proposed model was used in a telegram bot to detect cyberbullying.

The dataset was pre-processed using Python's Natural Language Toolkit library. LSTM is the final architecture that obtains word embedding for each token. The performance metrics evaluated were precision (76%), recall (31%), ROC area under the Receiver Operating Characteristic (ROC) curve (97%), F1-score (44%), and accuracy (95.2%). Pre-processing techniques were not used to eliminate symbols.

Bu et al. (2018) proposed a hybrid DL model of CNN and long-term recurrent convolutional networks (LRCN) to detect cyberbullying in social network comments. CNN and LSTM were used to study the features of the characters.

The word-level LRCN was utilized to capture all high-level semantic data from a sequence of words, whereas a character-level CNN was used to study all low-level syntactic data from a sequence of characters. A Kaggle dataset was used to train the proposed model.

The performance metric accuracy was evaluated to check the performance level of the model. This method does not yield accurate results, and only a single parameter is used for the result obtained.

Reviewing the relevant literature revealed that most of the authors used Python, TensorFlow, Keras, and Google Colab platforms for the implementation of their models.

18.5 RESEARCH GAP ANALYSIS

In this section, the most common issues related to SBD and cyberbullying detection in OSNs, as well as future research directions, are discussed.

- The collection and analysis of huge amounts of data and unbiased datasets with user profiles and other content for real-time SBD techniques are challenging.
- Scalability is another challenge in validating the effectiveness of DL models because of the growing rate of OSN data.
- There are many OSNs; however, researchers have proposed models only for commonly used social networks.
- Using an existing dataset is a challenge.
- Another important challenge is the lack of a security system to protect user activities in most spam-attacked networks (Khang et al., 2022).
- All countries use OSN of different languages that also comprise tweets or spam bots of different languages. These tweets of different languages should also be detected using various techniques.
- Researchers have also noted that spam bots can communicate with other accounts, post tweets on various topics, and behave like humans.
- It is noticed from the literature review that the interaction of users in OSNs is updated because of the use of different slang and languages.
- In noisy text, SBD is a challenging task because of sparseness issues and irresponsibility for language use on OSNs.
- Analysis of the computational time and time complexity of the proposed method is a major challenge when considering the method to be more effective in detecting spam bots and cyberbullying.

18.6 OUTCOME OF LITERATURE SURVEY

The proposed systematic review provides a clear explanation for spam bot and cyberbullying detection in OSNs using various DL techniques.

This systematic review guides researchers to innovate novel DL techniques to detect spam bots and cyberbullying. This section provides the overall outcome of the systematic review. The first section of this review contains a basic introduction to spam bot and cyberbullying detection.

The section is composed of types of spam and cyberbullying, different methods to detect spam bots and cyberbullying, and the stages involved in detecting spam bots and cyberbullying using DL.

A brief description of the key contributions of this review is given with a graphical representation of the total number of studies reviewed from 2018 to 2022. The next section contains the review methodology.

This section comprises the planning, identifying, and reporting stages. The key words used to select literature, inclusion and exclusion criteria, and literature-searched platforms are given in detail. Subsequently, performance metrics are used to validate the proposed methods for spam bot and cyberbullying detection.

The next part contains the SBD in OSN using DL techniques. Deep learning models, such as deep forest, DNN, CNN, BiLSTM, GNN, DGNN, RNN, LSTM, RGA, BGSRD, and GAN, are used to detect spam bots in OSNs.

The main aim of these models is to detect spam bots from user profiles, posts, comments, hashtags, SMS, interactions, and accounts. The datasets used by the authors to train and to validate the models are honeypot, X/Twitter, Kaggle, Weibo, and Cresci datasets.

The major drawback of all the methods of SBD using DL techniques is that they do not choose the optimal features for SBD. Moreover, all the models consume more time to perform a particular task.

The computational complexity, cost, and structure complexity are also main drawbacks of the SBD methods.

The next section contains the SBD in OSN using DL techniques, particularly on the X/Twitter platform. One of the globally used OSN is the X/Twitter platform, with 330 million users actively using X/Twitter.

The drawbacks of these techniques are similar to those of the aforementioned technique. Many authors have attempted to make their methods superior by providing various results in terms of accuracy, F-measure recall, and precision.

However, these techniques failed to show their effectiveness by providing computational and time complexities. Therefore, methods that use DL or ML techniques must provide an analysis of the computational and time complexities to reveal the effectiveness of the proposed method.

The next section describes cyberbullying detection in OSNs using DL techniques. Most authors have used only the models to classify a tweet or an account as non-bullying or bullying. To accurately classify tweets or accounts into two classes, it is important to perform pre-processing, feature extraction, and feature selection.

Many authors have used pre-processing, feature extraction, and classification, but they have refused to add feature selection. Feature selection using bio-inspired optimization algorithms allows the model to classify tweets or accounts into two classes more accurately. These are few outcomes of the systematic review on spam bot and cyberbullying detection in OSNs.

18.7   CONCLUSION

Online social networks are energetic tools used by people worldwide to communicate with one another. This systematic review provides online social SBD using DL models from 2018 to 2022. Basic information about the botnet, spam and its types, and social media bots was provided. The steps involved in detecting spam bots in OSN using DL were also discussed.

X/Twitter is considered the most accessible spam bot social network. Therefore, the technique used to detect spam bots on the X/Twitter platform was also analyzed. There are many OSNs, but SBD techniques have been proposed only for rare social networks, such as Facebook, Instagram, YouTube, and X/ Twitter.

A plethora of studies have centered around the X/Twitter platform. Prior research has mainly concentrated on detecting spam bots and cyberbullying through anomaly-based methods, using single feature sets and simple relation structures to identify anomalous messages or accounts.

Anomaly detection techniques that rely on a single feature set have faced more difficulties in accurately detecting anomalous events, users, and spam bots. This chapter has reviewed the most significant challenges in this area and provided solutions to address these challenges (Rana et al., 2021).

18.8   FUTURE SCOPE

To protect OSN users' data and profiles from potential spam attacks, it is crucial to implement decentralized blockchain techniques in social networks. This will ensure that user information is secure and not accessible to third-party authentication (Khanh & Khang, 2021).

Given the ever-increasing size of OSN data, existing methods are no longer adequate to manage these data. Therefore, big data techniques can be employed to develop more effective ways of handling such data. Because the existing database contains outdated information, it can be archived, deleted, or marked as private. In preparation for future research, a new database can be created.

A smaller number of hybrid techniques are being used to detect spam and cyberbullying in OSNs. Therefore, there is a need to encourage the use of hybrid techniques for detecting spam bots and cyberbullying among researchers. Researchers are primarily focused on developing new approaches to detect spam bots and cyberbullying in popular social networks, such as Facebook, X/ Twitter, and email.

In addition to these OSNs, there are other social networks, such as LinkedIn, Pinterest, and YouTube, that are also susceptible to spam bots and cyberbullying. It is essential to employ proper techniques to detect and prevent these malicious activities in the future.

The detection of multilingual cyberbullying is a challenging task that requires further research. Developing an optimization algorithm to prolong the lifespan of techniques used for spam bot and cyberbullying detection is crucial for the future (Khang et al., 2023c).

REFERENCES

Abarna S., Sheeba J. I., Jayasrilakshmi S., Pradeep Devaneyan S., (2022). "Identification of cyber harassment and intention of target users on social media platforms," *Engineering Applications of Artificial Intelligence*, 115, 105283. https://doi.org/10.1016/j.engappai.2022.105283

Agrawal S., Amit A., (2018). "Deep learning for detecting cyberbullying across multiple social media platforms," In *European Conference on Information Retrieval*, pp. 141–153. Springer. https://doi.org/10.1007/978-3-319-76941-7_11.

Ahmed H., Issa T., Sherif S., (2018). "Detecting opinion spams and fake news using text classification," *Security and Privacy*, 1(1), e9. https://doi.org/10.1002/spy2.9

Ahmed T., Shahriar I., Mohsinul K., Hasan M., Kamrul H., (2022). "Performance analysis of transformer-based architectures and their ensembles to detect trait-based cyberbullying," *Social Network Analysis and Mining*, 12(1), 1–17. https://doi.org/10.1007/s13278-022-00934-4

Al-Ajlan M. A., Mourad Y., (2018). "Optimized twitter cyberbullying detection based on deep learning," In *2018 21st Saudi Computer Society National Computer Conference (NCC)*, pp. 1–5. IEEE. https://doi.org/10.1109/NCG.2018.8593146

Alauthman M., (2020). "Botnet spam e-mail detection using deep recurrent neural network," *International Journal*, 8(5). https://doi.org/10.30534/ijeter/2020/83852020

Alhassun A. S., Murad A. R., (2022). "A combined text-based and metadata-based deep-learning framework for the detection of spam accounts on the social media platform Twitter," *Processes* 10 (3), 439. https://doi.org/10.3390/pr10030439

Ali Alhosseini S., Raad Bin T., Pejman N., Christoph M., (2019). "Detect me if you can: Spam bot detection using inductive representation learning," *Companion Proceedings of the 2019 World Wide Web Conference*, 148–153. https://doi.org/10.1145/3308560.3316504

Alkhamees M., Saleh A., Al-Qurishi M., Al-Rubaian M., Amir H., (2021). "User trustworthiness in online social networks: A systematic review," *Applied Soft Computing*, 103, 107159. https://doi.org/10.1016/j.asoc.2021.107159

Almadhoor L., (2021). "Social media and cybercrimes," *Turkish Journal of Computer and Mathematics Education (TURCOMAT)*, 12(10), 2972–2978. https://doi.org/10.17762/turcomat.v12i10.4947

Alom Z., Barbara C., Elena F., (2020). "A deep learning model for Twitter spam detection," *Online Social Networks and Media*, 18, 100079. https://doi.org/10.1016/j.osnem.2020.100079

Alotaibi M., Bandar A., Abdul R., (2021). "A multichannel deep learning framework for cyberbullying detection on social media," *Electronics*, 10(21), 2664. https://doi.org/10.3390/electronics10212664.

Aroyo L., Lucas D., Nithum T., Olivia R., Rachel R., (2019). "Crowdsourcing subjective tasks: The case study of understanding toxicity in online discussions," *Companion Proceedings of the 2019 World Wide Web Conference*, 1100–1105. https://doi.org/10.1145/3308560.3317083

Atoum J. O., (2021). "Cyberbullying detection neural networks using sentiment analysis," In *2021 International Conference on Computational Science and Computational Intelligence (CSCI)*, pp. 158–164. IEEE. https://doi.org/10.1109/CSCI54926.2021.00098

Ban X., Chao C., Shigang L., Yu W., Jun Z., (2018). "Deep-learnt features for Twitter spam detection," In *2018 International Symposium on Security and Privacy in Social Networks and Big Data (SocialSec)*, pp. 208–212. IEEE. https://doi.org/10.1109/SocialSec.2018.8760377

Banerjee V., Jui T., Pooja G., Pallavi V., (2019). "Detection of cyberbullying using deep neural network," In *2019 5th International Conference on Advanced Computing & Communication Systems (ICACCS)*, pp. 604–607. IEEE. https://doi.org/10.1109/ICACCS.2019.8728378

Behzadi M., Ian G. H., Ali D., (2021). "Rapid cyberbullying detection method using compact BERT models," In *2021 IEEE 15th International Conference on Semantic Computing (ICSC)*, pp. 199–202. IEEE. https://doi.org/10.1109/ICSC50631.2021.00042

Birunda S. S., Kanniga Devi R., (2021). "A novel score-based multi-source fake news detection using gradient boosting algorithm," In *2021 International Conference on Artificial Intelligence and Smart Systems (ICAIS)*, pp. 406–414. IEEE.

Bozyiğit A., Semih U., Efendi N., (2021). "Cyberbullying detection: Utilizing social media features," *Expert Systems with Applications*, 179, 115001. https://doi.org/10.1016/j.eswa.2021.115001

Castellanos A., Ortega-Ruipérez B., David A., (2021). "Teachers' perspectives on cyberbullying: A cross-cultural study," *International Journal of Environmental Research and Public Health*, 19(1), 257. https://doi.org/10.3390/ijerph19010257

Chandra N., Sunil Kumar K., Subhranil S., (2018). "Cyberbullying detection using recursive neural network through offline repository," In *2018 7th International Conference on Reliability, Infocom Technologies and Optimization (Trends and Future Directions) (ICRITO)*, pp. 748–754. IEEE. https://doi.org/10.1109/ICRITO.2018.8748570

Chang Ho-Chun H., Emily C., Meiqing Z., Goran M., Emilio F., (2021). "Social bots and social media manipulation in 2020: The year in review," *arXiv preprint arXiv:2102.08436*. https://doi.org/10.48550/arXiv.2102.08436

Chen, Hsin-Yu, and Cheng-Te Li, "HENIN: Learning Heterogeneous Neural Interaction Networks for Explainable Cyberbullying Detection on Social Media," *arXiv preprint arXiv:2010.04576* (2020), DOI: 10.48550/arXiv.2010.04576

Cheng Y., Dan L., Zhiyuan G., Binyao J., Jiaxin L., Xi F., Jinkun G., et al., (2019). "Dlbooster: Boosting end-to-end deep learning workflows with offloading data preprocessing pipelines," *Proceedings of the 48th International Conference on Parallel Processing*, 1–11. https://doi.org/10.1145/3337821.3337892

Chia Zheng L., Michal P., Fumito M., Gniewosz L., Michal W., (2021). "Machine learning and feature engineering-based study into sarcasm and irony classification with application to cyberbullying detection," *Information Processing & Management*, 58(4), 102600. https://doi.org/10.1016/j.ipm.2021.102600

Chowdhury R., Kumar Gourav D., Banani S., Samir Kumar B., (2020). "A method based on nlp for Twitter spam detection," Preprints 2020, 2020070648. https://doi.org/10.20944/preprints202007.0648.v1

Cresci S., Fabrizio L., Daniele R., Serena T., Maurizio T., (2019). "Cashtag piggybacking: Uncovering spam and bot activity in stock microblogs on Twitter," *ACM Transactions on the Web (TWEB)*, 13(2), 1–27. https://doi.org/10.1145/3313184

Dadvar M., Kai E., (2020). "Cyberbullying detection in social networks using deep learning based models," In *International Conference on Big Data Analytics and Knowledge Discovery*, pp. 245–255. Springer. https://doi.org/10.1007/978-3-03059065-9_20

Daouadi K. E., Rim Zghal R., Ikram A., (2019). "Bot detection on online social networks using deep forest," In *Computer Science On-line Conference*, pp. 307–315. Springer. https://doi.org/10.1007/978-3-030-19810-7_30

Fan R., Oleksandr T., Vu T., (2020). "Social media bots and stock markets," *European Financial Management*, 26(3), 753–777. https://doi.org/10.1111/eufm.12245

Fang Y., Shaoshuai Y., Bin Z., Cheng H., (2021). "Cyberbullying detection in social networks using Bi-gru with self-attention mechanism," *Information*, 12(4), 171. https://doi.org/10.3390/info12040171

Fazil M., Amit Kumar S., Muhammad A., (2021). "DeepSBD: A deep neural network model with attention mechanism for socialbot detection," *IEEE Transactions on Information Forensics and Security*, 16, 4211–4223. https://doi.org/10.1109/TIFS.2021.3102498

Gada M., Kaustubh D., Smita S., (2021). "Cyberbullying detection using LSTM-CNN architecture and its applications," In *2021 International Conference on Computer Communication and Informatics (ICCCI)*, pp. 1–6. IEEE. https://doi.org/10.1109/ICCCI50826.2021.9402412

Guo Q., Haiyong X., Yangyang L., Wen M., Chao Z., (2021a). "Social bots detection via fusing BERT and graph convolutional networks," *Symmetry*, 14(1), 30. https://doi.org/10.3390/sym14010030

Guo Z., Lianggui T., Tan G., Keping Y., Mamoun A., Andrii S., (2021b). "Deep graph neural network-based spammer detection under the perspective of heterogeneous cyberspace," *Future Generation Computer Systems*, 117, 205–218. https://doi.org/10.1016/j.future.2020.11.028

Hang, Ong Chee, and Halina Mohamed Dahlan, "Cyberbullying lexicon for social media," In *2019 6th International Conference on Research and Innovation in Information Systems (ICRIIS)*, pp. 1–6. IEEE (2019): 2319–2323, DOI: 10.1109/ICRIIS48246.2019.9073679

Harrigan P., Timothy M. D., Kristof C., Julie A. L., Geoffrey N. S., Uwana E., (2021). "Identifying influencers on social media," *International Journal of Information Management*, 56, 102246. https://doi.org/10.1016/j.ijinfomgt.2020.102246

Herath Thilini B. G., Prashant K., Monjur A., (2022). "Cybersecurity practices for social media users: A systematic literature review," *Journal of Cybersecurity and Privacy*, 2(1), 1–18. https://doi.org/10.3390/jcp2010001

Hung-Wei Y., Li-Chin H., Min-Shiang H., (2021). "Research on detection and prevention of mobile device botnet in cloud service systems," *International Journal of Network Security*, 23(3), 371–378. https://doi.org/10.6633/IJNS.202105

Iwendi C., Gautam S., Suleman K., Praveen Kumar R. M., (2020). "Cyberbullying detection solutions based on deep learning architectures," *Multimedia Systems*, 1–14. https://doi.org/10.1007/s00530-020-00701-5

Jogin Manjunath M., Madhulika S., Divya G. D., Meghana R. K., Apoorva S., (2018). "Feature extraction using convolution neural networks (CNN) and deep learning," In *2018 3rd IEEE International Conference on Recent Trends in Electronics, Information & Communication technology (RTEICT)*, pp. 2319–2323. IEEE. https://doi.org/10.1109/RTEICT42901.2018.9012507

Kaur P., Amandeep D., Anushree T., Ebtesam A. A., Abeer Ahmed A., (2021). "A systematic literature review on cyberstalking. An analysis of past achievements and future promises," *Technological Forecasting and Social Change*, 163, 120426. https://doi.org/10.1016/j.techfore.2020.120426

Khang A., (2021). "Material4Studies," *Material of Computer Science, Artificial Intelligence, Data Science, IoT, Blockchain, Cloud, Metaverse, Cybersecurity for Studies*. www.researchgate.net/publication/370156102_Material4Studies

Khang A., Gupta S. K., Shah V., Misra A., (Eds.). (2023a). *AI-Aided IoT Technologies and Applications in the Smart Business and Production*. CRC Press. https://doi.org/10.1201/9781003392224

Khang A., Gupta S. K., Hajimahmud V. A., Babasaheb J., Morris G., (2023b). *AI-Centric Modelling and Analytics: Concepts, Designs, Technologies, and Applications* (1st ed.). CRC Press. https://doi.org/10.1201/9781003400110

Khang A., Vrushank S., Rani S., (2023c). *AI-Based Technologies and Applications in the Era of the Metaverse* (1st ed.). IGI Global Press. https://doi.org/10.4018/9781668488515

Khang A., Hahanov V., Abbas G. L., Hajimahmud V. A., (2022). "Cyber-physical-social system and incident management," In *AI-Centric Smart City Ecosystems: Technologies, Design and Implementation*, Vol. 2, No. 15 (1st ed.). CRC Press. https://doi.org/10.1201/9781003252542-2

Khanh H. H., Khang A., (2021). "The role of artificial intelligence in blockchain applications," *Reinventing Manufacturing and Business Processes through Artificial Intelligence*, 2(20–40) (CRC Press). https://doi.org/10.1201/9781003145011-2

Kumar A., Nitin S., (2020). "Multi-input integrative learning using deep neural networks and transfer learning for cyberbullying detection in real-time code-mix data," *Multimedia Systems*, 1–15. https://doi.org/10.1007/s00530-020-00672-7

Kumar A., Nitin S., (2021a). "A Bi-GRU with attention and CapsNet hybrid model for cyber-bullying detection on social media," *World Wide Web*, 1–14. https://doi.org/10.1007/s11280021–00920–4.

Kumar A., Nitin S., (2021b). "Multimodal cyberbullying detection using capsule network with dynamic routing and deep convolutional neural network," *Multimedia Systems*, 1–10. https://doi.org/10.1007/s00530-020-00747-5

Kumar G. D., Kameswara Rao M., Premnath K., (2020). "A recurrent neural network model for spam message detection," In *2020 5th International Conference on Communication and Electronics Systems (ICCES)*, pp. 1042–1045. IEEE. https://doi.org/10.1109/ICCES48766.2020.9137940

Kumar S., Shivang G., Yatharth V., Anil Singh P., (2021). "Content based bot detection using bot language model and BERT embeddings," In *2021 5th International Conference on Computer, Communication and Signal Processing (ICCCSP)*, pp. 285–289. IEEE. https://doi.org/10.1109/ICCCSP52374.2021.9465506

Kumari K., Jyoti Prakash S., (2021). "Identification of cyberbullying on multi modal social media posts using genetic algorithm," *Transactions on Emerging Telecommunications Technologies*, 32(2), e3907. https://doi.org/10.1002/ett.3907

Latah M., (2019). "The art of social bots: A review and a refined taxonomy," *arXiv preprint arXiv:1905.03240*. https://doi.org/10.48550/arXiv.1905.03240

Latah M., (2020). "Detection of malicious social bots: A survey and a refined taxonomy," *Expert Systems with Applications*, 151, 113383. https://doi.org/10.1016/j.eswa.2020.113383

Lingam G., Rashmi Ranjan R., Durvasula V. L. N. S., (2019). "Deep Q-learning and particle swarm optimization for bot detection in online social networks," In *2019 10th International Conference on Computing, Communication and Networking Technologies (ICCCNT)*, pp. 1–6. IEEE. https://doi.org/10.1109/ICCCNT45670.2019.8944493

Lingam G., Rashmi Ranjan R., Durvasula V. L. N. S., Sajal K. D., (2020). "Social botnet community detection: A novel approach based on behavioral similarity in Twitter network using deep learning," In *Proceedings of the 15th ACM Asia Conference on Computer and Communications Security*, 708–718. https://doi.org/10.1145/3320269.3384770

Luo L., Xiaofeng Z., Xiaofei Y., Weihuang Y., (2020). "Deepbot: A deep neural network based approach for detecting Twitter bots." In *IOP Conference Series: Materials Science and Engineering*, 719(1), 012063. IOP Publishing. https://doi.org/10.1088/1757–899X/719/1/012063

Madisetty S., Maunendra Sankar D., (2018). "A neural network-based ensemble approach for spam detection in Twitter," *IEEE Transactions on Computational Social Systems*, 5(4), 973–984. https://doi.org/10.1109/TCSS.2018.2878852

Mahat M., (2021). "Detecting cyberbullying across multiple social media platforms using deep learning," In *2021 International Conference on Advance Computing and Innovative Technologies in Engineering (ICACITE)*, pp. 299–301. IEEE. https://doi.org/10.1109/ICACITE51222.2021.9404736.

Mahbub S., Eric P., Kayes A. S. M., (2021). "Detection of harassment type of cyberbullying: A dictionary of approach words and its impact," *Security and Communication Networks*, 2021. https://doi.org/10.1155/2021/5594175

Mariappan M. B., Kanniga D., Yegnanarayanan V., Ming K. L., Panneerselvam T., (2022a). "Using AI and ML to predict shipment times of therapeutics, diagnostics and vaccines in e-pharmacy supply chains during COVID-19 pandemic," *International Journal of Logistics Management*.

Mariappan M. B., Kanniga D., Yegnanarayanan V., Samuel F. W., (2022b). "A large-scale realworld comparative study using pre-COVID lockdown and postCOVID lockdown data on predicting shipment times of therapeutics in e-pharmacy supply chains," *International Journal of Physical Distribution & Logistics Management* (ahead-of-print), 52(7), https://www.emerald.com/insight/content/doi/10.1108/IJPDLM-05-2021-0192/full/html

Masood F., Ahmad A. N., Assad A., Hasan A. K., Ikram U. D., Mohsen G., Mansour Z., (2019). "Spammer detection and fake user identification on social networks," *IEEE Access*, 7, 68140–68152. https://doi.org/10.1109/ACCESS.2019.2918196

Murshed B. A. H., Jemal A., Suresha M., Mufeed A. N. S., Hasib Daowd E. Al-Ariki, (2022). "DEA-RNN: A hybrid deep learning approach for cyberbullying detection in Twitter social media platform," *IEEE Access*, 10, 25857–25871. https://doi.org/10.1109/ACCESS.2022.3153675

Najari S., Mostafa S., Reza F., (2022). "GANBOT: A GAN-based framework for social bot detection," *Social Network Analysis and Mining*, 12(1), 1–11. https://doi.org/10.1007/s13278-021-00800-9

Noekhah S., Naomie Binti S., Nor Hawaniah Z., (2020). "Opinion spam detection: Using multi-iterative graph-based model," *Information Processing & Management*, 57(1), 102140. https://doi.org/10.1016/j.ipm.2019.102140.

Orabi M., Djedjiga M., Zaher Al A., Ibrahim K., (2020). "Detection of bots in social media: A systematic review," *Information Processing & Management*, 57(4), 102250. https://doi.org/10.1016/j.ipm.2020.102250

Ortiz-Marcos J. M., Tomé-Fernández M., Fernández-Leyva C., (2021). "Cyberbullying analysis in intercultural educational environments using binary logistic regressions," *Future Internet*, 13(1), 15. https://doi.org/10.3390/fi13010015

Oskouei M. D., Seyed Naser R., (2018). "An ensemble feature selection method to detect web spam," *Asia-Pacific Journal of Information Technology and Multimedia*, 7(2), 99–113. https://doi.org/10.17576/apjitm-2018–0702–08

Paul S., Sriparna S., (2020). "CyberBERT: BERT for cyberbullying identification," *Multimedia Systems*, 1–8. https://doi.org/10.1007/s00530-020-00710-4.

Peiling Y., and Zubiaga A., "Cyberbullying detection across social media platforms via platform-aware adversarial encoding," In *Proceedings of the International AAAI Conference on Web and Social Media* 16 (2022): 1430–1434. https://ojs.aaai.org/index.php/ICWSM/article/view/19401

Ping H., Sujuan Q., (2018). "A social bots detection model based on deep learning algorithm," In *2018 IEEE 18th International Conference on Communication Technology (ICCT)*, pp. 1435–1439. IEEE. https://doi.org/10.1109/ICCT.2018.8600029

Rahman S., Kamrul Hasan T., Sabia Khatun M., (2021). "An empirical study to detect cyberbullying with TFIDF and machine learning algorithms," In *2021 International Conference on Electronics, Communications and Information Technology (ICECIT)*, pp. 1–4. IEEE. https://doi.org/10.1109/ICECIT54077.2021.9641251

Rana G., Khang A., Sharma R., Goel A. K., Dubey A. K., (Eds.). (2021). *Reinventing Manufacturing and Business Processes through Artificial Intelligence*. CRC Press. https://doi.org/10.1201/9781003145011

Risch J., Ralf K., (2020). "Toxic comment detection in online discussions," In *Deep Learning-based Approaches for Sentiment Analysis*, pp. 85–109. Springer. https://doi.org/10.1007/978981-15-1216-2_4.

Rodríguez-Ruiz J., Mata-Sánchez J. S., Raúl M., Loyola-González O., López-Cuevas A., (2020). "A oneclass classification approach for bot detection on Twitter," *Computers & Security*, 91, 101715. https://doi.org/10.1016/j.cose.2020.101715

Roy P. K., Jyoti Prakash S., Snehasish B., (2020). "Deep learning to filter SMS spam," *Future Generation Computer Systems*, 102, 524–533. https://doi.org/10.1016/j.future.2019.09.001

San Biagio M., Roberto A., Valentina M., Ernesto La M., Vito M., (2021). "A new SOCMINT framework for threat intelligence identification," In *2021 International Conference on Computational Science and Computational Intelligence (CSCI)*, pp. 692–697. IEEE. https://doi.org/10.1109/CSCI54926.2021.00180

Seok-Jun B., Sung-Bae C., (2018). "A hybrid deep learning system of CNN and LRCN to detect cyberbullying from SNS comments," In *International Conference on Hybrid Artificial Intelligence Systems*, pp. 561–572. Springer. https://doi.org/10.1007/978-3-319-92639-1_47

Shahzad A., Nazri Mohd N., Muhammad Zubair R., Abdullah K., (2021). "An improved framework for content-and link-based web-spam detection: A combined approach," *Complexity*, 2021. https://doi.org/10.1155/2021/6625739.

Shannag F., Bassam H. H., Hossam F., (2022). "The design, construction and evaluation of annotated Arabic cyberbullying corpus," *Education and Information Technologies*, 1–47. https://doi.org/10.1007/s10639–022–11056-x

Sharma P., Uma B., (2018). "Machine learning based spam email detection," *International Journal of Intelligent Engineering and Systems*, 11(3), 1–10. https://doi.org/10.22266/ijies2018.0630.01

Sharma S., Manu S., (2021). "Exploring feature selection technique in detecting sybil accounts in a social network," In *International Conference on Innovative Computing and Communications*, pp. 695–708. Springer. https://doi.org/10.1007/978-981-15-5148-2_61

Shetty N. P., Balachandra M., Arshia A., Sushant K., (2022). "An enhanced sybil guard to detect bots in online social networks," *Journal of Cyber Security and Mobility*, 105–126. https://doi.org/10.13052/jcsm2245–1439.1115

Song, Tae-Min, and Juyoung Song, "Prediction of risk factors of cyberbullying-related words in Korea: Application of data mining using social big data," *Telematics and Informatics* 58 (2021): 101524. DOI: 10.1016/j.tele.2020.101524

Tae-Min S., Juyoung S., (2021). "Prediction of risk factors of cyberbullying-related words in Korea: Application of data mining using social big data," *Telematics and Informatics*, 58, 101524. https://doi.org/10.1016/j.tele.2020.101524

Vengatesan K., Abhishek K., Radhakrishana N., Deepak Kumar V., (2018). "Anomaly based novel intrusion detection system for network traffic reduction," In *2018 2nd International Conference on I-SMAC (IoT in Social, Mobile, Analytics and Cloud)(I-SMAC) I-SMAC (IoT in Social, Mobile, Analytics and Cloud)(I-SMAC), 2018 2nd International Conference on, IEEE*, 688–690. IEEE. https://doi.org/10.1109/I-SMAC.2018.8653735

Venkateswarlu B., Viswanath S., (2021). "Optimized generative adversarial network with fractional calculus based feature fusion using Twitter stream for spam detection," Information Security Journal: A Global Perspective. Volume 31, 2022 - Issue 5 https://doi.org/1 0.1080/19393555.2021.1956024.

Wang T., Zhihan C., Shuo W., Jianhuang W., Lianyong Q., Anfeng L., Mande X., Xiaolong L., (2019). "Privacy-enhanced data collection based on deep learning for Internet of vehicles," *IEEE Transactions on Industrial Informatics*, 16(10), 6663–6672. https://doi.org/10.1109/TII.2019.2962844

Wei F., Trang N., (2020). "A lightweight deep neural model for SMS spam detection," In *2020 International Symposium on Networks, Computers and Communications (ISNCC)*, pp. 1–6. IEEE. https://doi.org/10.1109/ISNCC49221.2020.9297350

Wei F., Uyen Trang N., (2019). "Twitter bot detection using bidirectional long short-term memory neural networks and word embeddings," In *2019 First IEEE International Conference on Trust, Privacy and Security in Intelligent Systems and Applications (TPS-ISA)*, pp. 101–109. IEEE. https://doi.org/10.1109/TPS-ISA48467.2019.00021

Wu Y., Yuzhou F., Shuaikang S., Jing J., Lai W., Haizhou W., (2021). "A novel framework for detecting social bots with deep neural networks and active learning," *Knowledge Based Systems*, 211, 106525. https://doi.org/10.1016/j.knosys.2020.106525

Xu Y., Paula T., (2021). "Towards descriptive adequacy of cyberbullying: Interdisciplinary studies on features, cases and legislative concerns of cyberbullying," *International Journal for the Semiotics of Law-Revue internationale de Sémiotique Juridique*, 34(4), 929–943. https://doi.org/10.1007/s11196–02109856–4

Yadav J., Devesh K., Dheeraj C., (2020). "Cyberbullying detection using pre-trained BERT model," In *2020 International Conference on Electronics and Sustainable Communication Systems (ICESC). IEEE*, pp. 1096–1100. https://doi.org/10.1109/ICESC48915.2020.9155700.

Yao M., Charalampos C., Daphney-Stavroula Z., (2019). "Cyberbullying ends here: Towards robust detection of cyberbullying in social media," *The World Wide Web Conference*, 3427–3433. https://doi.org/10.1145/3308558.3313462

Yuvaraj N., Victor C., Balasubramanian G., Arulprakash P., Srihari K., Gaurav D., Arsath R. R., (2021). "Automatic detection of cyberbullying using multi-feature based artificial intelligence with deep decision tree classification," *Computers & Electrical Engineering*, 92, 107186. https://doi.org/10.1016/j.compeleceng.2021.107186

Zaib M. H., Faisal B., Kashif Naseer Q., Sumaira K., Muhammad R., Gwanggil J., (2021). "Deep learning based cyber bullying early detection using distributed denial of service flow," *Multimedia Systems*, 1–20. https://doi.org/10.1007/s00530-021-00771-z

Zhang S., (2021). "From flaming to incited crime: recognising cyberbullying on Chinese WeChat account," *International Journal for the Semiotics of Law-Revue internationale de Sémiotique Juridique*, 34(4), 1093–1116. https://doi.org/10.1007/s11196020-09790-x

Zhao C., Yang X., Xuefeng L., Hongliang Z., Yixian Y., Yuling C., (2020). "An attention-based graph neural network for spam bot detection in social networks," *Applied Sciences*, 10(22), 8160. https://doi.org/10.3390/app10228160

Index

For Product Safety Concerns and Information please contact our EU
representative GPSR@taylorandfrancis.com
Taylor & Francis Verlag GmbH, Kaufingerstraße 24, 80331 München, Germany